THE 'CABINET' OF THE BROTHERS GERARD AND JAN REYNST

KONINKLIJKE NEDERLANDSE AKADEMIE VAN WETENSCHAPPEN
VERHANDELINGEN AFDELING LETTERKUNDE, NIEUWE REEKS, DEEL 99

ANNE-MARIE S. LOGAN

THE 'CABINET' OF THE BROTHERS GERARD AND JAN REYNST

NORTH-HOLLAND PUBLISHING COMPANY
AMSTERDAM/OXFORD/NEW YORK 1979

I S B N 0 7204 8342 5

Communicated at the meeting of June 9, 1975 by Prof. Dr. J. G. van Gelder and Prof. Dr. H. K. Gerson †

Published with financial assistance from:
Mr. S. Nijstad,
Dr. Elisabeth Reynst
and Mrs. Marguerite Seagesser

CONTENTS

ACKNOWLEDGEMENTS 7

I. INTRODUCTION 9

II. THE REYNST FAMILY
1. Ancestors 13
2. Gerrit Reynst and Margaretha Nicquet 13
3. Gerard Reynst and Anna Schuyt 19
4. Jan Reynst 29

III. THE COLLECTIONS OF GERARD AND JAN REYNST
A. Their contents 37
1. Caelaturae 38
2. Icones 45
3. Descriptions from visitors to the Reynst collections
   Amsterdam: Aernout van Buchell (1639) 55
   Amsterdam: Barthold Neuhauser (ca. 1650) 59
   Amsterdam: Christian Knorr von Rosenroth (1663) 63
   Venice: Carlo Ridolfi (1646) 66
4. The collection of Andrea Vendramin 67
5. The 'Dutch Gift' 75

B. Their nature 87
1. Formation 87
2. Character of the collections 98

IV. CONCLUSION 102

CATALOGUE OF PAINTINGS AND SCULPTURES IN THE COLLECTION OF GERARD AND JAN REYNST

Abbreviations used in the catalogue 108

Royal inventories 109

Italian paintings in the collection of Gerard Reynst, partly formerly in the collection of Jan Reynst 110

Dutch paintings in the collection of Gerard Reynst 160

Paintings in the collection of Jan Reynst, mentioned by Ridolfi 165

Painting in the collection of Gerard Reynst, appraised on 12 May, 1672 169

Drawings in the collection of Gerard and Jan Reynst   170

Paintings formerly in the collection of Andrea Vendramin, Venice, probably bought *en bloc* by Gerard Reynst (based on Borenius)   171

Sculptures in the collection of Gerard Reynst, reproduced in the ICONES   175

Sculptures presented to Charles II as part of the 'Dutch Gift' in 1660   214

Sculpture illustrated in the Vendramin catalogue DE SCULPTURIS, but not engraved for the ICONES   218

Sculptures in the Rijksmuseum van Oudheden, Leiden, formerly in the collection of Andrea Vendramin (based on Oudendorp)   219

Tomb monuments formerly in the collection of Gerard Reynst, reproduced in the Vendramin catalogue 'De Antiquorum Tumulis', Venice, 1627   221

APPENDIX

Dutch and Flemish paintings and engravings associated incorrectly with the collection of Gerard Reynst   228

French and Italian paintings and engravings associated incorrectly with the collection of Gerard Reynst   234

List of paintings in the collection of Gerard Reynst   236

List of paintings in the collection of Jan Reynst   240

List of paintings included in the 'Dutch Gift'   242

List of paintings in the English Royal collections formerly with Reynst   243

Concordance of the Vendramin catalogue DE SCULPTURIS with the ICONES   244

Inventory of Jacques Nicquet   248

Bibliography   269

Exhibitions   273

Index   274

List of text figures, plates and credit for photographs   287

Colophon   293

ACKNOWLEDGMENTS

In the course of this study I have received assistance in numerous ways from kind people helping me trace archival material, works of art and copies of the CAELATURAE and the ICONES.

Special thanks are due to Dr. I.H. van Eeghen and Mr. S.A.C. Dudok van Heel from the Gemeentelijke Archiefdienst in Amsterdam, who brought much new data on the Reynst family to my attention. Mr. Dudok van Heel in particular found important new facts about Jan Reynst, his residence in Venice and his designation of Gerard Reynst as the universal heir. Thanks to his efforts we also know the inventory of Jacques Nicquet which is of significance for the Reynst brothers as collectors. Mr. Dudok van Heel generously offered to transcribe it and allowed me to include it in the present study (see pp. 248–268). This offer deserves all the more thanks since the Nicquet inventory is so severely damaged by fire that it can no longer be consulted publicly. Mr. G. W. van der Meiden made the extensive correspondence of Jan Reynst at the Algemeen Rijksarchief in The Hague available for which I am most grateful. Mr. S. de Tuoni, Fondazione Cini, Venice, kindly searched in the archives of Venice for material concerning Jan Reynst, and Mr. F. Ludwigs, Riksarkivet, Stockholm, assisted me in identifying letters by Michel Le Blon.

Dr. H.-E. Teitge, Deutsche Staatsbibliothek, East Berlin, made it possible for me to study the papers on the Reynst collection, left by Emil Jacobs. I would like to thank Sir Oliver Millar for allowing me to consult the English royal inventories. Miss Jennifer Sherwood, Department of Drawings, Windsor Castle, kindly assisted me in tracing references to sculptures in the English royal collections that were part of the 'Dutch Gift'.

Many identifications of the antique sculptures are due to help received from Professor Sheldon A. Nodleman. I would also like to thank Professor H. Brunsting, Leiden, for sharing his research with me on those sculptures, tomb monuments and reliefs in the Rijksmuseum van Oudheden that formerly belonged to Gerard Reynst, and Dr. P.J.J. Stuart for allowing me to study the originals. Dr. Gerald Heres, Staatliche Museen, East Berlin, kindly assisted me in retrieving information on the sculptures formerly in the collection of the Elector of Brandenburg. Thanks are also due to Professor and Mrs. M. Raumschüssel, Skulpturensammlung, Dresden, for their assistance in collecting data on the sculptures from the Reynst collection that came to Dresden in the early 18th century.

I would like to express my deepest gratitude for continuous encouragement to Professor E.H. Begemann and Professor J.G. van Gelder. Professor Begemann has given me much constructive advice and valuable assistance over the years. From Professor Van Gelder I have received many enlightening and most useful suggestions for which I am very grateful. Thanks to his generous, continuous support the manuscript was accepted by the Dutch Royal Academy for publication. Special words of thanks are also due to Dr. I.I.E. van Gelder-Jost for helpful suggestions and information on the subject, especially with regard to the antique sculptures.

Finally, I would like to express my thanks to Dr. Elisabeth Reynst, whose initial work on the collection of the brothers Gerard and Jan Reynst was the basis for this expanded study. Her support was most welcome and appreciated.

I. INTRODUCTION

The works of art assembled by the brothers Gerard and Jan Reynst, both successful merchants, grew into one of the largest, most important and most special collections of its kind of Italian paintings, antique sculptures, *naturalia* and various other collector's items in the Netherlands of the middle of the 17th century.[1]

By 1638, the collection was so well known that Princess Amalia van Solms, the wife of stadholder Frederick Henry, expressed interest in seeing it. She was most taken by a statue of *Cleopatra* which Gerard Reynst eventually presented to her.[2] Between 1639 and 1668, the house on the Keizersgracht in Amsterdam was also visited by a number of Dutch and foreign travellers who fortunately left descriptions of the collection.

The greatest tribute to the brothers Reynst as collectors came in 1660, when the States of Holland and West Friesland selected twenty-four paintings and twelve antique sculptures which they presented as the so-called 'Dutch Gift' to king Charles II of England. Both Gerard and Jan would undoubtedly have been pleased to see part of their paintings and sculptures included in the English royal collections.

Another honor for the two collectors was Ridolfi's dedication of the first part of his *Le Maraviglie dell' Arte* in June, 1646, to 'gl' Illustrissimi Signori Fratelli Reinst, IL CAVALIER GIOVANNI SIGN. DI NIEL, E Commissario appresso la Maestà Christianissima per li

---

[1] Much of the basic data was assembled by Elisabeth Reynst, *Sammlung Gerrit und Jan Reynst* (typewritten manuscript, 1964, 40 pp.).
Fundamental for the reconstruction of the Reynst collection are the following articles and books (listed chronologically): Franciscus Oudendorp, *Brevis monumentorum ab amplissimo viro Gerardo Papenbroekio academiae Lugduno-Bataviae legatorum*, Leiden, 1746; Carlo Ridolfi, *Le Maraviglie dell' Arte, Ouero le vite de gl' illustri pittori veneti e dello stato*, 2 parts, Venice, 1648; edited by Detlev von Hadeln, Berlin, 1914, 1924; A. Bredius, 'Italiaansche Schilderijen in 1672 door Amsterdamsche en Haagsche schilders beoordeeld,' *Oud Holland*, 4, 1886, pp. 41–46, 278–80; 34, 1916, pp. 88–93; J. C. Breen and Fuchs, 'Aus dem 'Itinerarium' des Christian Knorr von Rosenroth', *Jaarboek van het Genootschap Amstelodamum*, XIV, 1916, pp. 239–45; G. J. Hoogewerff, *De twee reizen van Cosimo de' Medici, Prins van Toscane, door de Nederlanden (1667–1669)*, Utrecht, 1919, pp. 77–78; Tancred Borenius, *The Picture Gallery of Andrea Vendramin*, London, 1923; Emil Jacobs, 'Das Museo Vendramin und die Sammlung Reynst,' *Repertorium für Kunstwissenschaft*, XLVI, 1925, pp. 15–38; G. J. Hoogewerff and J. Q. van Regteren Altena, *Arnoldus Buchelius*, 'Res Pictoriae' (1583–1639), (Quellenstudien zur Holländischen Kunstgeschichte, XV), 's-Gravenhage, 1928, pp. 96–100; Tancred Borenius, 'More about the Andrea Vendramin Collection,' *Burlington Magazine*, LX, 1932, pp. 140–45; Frits Lugt, 'Italiaansche kunstwerken in Nederlandsche verzamelingen van vroeger tijden,' *Oud Holland*, LIII, 1936, pp. 97–135; J. W. C. van Campen, ed., *Notae quotidianae van Aernout van Buchell* (Werken uitgegeven door het historisch genootschap, III, no. 70), Utrecht, 1940, pp. 77–78 and pp. 94–96; B. H. Stricker, 'De Verzameling Reynst, Egyptische Antiquiteiten,' *Vooraziatisch-egyptisch Gezelschap 'Ex Oriente Lux'*, *Mededeelingen en Verhandelingen*, VII, 1947, pp. 255–67;

Potentissimi Stati delle Provincie Vnite, E GERARDO SENATORE D'AMSTERDAMO' (text fig. 7).³

Ridolfi, who was the only contemporary Italian author to refer to the two brothers as collectors,⁴ probably had a good reason for his dedication. In his introductory remarks he stressed the generosity of the brothers Reynst towards living artists and we may therefore assume that he himself had received a certain amount of money, most likely through Jan, whom he must have known. The title page,⁵ designed by Ridolfi himself, prominently displays the coat of arms of the Reynst family: three silver jars (against a blue ground) with a six-pointed (golden) star inserted at the top.⁶ Two putti are supporting it above a monument on which the allegorical figure of history is writing LE MARAVIGLIE DELL' ARTE. Father Time, holding an hourglass and a scythe, is hovering in the air in the distance. Ridolfi modelled his design after Cesare Ripa's description of the figure of *Historia*,⁷ alluding to the concept of Time versus History, of History surviving and preserving the works of art described in the book from the wrath of Time. Thus, he indicated that his *Maraviglie* and through it the contributions of the Reynst brothers would outlast time. The scroll held by a putto at the lower left, inscribed VIVIMVS MORITURI, MORIMVR VICTVRI (we will live to die, we will die to win) also stresses this element of transitoriness of earthly things.

Denis Mahon, 'Notes on the 'Dutch Gift' to Charles II,' *Burlington Magazine*, XCI, 1949, pp. 303–05; 349–50; XCII, 1950, pp. 12–18; Simona Savini-Branca, *Il Collezionismo veneziano nel '600* (Università di Padova, Pubblicazioni della Facoltà di lettere e filosofia, XLI), Padua, 1965, especially pp. 55, 70–71, 264–70; R.W. Scheller, 'Rembrandt en de encyclopedische verzameling,' *Oud Holland*, LXXXIV, 1969, pp. 116–17.
The Reynst collection is also mentioned briefly in Christiaan Kramm, *De levens en werken der hollandsche en vlaamsche kunstschilders*, V, Amsterdam, 1861, pp. 1363–65; A. Bredius, 'De Kunsthandel te Amsterdam in de XVIIe eeuw,' *Amsterdamsch Jaarboekje*, 1891, p. 55; Hanns Floerke, *Studien zur niederländischen Kunst- und Kulturgeschichte*, Munich/Leipzig, 1905, p. 105 and pp. 173–74; A. Donath, *Psychologie des Kunstsammelns*, Berlin, 1911, p. 52; Frits Lugt, *Mit Rembrandt in Amsterdam*, Berlin, 1920, pp. 65–66; A. Heppner, in A.E. d'Ailly ed., 'Amsterdamsche Verzamelaars,' in *Zeven Eeuwen Amsterdam*, VI, Amsterdam (s.a.), pp. 240–41; Francis Henry Taylor, *The Taste of Angels*, Boston, 1948, pp. 238–39, 267–68; H.E. van Gelder, 'Een praatje over de kunsthandel en het verzamelen in Nederland,' *Oude Kunst- en Antiekbeurs Delft* (exh.cat.), Delft (Museum 'Het Prinsenhof'), 1956, p. 22; Ernst Brochhagen, *Karel Dujardin*, Diss. Cologne, 1957, p. 6; Th.H. Lunsingh Scheurleer, *Sprekend verleden*, Amsterdam, 1959, pp. 14–15; Gisela Thieme, *Der Kunsthandel in den Niederlanden im siebzehnten Jahrhundert*, Cologne, 1959, p. 11; Niels von Holst, *Künstler, Sammler, Publikum*, Darmstadt, 1960, p. 106; J.Q. van Regteren Altena and P.J.J. van Thiel, *De portret-galerij van de Universiteit van Amsterdam en haar stichter Gerard van Papenbroeck 1673–1743*, Amsterdam, 1964, p. 34; Hannelore Sachs, *Sammler und Mäzene*, Leipzig, 1971, p. 83.

² Mentioned by Van Buchell in the *Notae Quotidianae* of his visit to the house of Gerard Reynst on September 4, 1639. Hoogewerff

In a four page introductory essay Ridolfi furthermore praised the brothers highly for having erected in Venice and Amsterdam at considerable expense two famous galleries where one admired many works by Raphael, Giovanni Bellini, Correggio, Parmegianino, Titian, Tintoretto, Veronese and other famous painters ('*hauendo elleno con generoso dispendio erette in Venetia, & in Amsterdamo due famose Galerie, oue si ammirano opere molte di Raffaelo, di Gio Bellino, del Correggio, del Parmegiano, di Titiano, del Tintoretto, di Paolo, e di qual si voglia altro insigne Pittore : onde vengono frequentate da Personaggi, e da begli ingegni ...*') (text fig. 7a–d).

As will be shown later, the core of the Reynst collection was bought from the Venetian nobleman Andrea Vendramin or his heirs and shipped to Amsterdam, where it was seen in the house of Gerard Reynst in the late 1630's. The most original part of the collection assembled by Gerard and Jan, however, were the paintings and sculptures acquired from other sources. These works generally were of higher quality. Only paintings from this group were engraved for the CAELATURAE. Forty-five of the one hundred and ten sculptures reproduced in the ICONES also came from sources other than Vendramin.

Jan seems to have been the driving force behind the formation of the collection. He was the one who bought the works of art. He

and Van Regteren Altena, *op.cit.*, pp. 97–98 and Van Campen, *op.cit.*, pp. 78 and 94 identified the sculpture with a version of the *Sleeping Ariadne* in the Vatican. The inventory drawn up for Amalia van Solms between 1654 and 1668 lists two sculptures of a *Cleopatra*, however, one reclining, one sitting. This earlier identification, therefore, is not that certain ; see p 58. I would like to thank Professor J.G. van Gelder for this reference.

³ Ridolfi, *op.cit.*, I, pp. 3–6.

⁴ The only other contemporary Italian source that mentioned the brothers in this context was Carlo Cesare Malvasia, who stated that they owned a large number of Carracci drawings (*Felsina Pittrice ...*, I, Bologna, 1678, p. 467).

⁵ Engraving, 194 × 142 mm. Signed in the plate at bottom left : *Eq. Car. Rodulphivs inu.* ; inscribed at bottom center : *Jacob. Picinus sculpsit Venet.*

⁶ J.E. Elias, *De Vroedschap van Amsterdam, 1578–1795*, II, Haarlem, 1905, p. 1098, App. c, no. 153 ; illustrated in color in M.A. van Rhede van der Kloot, *De Gouverneurs-Generaal van Nederlandsch-Indië 1610–1688*, 's-Gravenhage, 1891, p. 297, and plate.

⁷ Cesare Ripa, *Iconologia*, Rome, 1603, p. 218.

also was the one who acquired the Vendramin collection, at great expense according to Van Buchell, and had it shipped to Amsterdam sometime before 1639, with the permission of the Republic of Venice. Gerard, as Jan's universal heir, inherited his brother's collection in 1646 and had it moved to Amsterdam.[8] What traditionally is called 'Reynst collection', therefore, is the collection of Gerard Reynst in Amsterdam that included the works of art assembled by Jan as well. The 'cabinet' of Reynst was sold in Amsterdam in late May, 1670.[9] Documents that would firmly establish its contents are not known. The reconstruction, therefore, had to be based on a number of sources such as descriptions from visitors and the two volumes of prints after paintings and sculptures owned by Gerard Reynst. A first effort at gathering this vastly scattered material was made by Dr. Elisabeth Reynst,[10] whose husband, François Henri Reynst (1909–1958), was a distant relative of the two collectors.

Since the formation of the Reynst collection is closely linked to the lives of the two brothers Gerard (1599–1658) and Jan (1601–1646), the older one living in Amsterdam, the younger one in Venice, a brief history of the Reynst family is to follow.

---

[8] See p. 35.

[9] See pp. 90–91.

[10] See beginning of note 1.

II. THE REYNST FAMILY

1. ANCESTORS

The ancestors of Gerard and Jan Reynst can be traced back to *Reynst Jacobz* who was a bargemaster of a boat sailing the Dutch waters. In 1536, he was recorded living on the Dam in Amsterdam.[1] One of his sons, *Pieter Reynst*, born ca. 1510, was a merchant in ashes for the soap industry. He lived on the Nieuwe Brugsteeg in the *Lybaert* (now no. 11), with the sign of the 'Three Watering Cans' (*Drie Gieters*).[2]

2. GERRIT REYNST AND MARGARETHA NICQUET

The most prominent of Pieter Reynst's seven children was *Gerrit (Gerard) Reynst*, born during the second half of the sixteenth century. In 1604, he was living on the Singel in Amsterdam, in a large house called the 'Flower Pot' (*Bloempot*),[3] actually Singel number 172. Reynst sold the house in the summer of 1612.[4]

As a successful merchant and shipowner, Reynst took a most active part in the opening of trade routes with the East as far as China. In 1599, he was one of the co-founders and directors of the New or Brabant Company (*Nieuwe/Brabantsche Compagnie*) which was transformed in 1600 into the United Company of Amsterdam (*Vereenigde Compagnie van Amsterdam*) and in 1602 expanded into the Dutch East India Company (*Vereenigde Nederlandsche Geoctroyeerde Oostindische Compagnie*),[5] the most important of a number of such trade organisations founded around the turn of the century. Gerrit Reynst's share in it amounted to 12,000 guilders.[6]

---

[1] J.E. Elias, *De Vroedschap van Amsterdam, 1578–1795*, I, Haarlem, 1903, p. 372, under no. 122.

[2] *Ibidem*, p. 327, under no. 122 c. The actual house number was established by S.A.C. Dudok van Heel (Kohier 1569, oude zijde, fol. 3). He also found that Pieter Reynst owned at least three additional houses in Amsterdam, one Nieuwebrugsteeg 9 which he rented for £ 115 (Kohier 1569, o.z., fol. 3), another one on the Nieuwendijk which was rented for £ 32 to fugitives (Kohier 1562, p. 332) and one on the Westside of the O.Z. Achterburgwal which he rented for £ 40 (Kohier 1562, p. 250).

[3] Elias, *op.cit.*, I, p. 373, no. 122 e.

[4] *Kwijtschelding* no. 36, p. 137 v of 27 July 1612. I would like to thank S.A.C. Dudok van Heel for this information.

[5] F.W. Stapel, *De Gouverneurs-Generaal van Nederlandsch-Indië in beeld en woord*, The Hague, 1941, no. II.

[6] J.G. van Dillen, *Het oudste aandeelhoudersregister van de Kamer Amsterdam der Oost-Indische Compagnie*, 's-Gravenhage, 1958, p. 115. Kindly pointed out by S.A.C. Dudok van Heel.

On March 20, 1602, this East India Company received the exclusive rights to trade with the Far East.[7] The success and expansion of commerce in the islands of the East Indies was closely linked to the flourishing of the city of Amsterdam. In order to raise the capital to finance these long expeditions, the company was issuing shares at various prices to distribute the losses as well as the gains of these often hazardous voyages around the Cape among the shareholders. The company was well established by 1627, primarily thanks to the efforts of its fourth governor-general, Jan Pietersz Coen (1587–1629).[8] This successful expansion of the trade turned the United Provinces into the richest nation of Europe with Amsterdam the money market of the world.[9]

Gerrit Reynst was one of seventeen directors (*bewindhebbers*) in charge of the East India Company and thus must have been most influential. He apparently travelled extensively, because in 1607, he visited the river Congo in Africa, together with Lucas van de Venne, and received letters from the 'Grave van Songes' who lived there.[10] On February 20, 1613, Gerrit Reynst was elected and accredited by the States General to serve as the second governor-general of the Dutch East Indies, roughly today's Indonesia.[11] As governor-general, Reynst was required to live in the Far East for at least five consecutive years, excluding the travelling time to

[7] Van Rhede van der Kloot, *op.cit.*, p. 1.

[8] Petrus Johannes Blok, *History of the People of the Netherlands*, New York and London, 1907, pp. 34–35.

[9] For a brief, basic survey of the commerce in Amsterdam during the 17th century see Violet Barbour, *Capitalism in Amsterdam in the 17th Century*, Ann Arbor, 1963.

[10] Res. Staten Generaal, 24 August 1607. Information kindly furnished by S.A.C. Dudok van Heel.

[11] Van Rhede van der Kloot, *op.cit.*, pp. 26–28 and Stapel, *loc.cit.* The portrait of Gerrit Reynst of ca. 1614–15 by a Dutch painter in Batavia was transferred in 1950 from Rijswijk Palace in Weltevreden on Java to the Rijksmuseum in Amsterdam, inv. no. A 3756; panel, 98.8 × 78 cm. A copy after it, inv. no. A 4526, is also there. See P.J.J. van Thiel et al., *Catalogue ...*, Amsterdam, 1976, pp. 711–12, ill. (I would like to thank Dr. van Thiel for this information). Also mentioned in J. de Loos-Haaxman, *De Landsverzameling Schilderijen in Batavia*, Leiden, 1941, pp. 19–21, 28–29, fig. 3. For Gerrit Reynst see furthermore A.J. van der Aa, *Biographisch Woordenboek der Nederlanden*, x, Haarlem, 1874, p. 92; E.W. Moes, *Iconographia Batava*, II, Amsterdam, 1905, p. 272, no. 6403; and P.C. Molhuysen and P.J. Blok, *Nieuw Nederlandsch Biografisch Woordenboek*, IV, Leiden, 1918, p. 1147.

and from the Islands. Reynst, therefore, gave the power of attorney to his brother-in-law Jacques Nicquet to assure continuous management of his affairs in Italy, Spain, Portugal, France, England and the Baltic States ('Oostland') during his absence.[12] In early June 1613, he sailed a fleet of nine ships towards the East Indies,[13] where he was to protect and secure the trade in spices, especially cloves, against strong English competition. Reynst did not arrive in Bantam (Western Java) until late in 1614. His tenure as governor general was cut short by his death in Djakarta on December 7, 1615.[14]

Gerrit Reynst was married to *Margaretha Nicquet*, who died at childbirth and was buried in the Nieuwe Kerk on October 23, 1603. Therefore, she did not accompany her husband to Indonesia, as was stated in the earlier literature.[15] Margaretha was the daughter of the well-to-do merchant *Jan Nicquet* (1539–1608), who had immigrated from Antwerp to Amsterdam during the latter part of the sixteenth century.[16]

The likeness of Jan Nicquet is known from a drawing by Hendrick Goltzius, made in preparation for the engraving of 1595 (B. 177; H. 202).[17] Nicquet was one of the Dutch merchants with very close ties to Venice. Jan was also a well-known collector.

---

[12] Not. Publ. J.F.Bruyningh, N.A.A. no. 131, pp. 115–17, dated 30 May 1613. I owe this reference to S.A.C.Dudok van Heel.

[13] F.W. Stapel, *Pieter van Dam, Beschryvinghe van de Oostindische Compagnie*, book I, part II (*Rijks Geschiedkundige Publicatiën*, 68), 's-Gravenhage, 1929, p. 523.

[14] Van Rhede van der Kloot, *loc.cit.*; Elias, *loc.cit.*; W.Wijnandts van Resandt, 'Oude Indische Families, v, Het Geslacht Reijnst, (In Indië van ± 1755 tot heden),' *Maandblad van het genealogisch-heraldiek genootschap, De Nederlandsche Leeuw*, XXV, 1907, column 139.

[15] Buried on the 'hoogkoor'. Recorded living in the 'Nieuwstad'. D.T.B. no. 1053, p. 11. Information kindly provided by S.A.C. Dudok van Heel. Elias, *loc.cit.*, Van Rhede van der Kloot, *loc.cit*, and (B. van Treslong Prins ?), 'Aanvullingen en aanteekeningen uit het Landsarchief te Batavia,' *De Navorscher*, 1929, p. 85, thought that Margaretha died in Djakarta sometime before 1619, or in 1616 respectively. This was questioned by W. Ph. Coolhaas, 'Aanvullingen en verbeteringen op Van Rhede van der Kloot's Gouverneurs-Generaal en Commissarissen-Generaal van Nederlandsch-Indië (1610–1888),' *De Nederlandsche Leeuw*, LXXIII, 1956, columns 339–40, where he pointed out that no archival material on Margaretha Nicquet was found in Djakarta (formerly Batavia), and that she, therefore, probably stayed behind.

[16] See Elias, *op.cit.*, I, pp. 307–08, under no. 97.

[17] E.K.J. Reznicek, *Hendrick Goltzius, Zeichnungen*, I, Utrecht, 1961, pp. 96, 364–65; fig. A. 246. The drawing is in the Rijksprentenkabinet, Amsterdam (Inv.no. 1884-A 335; graphite on ivory-colored prepared board, 96×78 mm.). For a reproduction of the engraving see *Goltzius en zijn school* (exh. cat.), Rotterdam (Museum Boymans-van Beuningen), 1972, no. 59.

The inventory drawn up after the death of his wife, *Margaretha Bosmans*, in December, 1612, lists sixty paintings by Dutch and some Flemish artists, among them five by Carel van Mander, five hundred and fifty prints, a copper statue of Venus and Cupid, a figure in marble and another one in copper, tapestries, and some books, among them on Roman antiquities.[18] Neither the paintings nor the sculptures figure in the collections of the grandsons, Gerard and Jan Reynst.

Nicquet's son Jacques is more important for the Reynst brothers in this respect. Jacques not only designated Gerard Reynst one of his principal heirs but specifically indicated a number of books containing art ('kunstboecken') should go to his nephews, Gerard and Jan. According to the inventory, drawn up in 1642, Jacques owned some thirty-seven paintings. Most of them were landscapes or still lifes and not one of them was listed with an artist's name. None could be traced to the collections of Gerard and Jan either. Nicquet also owned several books of prints, among them by Dürer, Aldegrever, Schongauer, Sadeler, an illustrated bible and Heijns' *Mirror of the World*. Furthermore, he had a number of books on Roman antiquities. Jacques Nicquet's library also included works by Italian, French, and Dutch authors.

The copper plate by Goltzius with prints probably referred to Goltzius' portrait of Jan Nicquet.[19]

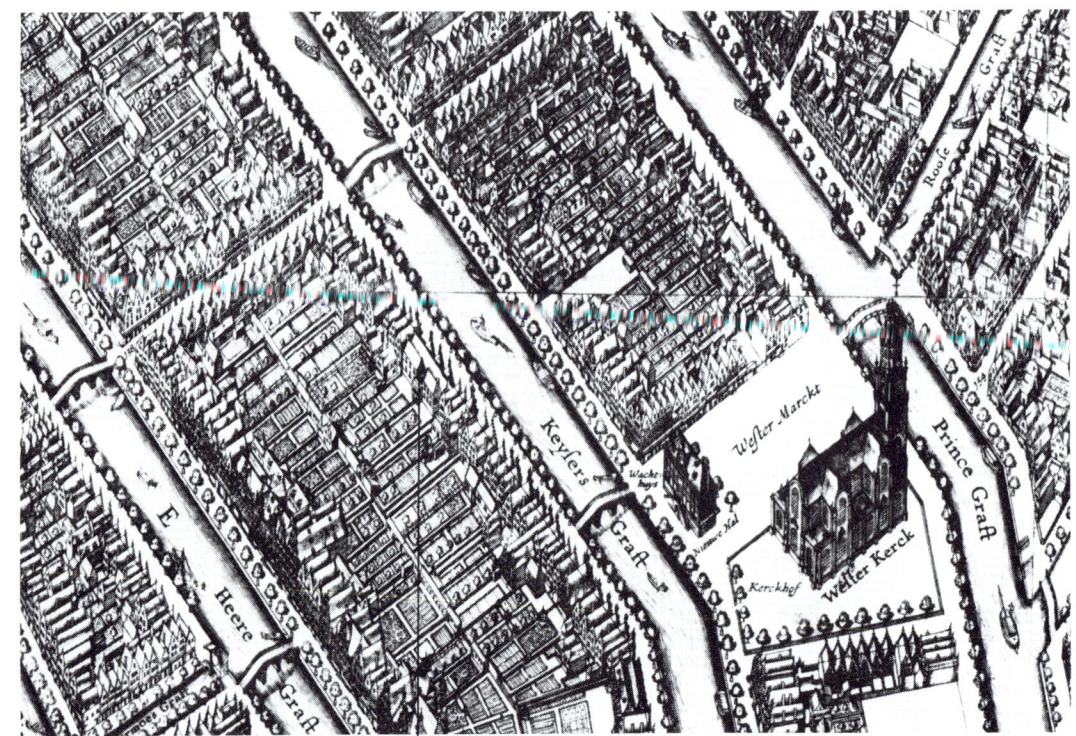

1

The house De Hoop on the Keizersgracht no. 209

Mr. S.A.C. Dudok van Heel not only brought this inventory to my attention but he also transcribed it, since the original can no longer be consulted, and allowed me to include it in the present study for which I am most grateful (see pp. 248–268).

Both the father Jan Nicquet as well as his son Jacques seem to have set an example for Gerard and Jan Reynst as collectors. *Jacques Nicquet*[20] in particular needs to be discussed briefly, since he played a rather dominant role in the lives of his two nephews. Jacques was born in Antwerp, in ca. 1573. He lived in Venice by 1595[21] and was negotiating in March, 1601, with the Venetian government about the restitution of sequestered Dutch ships and goods.[22] By 1603, Jacques was in Amsterdam, where he married Clara de Haze.[23] As mentioned earlier, Jacques Nicquet received the power of attorney from Gerrit Reynst in 1613, when the latter left for Indonesia. Jacques was also designated first guardian of Gerrit's children.[24] In 1620, Nicquet was re-married to Geertruyd Jacobs Hinlopen, a member of the Hinlopen family of collectors.[25] Five years later, Jacques signed a contract on behalf of Jan (Giovanni) Reynst who was referred to as 'tot Venetien', or 'in Venice'. Jacques Nicquet is mentioned again in 1631, when he accompanied Gerrit's eldest son, Gerard Reynst, for the latter's publication of his bann to marry Anna Schuyt. Nicquet was buried

---

[18] Not. Publ. J.F. Bruyningh, N.A.A. no. 197, pp. 436-62 ; December 14, 1612–January 19, 1613. The inventory also lists several very expensive diamonds. Published in parts by A. Bredius, *Künstler-Inventare*, II, The Hague, 1916, pp. 394–97.

[19] Not. Publ. B. Baddel, N.A.A. no. 957, omslag C ; March 29–June 23, 1642.

[20] Not listed in Elias, *op.cit*. All the relevant information on Jacques Nicquet was provided by S.A.C. Dudok van Heel.

[21] Wilfrid Brulez, *Marchands flamands à Venise (1568–1605)*, I, Brussels/Rome, 1965, p. 202, no. 597. Brulez published a large number of other documents concerning especially Jacques, but also Jan Nicquet. See also K. Heeringa, *Bronnen tot de geschiedenis van den Levantschen handel* (*Rijks Geschiedkundige Publicatiën*, 9, 10), 's-Gravenhage, 1910, pp. 33, 42, 67, 803 for additional material. A good survey on Holland's relationship with Venice was given by J. C. de Jonge, *Nederland en Venetie*, 's Gravenhage, 1852, especially pp. 283–311 with regard to the trade. See also J. H. Kernkamp, "Scheepvaart- en Handelsbetrekkingen met Italië tijdens de opkomst der Republiek," *Nederlandsch Historisch Instituut te Rome, Mededeelingen*, II, series, VI, 1936, pp. 53–85.

[22] P. J. Blok, *Relazioni Veneziane, Venetiaansche Berichten over de Vereenigde Nederlanden van 1600–1795* (*Rijks Geschiedkundige Publicatiën*, 7), 's-Gravenhage, 1909, pp. 42–43, no. 22 : letter from Jacques Nichetti in Venice to his father Giovanni Nichetti in Amsterdam, 16 March 1601 (R. A., Arch. Holland 2634 d). We know from a document that the 'natione fiamenga' comprised in 1596 at least twenty-one Flemish merchants residing in Venice, among them Jan and Jacques Nicquet (see Brulez, *op.cit*., p. XIX, pl. 1). These Flemish merchants worked in close cooperation with the Dutch and interested them more and more in trading with the

in the Nieuwe Kerk in Amsterdam on March 28, 1642 ('koor'). At that time he was living on the Keizersgracht, in the house *De Oranjeappel*, today's number 345, which he rented. His brother-in-law Samuel Bloemart rented the house next door, number 343. Both residences were situated opposite the Amsterdam theater.[26] Jacques was also acquainted with Constantijn Huygens, who wrote to his parents from Venice in 1620 that the introductions received from Nicquet were most helpful.[27]

Nothing specific is known about Gerrit Reynst's possessions, but from the little we do know, it appears that he was interested in collecting, because his name is mentioned twice in connection with the sale of an art object. Thus, on August 30, 1612, his son-in-law Samuel Bloemaert acquired for him from the estate of Claes Rauwart a painting representing *Holofernes*, for the sum of twenty guilders.[28] The second reference has to do with a special gift of about 20,000 guilders on behalf of the States General to Sultan Ahmed I, in appreciation of a favorable trade treaty with Turkey, signed on July 6, 1612. Among the many objects included in this gift was an enameled harness which was bought from Gerrit Reynst for the sum of two hundred guilders.[29] This was the first known instance that a special item was purchased from the Reynst family to be presented officially to a head of state.

Mediterranean region. Much of the Dutch trade initially was financed by capital from Flemish immigrants (Brulez, *op.cit.*, p. xv).

[23] Born Antwerp, ca. 1585; publication of the banns on July 4, 1603 (D.T.B. no. 410, p. 484). Clara de Haze was buried in the Nieuwe Kerk, Amsterdam on May 8, 1618 ('hoogkoor'). She was living in the Oude Hoogstraat. Clara de Haze was not included either by Elias, *loc.cit.* She was the sister of Constantia de Haze, who was married to Jacques' brother Jan (see also footnote 34), and of Sara de Haze, married to Jan's business partner, Paulo de Wilhem (see footnote 81). For the De Haze family in general see Elias, *op.cit.*, II, pp. 599–605, no.236.

[24] Not. Publ. J. F. Bruyningh, N.A.A. no. 131, p. 115.

[25] Born ca. 1587; publication of the banns on February 7, 1620 (D.T.B. no. 424, p. 142). Geertruyd Jacobs Hinlopen was buried in the Nieuwe Kerk, 'koor', on July 12, 1622. She lived on the Herengracht. For the Hinlopen family as collectors see S.A.C. Dudok van Heel, in *Maandblad...Amstelodamum*, 56, 1969, pp. 233–37. See also Elias, *op.cit.*, I, p. 309.

[26] Thes. extr. ord. no. 284, p. 336 v (1647).

[27] Worp, *op.cit.*, I, pp. 47–50, no. 83. Letter of June 18, 1620.

[28] Bredius, *op.cit.*, V, The Hague, 1918, p. 1745.

[29] N. de Roever, 'Een vorstelijk geschenk,' *Oud Holland*, I, 1883, pp. 185–86. Heeringa, *op.cit.*, pp. 264–70 and especially p. 270 reads the name as *Gerrit Kemp* (?) in which case the identification with Reynst would be mistaken.

3. GERARD REYNST AND ANNA SCHUYT

Gerrit Reynst and Margaretha Nicquet had seven surviving children, three sons and four daughters, born between 1589 and 1603.[30] Their eldest son Pieter perished at sea on November 25, 1615, on his way to Java.[31] The two surviving sons, Gerard (Gerrit) and Jan Reynst continued the family tradition and became successful merchants, one residing in Amsterdam, the other one in Venice. *Gerard (Gerrit) Reynst* was born in Amsterdam, in 1599.[32] Nothing is known about his youth. He may possibly have travelled to Italy in the company of the Dutch painter Pieter van Laer (ca. 1592–1642) which would have been sometime before 1638, the year Van Laer returned to Haarlem for good.[33] On May 13, 1631, Gerard Reynst married Anna Schuyt, the daughter of Albert Gijsbertsz Schuyt and Anna Bernard.[34] Schuyt was an underwriter and merchant in Amsterdam, trading with Italy and the Levant, who had good connections with Venice.

On April 4, 1634, Gerard and Jan Reynst purchased together a house on the Keizersgracht, today number 209, called *De Hoop* (Hope).[35] This date probably marked a starting point and one may assume that both began collecting seriously around that time, installing the recently acquired works of art in their new residence. The house, as it appears today (text fig. 3) dates back only to 1738, when much of the original structure was altered. Initially, the *De*

---

[30] Their correct sequence is found in the last will of Gerrit Reynst (Not. Publ. J.F. Bruyningh, N.A.A. no. 181, pp. 110v–112; dated June 1, 1613), where they were designated his heirs: *Catharina*, baptized Oude Kerk, December 10, 1589; *Jan*, baptized Nieuwe Kerk, July 14, 1591 (died young); *Pieter*, baptized Oude Kerk, September 4, 1592; *Margryt*, baptized Oude Kerk, October 24, 1593 (died young); *Margaretha*, baptized Oude Kerk, October 7, 1598; *Gerard*, no baptismal records found, born in 1599; *Wijntje*, no baptismal records found, born in 1600. In April, 1620 she stated that she was 20 years old. *Jan*, baptized Oude Kerk, October 26, 1601; *Constantia*, baptized Nieuwe Kerk, October 21, 1603. (Her mother, Margaretha Nicquet, was buried there two days later). Two children of Gerrit Reynst and Margaretha Nicquet were buried in the Nieuwe Kerk on December 21, 1594 and January 4, 1595. The above information was kindly provided by S.A.C. Dudok van Heel.

[31] Wijnandts van Resandt, *op.cit.*, column 139.

[32] Elias, *op.cit.*, I, p. 447, no. 153; Moes, *op.cit.*, p. 272, no. 6404. Jan Reynst calls him 'Gerard' in a letter of 1646, and this spelling, therefore, is preferred. S.A.C. Dudok van Heel suggested that since Gerard and Wijntje apparently were born within a very short time span, the two may actually have been twins.

[33] Kramm, *op.cit.*, p. 1365. Axel Janeck does not mention Kramm's statement in his recent dissertation on Van Laer (1968).

[34] Elias, *op.cit.*, II, p. 1017, no. 451. After the death of Anna Bernard in 1604, Schuyt (1576–1632) remarried in 1605 Constantia de Haze, the widow of Jan Nicquet (1565–99). Jan was the brother of Margaretha Nicquet, thus the uncle of Gerard and Jan Reynst. In 1603, Jacques Nicquet, Jan's younger brother, had married Constantia's sister, Clara de Haze (see footnote 23). This shows, how the families of merchants trading with Italy were also closely related through marriage.

Albert Schuyt belonged to the same group of merchants as Jacques Nicquet and Paulo de Wilhem,

Hoop looked like Keizersgracht no. 213. (A suggested reconstruction is given in text fig. 2, top).

The Keizersgracht was believed to be one of the most beautiful sights in Amsterdam. Many successful and rich merchants lived along this canal and their houses were renowned for their opulent interiors.[36] Queen Marie de Medicis was brought especially to the Keizersgracht during her visit to Amsterdam in September, 1638, and her amazement at the sight was described extensively by Caspar van Baerle.[37] The history of the house *De Hoop* up to the present was investigated by B. Bijtelaar and published in a series of articles,[38] from where the following brief summary is culled, concentrating primarily on the time span when the house was in the Reynst family, i.e. the years 1634–86.

Keizersgracht 209 is situated on the eastern side of the canal, close to the Hartenstraat, a section that – up to the Leidse gracht – was added to the city of Amsterdam during its third expansion in 1611 (text fig. 1).[39] Between July 11 and August 22, 1619, Barent van Hoorn had bought three lots along that section of the gracht totaling 88 feet in width and 170 feet in length,[40] on which he built three houses, today's numbers 209 which he named *De Hoop*, number 211, called *De Liefde* (Love) and number 213, possibly once called *Het Geloof* (Faith).[41] On August 7, 1622, Van Hoorn's

Jan Reynst's business partner, because their respective signatures are found in the same document of 9 November 1618 (see Heeringa, *op.cit.*, I, p. 67, no. 50).

[35] *Kwijtscheldingen* 3C, fol 438. The purchase of the house by Gerard and Jan Reynst and the specific date are mentioned only indirectly and much later in an annotation on the bill of sales written on August 27, 1686, when the house *De Hoop* was sold to Joan Graefland. The annotation reads as follows: *...De brieve in dezen gemelt is den 4 april 1634 ten behoeve van Gerard en Joan Reynst verleden*; see B. Bijtelaar, 'Het Huis Keizersgracht 209,' *Ons Amsterdam*, XIII, no. 1, January 1961, p. 9 for this reference. The *kwijtscheldingen* for the year 1634 are no longer extant. The statement by A.E. d'Ailly, in *Zeven Eeuwen Amsterdam*, III, Amsterdam (s.a.), p. 217, that the house was built by Gerard Reynst needs to be corrected.

[36] Frits Lugt, *Mit Rembrandt in Amsterdam*, Berlin, 1920, pp. 59–70.

[37] Kasper van Baerle, *Blyde Inkomst der allerdoorluchtigste Koninginne, Maria de Medicis, t'Amsterdam*, Amsterdam, 1639, pp. 43–44: 'Zy stond verbaest toenze de Keizersgraft zagh, wiens weerga, zoo men vreemdelingen geloven magt, in geheel Europe niet te vinden is; het zy datze haar oogen liet weiden langs de lange streeck der huizen, of het gezicht sloegh op de gebouwen, tot pracht, schoonheid, cieraed, en gerief der inwoonderen gebouwt;...'

[38] B. Bijtelaar, 'Het Huis Keizersgracht 209,' *Ons Amsterdam*, XIII, no. 1, January 1961, pp. 6–11; no.2, February 1961, pp. 52–57; no.4, April 1961, pp. 104–11, with many illustrations.

[39] Jan Wagenaar, *Amsterdam in zijne opkomst...*, I, Amsterdam, 1760, pp. 46–48, especially p. 47: 'Door deeze uitlegging, werdt de stad aan de Westzyde vermeerderd met drie aanzienlyke graften, de Heeren-graft, de Keizers-graft en de Prinsen-graft, die van't Noorden naar 't zuiden liepen, en omtrent de plaats der Leidsche-graft eindigden.'

[40] *Kwijtscheldingen* 40, fol. 36 and 57; see Bijtelaar, *op.cit.*, p. 6 and notes 4, 5.

[41] Bijtelaar, *op.cit.*, p. 105.

2

3

widow sold the *De Hoop* to Henrick Dirck Boelensz (1586–1638),[42] who in turn resold it to Gerard and Jan Reynst in 1634.[43] The house was still singled out in 1662, when Melchior Fokkens described the sights of Amsterdam, calling it 'one of the most renowned, with many beautiful paintings and valuable objects gathered abroad, estimated to be worth about three tons of gold or more rather than less'.[44] In 1664, both Fokkens and Filip von Zesen praised the Keizersgracht as a most beautiful gracht in Amsterdam, where rich and powerful merchants lived and where the houses resembled royal palaces rather than merchants' residences.[45] The earliest description of works of art assembled by Gerard Reynst in his house on the Keizersgracht dates from 1639. On September 4, the lawyer and antiquary Aernout van Buchell (1565–1641) visited the collection and noted his impressions in his *Notae Quotidianae* which are preserved in the library of the University of Utrecht.[46] Since Van Buchell's remarks are of great significance for our understanding of the formation of the collection, they will be discussed *in extenso* in the following chapter. Gerard Reynst also owned a country house in Huizen (Huyzen) near Naarden to escape the city life which was mentioned in a poem from c. 1657 by Jan Vos on the homestead *Kommerrust* that belonged at that time to Jan Uittenbogaert.[46a] In this poem the goddess Minerva visits first *Kommerrust* and then

---

[42] *Kwijtscheldingen* 29, fol. 200; see Bijtelaar, *op.cit.*, p. 6 and note 7.

[43] See footnote 35 above.

[44] Melchior Fokkens, *Beschrijvingh der wijdtvermaarde koopstadt Amstelredam*, Amsterdam, 1662, p. 71: 'Dus gaat men de Ree-straat voorby, hier staat het Kostelyke Huys van de Heer C. (sic) Reynst zaliger, aan de Oost-zyde, dit is een van de vermaardste Huyzen van Amsterdam, aangaande de kostelyke schilderijen, en andere uytheemsch vergezochte kostelijkheydt, geschat deze vreemde rariteyten wel op drie tonnen gouts, of meerder en niet min.' A 'ton of gold' was about equivalent to 100,000 guilders.

[45] Fokkens, *op.cit.* second edition, p.395; F. von Zesen, *Beschreibung der Stadt Amsterdam*, Amsterdam, 1664, p. 357.

[46] Ms. 1827, *Notae Quotidianae* 1634–1641; published in parts and edited by Hoogewerff and Van Regteren Altena; additional material by J.W.C. van Campen, see p. 55, footnotes 44–45.

[46a] Jan Vos, *Alle Gedichten*, Amsterdam, 1726 (second edition), p. 233. The country house later belonged to Lucretia van Hoorn, who was the wife of Uittenbogaert and the sister of Simon van Hoorn who was well acquainted with Gerard Reynst. P.A.F. van Veen, *De soeticheydt des buytenlevens, vergheselschapt met de boucken*, (Diss. Leiden, The Hague 1960, pp. 35–36) dates *Kommerrust* between 1656 and 1658. I would like to thank Professor Van Gelder and S.A.C. Dudok van Heel for referring me to this poem by Jan Vos.

[47] Hans Bontemantel, *De Regeeringe van Amsterdam soo in't civiel als crimineel en militaire (1653–1672)*, edited by G.W. Kernkamp (*Werken uitgegeven door het historisch genootschap*, III, nos. 7, 8), 's-Gravenhage, 1897.

continues to see the houses of Cornelis de Graeff, Hinloopen, Grouwels, the widow Rensselaer and Reynst. The relevant passage reads as follows:

Nu quam zy met haar stoet in 't hof van Reinst te daalen :
Hij laat om deeze plaats zyn huis vol ruime zaalen,
Daar al de geesten van oudt Roomen en Atheen.
Haar kunst in zilver, goudt, kooraalen, marmersteen,
Yvoor, en andre stof, op't heerelijkst vertoogen.
Hy kan veel meer in 't loof dan in zyn kunst beoogen.
De wondren van 't gewasch verdooven alle kunst.

(Now Minerva came with her retinue and descended into Reynst's country estate. In favor of this place he left his house full of spacious rooms, where all the minds of classical Rome and Athens display most beautifully their art in silver, gold, coral, marble, ivory and other material. Here in nature he can keep his eyes on much more than art, for the wonders of nature numb all art).

From about 1646 on, Gerard Reynst took active part in the community life of Amsterdam. Most of the references to his public offices are found in Hans Bontemantel's account and notes of meetings and proceedings of the Amsterdam town council, published by Kernkamp,[47] and in Jan Wagenaar's description of

Amsterdam.[48] The existing documents, however, furnish little about Reynst's personality. Gerard Reynst was 'master of law' (the oldest reference to this dates from 1639). If he was *Jus Utriusque Doctor* it was rather normal to call him *Dr Gerret Reynst*, as some documents did.[49] Furthermore, he asked Van Baerle, the eminent professor of philosophy at the Amsterdam Athenaeum, to write a eulogy for his deceased brother Jan which indicates that Van Baerle was acquainted with the two brothers.[50]

Gerard must have been a Calvinist, since he voted with the Calvinist factions in the government.[51] Both he and Jan Reynst also carried the title of *Heer van Niel* which they had inherited from the husband of their sister Constantia, Joan Carlo Smissaert.[52] With Smissaert's death in 1644, his title first passed on to Jan Reynst and after the latter's death two years later, Gerard Reynst inherited it. This may be deduced from the fact that Ridolfi referred to Gerard Reynst in 1646 as 'senator' of Amsterdam, while Jan Reynst was called 'sign. di Niel'.[53]

In 1646, Gerard Reynst was listed for the first time as councillor (*raad*) in the town council of the city of Amsterdam (*vroedschap*), a post he held until his death in 1658.[54] As such he was one of thirty-six councillors who, together with a sheriff, four burgomasters and nine magistrates, were in charge of governing the city

5

[48] Jan Wagenaar, *Amsterdam* in *zyne opkomst*, 3 vols., Amsterdam, 1760–67.

[49] Kernkamp, *op.cit.*, II, p. 105; Resol. Raad II, 1, p. 57. According to S.A.C. Dudok van Heel, who provided the above facts on Reynst's background as a lawyer, Reynst apparently studied abroad (Italy? France?), since his name could not be found in any of the registers of Dutch universities.

[50] Letter by Van Baerle to Huygens, dated September 27, 1646; published by Worp, *op.cit.*, IV, pp. 353–54, no. 4461.

[51] On September 29, 1654, Gerard Reynst was among the magistrates who had assembled to elect a new

of Amsterdam. In 1650, and again in 1654, Reynst was elected magistrate (*schepen*),[55] and was in charge more specifically of the criminal and civil affairs within the city. During 1651, 1652, and 1654, he was one of five commissioners in charge of marital affairs (*commissaris van huwelijksche zaken*)[56] which also included questions about relationships between masters and servants and settlement of quarrels between these parties.

Gerard Reynst was a most successful, wealthy merchant who, for example, cleared 650,400 guilders through the city exchange in 1645.[57] His interest and close affiliation with the exchange bank (*wisselbank*) is further documented for the years 1653, 1655 until 1658, when he was one of its commissioners.[58] This city exchange bank was founded in January, 1609, as a direct result of the flourishing trade with foreign nations in order to convert foreign currencies. The exchange bank was governed by three elected commissioners who were either former magistrates or councillors. They had to be well versed in the trade and dealt primarily with the bills of exchange. Only the commissioners were allowed to deal in foreign coins and they established the current rates of exchange. Twice a year the clients of the city exchange had to check with a commissioner about their pending accounts or be fined twenty guilders.[59] The commissioners pledged to accept, safeguard and redeem the money of the city and its citizenry.

---

burgomaster in place of the deceased Cornelis Bicker. The election ended in a stalemate because the liberals supported Albert Dircksz. Pater while the Calvinists, among them Reynst, strongly favored Hendrick Dircksz Spiegel. A compromise was worked out only on the following day and Pater's nomination was confirmed; Kernkamp, *op.cit.*, I, pp. 93–94 and II, p. 111; Elias, *op.cit.*, I, pp. CIX–CX.

[52] Elias, *op.cit.*, I, p. 375, under Bijl. 2.

[53] First pointed out by Jacobs, *op.cit.*, p. 24, note 3.

[54] Wagenaar, *op.cit.*, III, p. 358, under 1646.
For a brief survey of the structure of the Amsterdam city council and the political functions of the individual members, see Peter Burke, *Venice and Amsterdam*, London, 1974, especially pp. 20–21, 40–44.

[55] Wagenaar, *op.cit.*, III, p. 342, under 1650 and 1654.

[56] Elias, *op.cit.*, II, p. 1339 under Gerrit Reynst. Further references to Gerard Reynst are found in Bontemantel under the following dates: 1649, lieutenant in the Kloveniersdoelen (Kernkamp, I, p. 198); 1652 and 1653, one of the governing members of the Kloveniersdoelen (Kernkamp, I, p. 175); 28 January 1653, listed as councillor (Kernkamp, II, p. 105); and on nominating list for new magistrates but not elected (Kernkamp, II, p. 108); 2 February 1653, listed as former magistrate (Kernkamp, II, p. 103); 10 September 1653, listed as absent during a meeting of the city council debating punishment of Franchoys Geesdorp (Kernkamp, I, p. 277); 28 January 1654, on nominating list for new magistrates and elected (Kernkamp, II, p. 110); 1 February 1656, present in meeting, listed among former magistrates (*oud-schepen*) (Kernkamp, II, p. 118).

[57] Wisselbank, Reg. 7 and 8; see Elias, *op.cit.*, I, p. 447, no. 153 and note c.

[58] Wagenaar, *op.cit.*, III, pp. 403–4, under 1653, 1655–1657, 1658, until his death on June, 29.

[59] Wagenaar, *op.cit.*, II, pp. 536–42.

With the growth of Amsterdam as a shipping center and a commodity market the exchange bank became more and more important and was for a long time the public bank in Northern Europe.

Gerard Reynst was also a member of the civic guard (*burgervendel*) which by the seventeenth century served primarily social rather than military purposes. In 1646 and 1649, he was lieutenant in the company of precinct 20, and by 1650, he had advanced to captain in precinct 41,[60] a prerequisite for a promotion to governing member of a guild. Reynst indeed was elected in 1652 and 1653,[61] to serve as one of the governors (*overlieden*) of the Kloveniersdoelen (arquebusiers). As such he was represented in Van der Helst's group portrait, dated 1655.[62] Reynst is sitting in front of the table, looking at the viewer and holding a glass of wine in his right hand. The other governors assembled for a meal of oysters are from left to right: Cornelis Witsen and Roelof Bicker, with Simon van Hoorn seated at the end of the table at the right (text fig. 4).[63]

Three years later, on June 29, 1658, Gerard Reynst fell into the Keizersgracht and drowned.[64] He was buried on July 4, in the Westerkerk in Amsterdam, not far from the house in which he had lived for the past twenty years. The four friends who escorted him to the place of burial were Hendrick Dircksz Spiegel, Joores

---

[60] J.A. Jochems, *Amsterdams oude Burgervendels (schutterij), 1580–1795*, Amsterdam, 1888, pp. 11, 40, 67; Kernkamp, *op.cit.*, I, p. 198. From 1620–50, the city of Amsterdam was divided into twenty precincts (*wijken*), after that date they were expanded to fifty-four. Each precinct had a company headed by a captain who was assisted by a lieutenant and three sergeants. Captains and lieutenants always were elected from members of government. See Kernkamp, *op.cit.*, I, pp. 194, 198, 200.

[61] Kernkamp, *op.cit.*, I, p. 175.

[62] J.J. de Gelder, *Bartholomeus van der Helst*, Rotterdam, 1921, pp. 101–02, 236–37, cat. no. 840, pl. XXIII. The second figure, identified by De Gelder (and Stricker, *op.cit.*, p. 257, plate) with Gerard Reynst, actually is Roelof Bicker, as may be seen in a comparison with Van der Helst's painting of the *Corporalship of Captain Roelof Bicker and Lieutenant Jan Michielsz Blaeuw* of 1643 (Amsterdam, Rijksmuseum, Inv. no. 1134; De Gelder, *op.cit.*, cat. no. 835, pl. VI).

[63] The order given by Bontemantel in listing the four governing members seems to be correct; see Kernkamp, *loc.cit.* This sequence was also taken over in the *Catalogue des tableaux, miniatures, pastels, dessins, encadres, etc. du Musée de l'Etat à Amsterdam*, Amsterdam, 1904, no. 1137, and by Bijtelaar, *op.cit.*, p. 7. Miss Bijtelaar (letter of 12 May 1974) pointed out further that Cornelis Witsen can be identified through Van der Helst's painting of the *Civic Guard under Captain Cornelis Witsen, Celebrating the Peace of Münster in 1648* (Amsterdam, Rijksmuseum, Inv. no. 1135; De Gelder, *op.cit.*, cat. no. 836, pl. XIV), and that the person at the very right must be the 37 year old Simon van Hoorn. I should like to thank Miss Bijtelaar for her kind assistance.

[64] Kernkamp, *op.cit.* II, p. 105: *Den vierden Julij is begraeven Dr Gerret Rynst, Raed en Outschepen. Den Heer Rynst viel op de Kysersgraft in het waeter en verdronck*. Elias, *op.cit.*, I, p. 447, no. 153 and note d. He was buried in tomb 123 in the center aisle.

[65] Kernkamp, *op.cit.*, II, p. 74. The same Simon van Hoorn was also represented in Van der Helst's painting. In 1660, he was one of the Dutch ambassadors who accom-

Backer, Bernard Schellinger and Simon van Hoorn.[65]

We may suppose from circumstantial evidence that Gerard Reynst's widow, Anna Schuyt, left the house on the Keizersgracht during 1668, and moved to the Nieuwe Heerengracht, corner Vijzelstraat, to live with her son, Joan Reynst.[66] She was buried in the Westerkerk on November 20, 1671.[67]

Gerard Reynst and Anna Schuyt had two married children, a son *Joan Reynst*,[68] born in Amsterdam in 1636, a daughter *Constancia Reynst*,[69] born in Amsterdam in 1638, and three unmarried sons *Gerard*, *Abraham* and *Albert*. All five children were mentioned in the protocol of notary Michiel Baers of August 7, 1673, where an agreement was reached between Pieter Schaep for Constancia Reynst and Joan Reynst on the one hand, and Gerard, Abraham and Albert on the other, on a certain amount of money Constancia and Joan had received at the time of their marriage in 1666 and 1667, respectively.[70]

Joan Reynst lived on the Herengracht, house number 498.[71]

In 1672, he bought the estate of Drakenstein near Lage Vuursche and obtained the courtesy title of Heer van Drakenstein and De Vuursche. The purchase possibly was made with funds inherited after his mother's death the year before. (The Slot Drakenstein is the official residence of Crown Princess Beatrix of the Netherlands). Houbraken[72] tells us that Joan Reynst travelled

---

panied the 'Dutch Gift' to London and described the reception at court in a letter to the States General (see footnote p. 83, 96). Furthermore, Vondel's poem *De Kunstkroon voor den koning van Groot-Britannië* was dedicated to him (see p. 76, text fig. 24).

[66] At the time of his marriage on June 27, 1667, Joan Reynst gave his address as Keizersgracht. The registers, recording the burial of Anna Schuyt, on the other hand, state that she came from the 'Nieuwe Heerengracht bij de Vijzelstraat' and 'Nieuwe Heerengraft' (D.T.B. 1101, 122). The latter document mentions her tomb as number 123 on the North side. I am most grateful to Dr. I.H. van Eeghen for sending me this information (letter of August 18, 1974).

[67] I.H. van Eeghen also kindly informed me (see preceding footnote) that Anna Schuyt's notary was I. van de Ven (N.A. 1136, 412 etc.), who died before her. The family archive Backer no. 570 includes several documents concerning the liquidation of the inheritance between the children. The one act signed by Michiel Baers is found there, dated August 7, 1673, which refers to the five children of Gerard Reynst. This same archive, furthermore, includes an account of how 133,000 guilders from the estate of Gerard Reynst and Anna Schuyt were to be divided among the children (June 1675), together with other smaller accounts. Both Joan Reynst and Pieter Schaep each received 1/5, Gerard and Abraham Reynst 3/5. 81,000 guilders worth of securities which were inherited by Gerard and Abraham Reynst, are listed as being deposited in Venice and Milan. Another list enumerates credit due to the various children amounting to over 50,000 guilders. Furthermore, there is a list mentioning the outstanding debts amounting to roughly 38,000 guilders and stocks deposited in Moscow under the name of Isaack Bernaerts. It also includes references to the ship *Aletta*, involved in an accident, and to the *Goude Leeuw*. Furthermore, we find accounts of debit and credit for Pieter Schaep for the years 1671–75 and for Joan Reynst, an account of interests received from various sources, as well as a receipt for 14,000 guilders for dealing in fruit in Italy (Milan and Venice).

to Rome accompanied by his friend and neighbor Karel Dujardin and that he went to visit other Italian cities as well, but returned home alone. This trip must have taken place in 1675. A portrait of *Joan Reynst* by Dujardin is preserved in the Rijksmuseum, Amsterdam.[73] With Joan Reynst, who was buried in the Westerkerk on May 9, 1695, the male succession of this branch of the Reynst family died out.

Constancia Reynst married Pieter Schaep on July 1, 1666, and went to live on the Southside of the Keizersgracht, in the house *Het derde vreede jaer*.[74] Pieter Schaep held among other posts the one of councillor from 1666 until 1672, of magistrate in 1664, and of captain in 1672. He was removed from office on September 10, 1672, with the advent of stadholder Willem III. Bartholomeus van der Helst [75] painted Constancia's portrait which was praised in a poem by Jan Vos, both mentioned by Houbraken.[76] Contrary to Elias, Constancia died in 1674 (not 1694) and was buried in the Oude Kerk in Amsterdam on January 5.[77] The following June, Pieter Schaep inherited from his wife, as his fifth part, the house on the Keizersgracht, valued at 27,000 guilders,[78] but he did not move there. The house *De Hoop* was sold on August 27, 1686, to Joan Graefland, magistrate of the town council, for the sum of 23,125 guilders[79] and thus left the Reynst family. None of the Reynst children continued to collect.

Albert Reynst is not mentioned, because he died in 1674 and was buried in the Wester Kerk on February 17. (Information kindly received from S.A.C. Dudok van Heel).
A family-tree from the 16th until the 18th century is included in the archive Backer no. 44.
For a list and brief description of the documents contained in the family archive Backer see I.H. van Eeghen, *Archief der Gemeente Amsterdam, Inventaris van het Familie-Archief Backer*, (Amsterdam), 1954, pp. 54, nos. 567, 570, 575.

[68] Elias, *op.cit.*, I, p. 447, under no. 153. The family archive Backer no. 575 includes a copy of the testament of Joan Reynst, Heer van Drakenstein and of Eva Hooftman, dated 29 November 1693. No works of art are mentioned.

The notary was Nicolaas Listingh and his acts date from 1653–83. The years 1667–80 which might have yielded material on the Reynst collection unfortunately are heavily damaged by fire. The same archive also includes documents concerning the processes over the inheritance of Eva de Wildt, granddaughter of Joan Reynst.

[69] Elias, *loc.cit.* and I, pp. 523–24, no. 195. The family archive Backer no. 567 contains the testament of Pieter Schaep and Constancia Reynst, dated 17 October (?) 1669. The only paintings mentioned are two portraits of ancestors. Furthermore, a golden crucifix is willed to the eldest son and there is reference to some illuminated manuscripts.

[70] N.A. 3756, Jan.-Dec. 1673,

Michiel Baers, folder B, p. 353. I should like to thank Dr. I.H. van Eeghen for this reference.

[71] Joan Reynst bought the ground in 1665. Kindly pointed out by S.A.C. Dudok van Heel.

[72] A. Houbraken, *De groote Schouburgh der Nederlantsche Konstschilders en Schilderessen*, ed. P.T.A. Swillens, III, Maastricht, 1953, pp. 46–47.

[73] *All the Paintings of the Rijksmuseum in Amsterdam*, Amsterdam, 1976, p. 204, no. A 191: canvas, 131 × 106 cm. Signed. Brochhagen, *op.cit.*, pp. 5–8, places Dujardin's second trip to Italy in the year 1675. Dujardin was a relative of Joan Reynst. The merchant, Nicolaes Du Gardyn, born in Emden in 1572, married in 1626 Maria Reynst (1588–1644), a sister of Joan's

4. JAN REYNST

Jan Reynst, Gerard's younger brother, was born in Amsterdam in 1601.[80] He is recorded living in Venice by January 20, 1625. On that day, his uncle Jacques Nicquet signed a contract in the name of *Gio Reijnst tot Venetien* that established a company, to be renewed after four years, between Paulo de Wilhem,[81] who owned the ships, and Gerard Smits on the one hand and Abraham Heijrmans and Jan Reynst on the other. De Wilhem participated with 50,000 guilders, including the 12,000 from Smits, Reynst and Heijrmans each with 25,000. The last two also were asked to reside in Venice, where the books were to be kept.[82]

An early reference to Jan is found in a letter by Cornelis Witsen, the consul at large of the United Provinces in Aleppo (Syria), to the States General of May 26, 1626, concerning a Dutch ship sent from Amsterdam to Alexandretta via Venice. The ship was consigned to Jan Reynst and made a stop in Venice, to take on additional cargo. According to Witsen, Jan Reynst saw to it that the Venetian merchants did not have to pay the customary fee to the consul in Aleppo. (A settlement between Witsen and the States General was arranged sometime later).[83] This incident shows that Jan Reynst had a strong enough position in Venice as a factor to stand up to Witsen.

We know from a receipt, dated 16 October 1632 that Jan and Gerard were business associates. A later receipt from 1645 shows Gerard participating with 25,000 guilders in Jan's affairs.[84]

grandfather, Gerrit Reynst. (Elias, *op.cit.*, I, pp. 372–73).

[74] J.C. Breen, 'Geschiedenis van het huis 'In 't DerdeVredejaer', Keizersgracht 604,' *Jaarboek van het Genootschap Amstelodamum*, XVII, 1919, pp. 73–79. Today, the house serves as museum of the Dutch Press (*Nederlandsch Persmuseum*).

[75] De Gelder, *op.cit.*, p. 172, cat. no. 120. Further portraits, also believed to represent Constancia Reynst, are listed under cat. nos. 8 and 539.

[76] Houbraken, *op.cit.*, p. 8 ; Moes, *op.cit.*, II, p. 271, no. 6399.

[77] I am grateful to Dr. I.H. van Eeghen for this information.

[78] Family archive Backer no. 570 : (June 1675) *aen de Heer Pieter schaep voor sijne 1/5 part in de scheijdinge aengenomen het huijs de hoop voor de somma van 27,000*. Reference kindly provided by Dr. I.H. van Eeghen.

[79] Elias, *op.cit.*, I, p. 525, under no. 196 ; Bijtelaar, *op.cit.*, p. 9.

[80] Elias, *op.cit.*, I, p. 374, no. 122 ee.

[81] Paulo de Wilhem (Hamburg 1581–Amsterdam 1648) was an older brother of David Le Leu de Wilhem, counselor of Prince Frederick Henry, who was a good friend of Jan Reynst (see p. 35). Paulo was a merchant trading with Italy and the Levant and lived on the Keizersgracht. In 1621, he became a factor in Amsterdam for king Christian IV of Denmark. He was a partner in the firm Abraham de Ligne and Paulo de Wilhem and figured in several documents that included Jacques Nicquet and Albert Schuyt as well. On September 2, 1614, he married Sara de Haze (1592–1635), a sister of Constancia de Haze, married to Jan Nicquet, and of Clara de Haze, married to Jacques Nicquet.

[82] Not. J. Warnaertsz, N.A.A. no. 659 B, pp. 121–124 v, dated January 20, 1625. The renewed contract is found *idem*, N.A.A. no. 663, pak 3, pp. 189–93, dated February 21, 1630. I am grateful to S.A.C. Dudok van Heel for the reference to these documents.

[83] Heeringa, *op.cit.*, I, p. 527, no.250.

[84] The 1632 document, Not. Publ. S. Ruttens, is lost. For the document dated July 17, 1645, see Not. Publ. J. Warnaertsz, N.A.A. no. 690, p. 169. Abraham Heijrmans was still Jan's business partner

As stated earlier, Jan Reynst purchased together with his brother Gerard the house *De Hoop* on the Keizersgracht, in 1634, where he kept some of his paintings according to Ridolfi.⁸⁵ Jan also owned a house in Venice, because Ridolfi specified in the introduction to his *Maraviglie dell'Arte* that Jan had built a 'famous gallery', where part of his paintings were kept. Neither the street nor the section of Venice is known, however, where Jan lived. Since Jan resided primarily in Italy, he always signed as *Gio(vanni)* rather than Jan Reynst.

From a letter by Caspar van Baerle to Constantijn Huygens we learn that Jan was well known to Dukes and republics in Italy as a merchant dealing in salt and grain.⁸⁶ Early in 1645, Jan left Venice for Holland. On March 23, he was in Antwerp and on June 16, 1645, he appeared at a meeting in Amsterdam, accompanied by Paulo de Wilhem, to discuss future negotiations with France to insure the safe passage of Dutch vessels in the Mediterranean.⁸⁷ Jan had emerged as one of the spokesmen for the merchants, shipowners and captains trading with the Levant, who were requesting the States General that foreign merchandise loaded unto Dutch ships was not to be taxed by the Dutch consuls stationed in the Levant, obviously a measure to attract more foreign freight. ⁸⁸ Subsequently, Jan Reynst was selected by these Dutch merchants trading in the Mediterranean to negotiate with

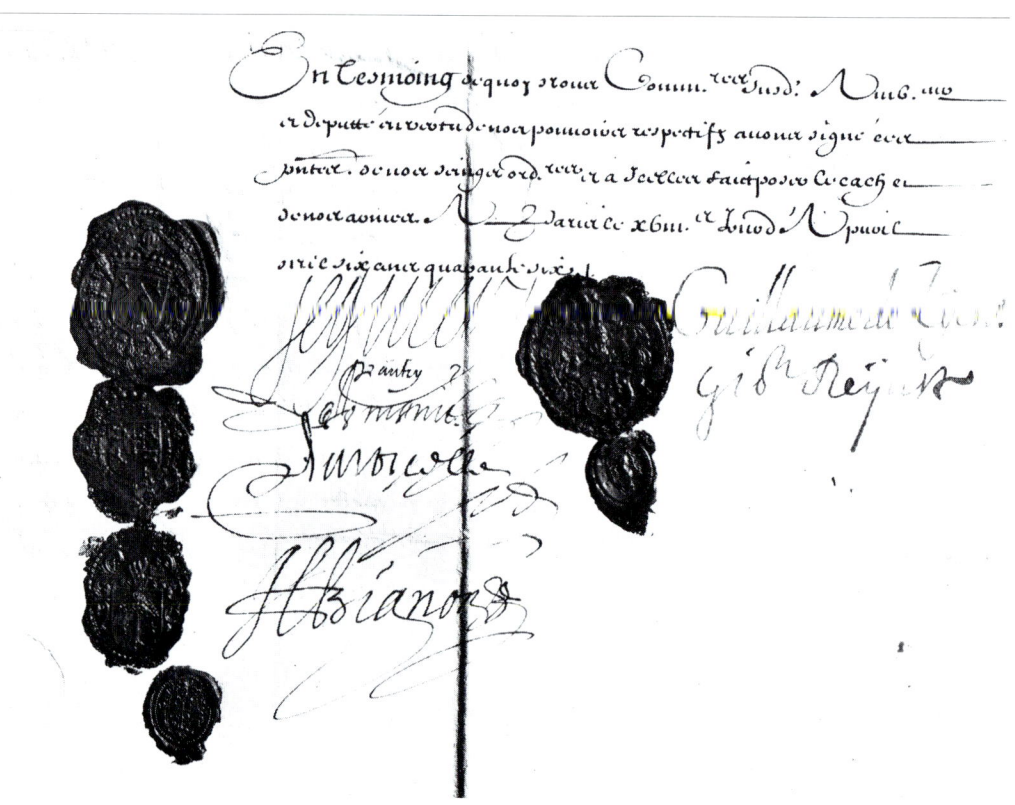

6

the French about keeping the waters open to the merchant fleets. Increasingly, Dutch cargo ships were harassed and searched by French war vessels that were patrolling the Mediterranean to insure that no war supplies of any kind reached their enemies, the Spaniards. The French suspected the Dutch of secretly assisting the Spaniards, a belief that was expressed several times by Cardinal Mazarin to Reynst during the negotiations in Paris.[89] The direct cause for Jan's mission was the confiscation of five Dutch merchant ships by French war vessels in the Mediterranean in February, 1645,[90] the sinking of one of them and the sale of the respective cargoes by the French.[91] Thus, on July 19, 1645, Jan Reynst was elected commissary and authorized by the States General to assist the Dutch ambassador to France, Willem van Lier, Heer van Oosterwijck, in negotiations at the French court for the United Provinces about the restitution of these five vessels with their freight and about reimbursement for the losses.[92]

The following brief summary of this mission is reconstructed from numerous documents preserved in the Algemeen Rijksarchief in The Hague, relating to Reynst's progress in his negotiations in Paris.[93] The lengthy report that Jan was requested by the States General to write about his mission is also found among these papers.[94]

---

in Venice. Information obtained from S.A.C. Dudok van Heel.

[85] Ridolfi, *op.cit.*, I, pp. 129, 232, 325 speaks of 'houses' in Amsterdam, where Jan Reynst kept his paintings.

[86] Worp, *op.cit.*, IV, p. 353, no. 4461.

[87] Heeringa, *op.cit.*, I, pp. 1071-73.

[88] *Ibidem*, p. 603, no. 288.

[89] Letters from Jan Reynst in Paris of January 13 and February 3, 1646, to States General and Directors of Levant Company; see note 93 below.

[90] Reynst wrote in his report to the States General in early May, 1646 that by the beginning of September, 1645, the five ships had been anchored off Toulon for seven months. The French finally decided that four of them should be returned, while the fifth ship, the Dolphin, was written off, since it sank during the fighting and its cargo was lost.

[91] Letter from Louis XIV of June 8, 1645, to his minister D'Inferville, in charge of the Police and Naval Finances, instructing him not to return anything from the five confiscated Dutch ships, to sell the grain before it spoiled, but not to hand over the proceeds to the Dutch until he gave the order to do so. (Algemeen Rijksarchief, The Hague; *Staten Generaal, LIAS-Frankrijk 6769, 1645-46*).

[92] Jan may have known Van Lier from the time the latter was ambassador of the United Provinces to Venice from 1627-1636, and had to intervene at times for Dutch merchants whose ships were seized by Venetians. See Heeringa, *op.cit.*, I, p. 76, note 3, when a ship of Albert Schuyt's, the future father-in-law of Gerard Reynst, was seized by the Venetians.

[93] I would like to thank G.W. van der Meiden, Deputy Keeper of the First Section, Algemeen Rijksarchief, The Hague, for the assistance I received in finding these documents. Jan Reynst's letters to the States General are in *Loketkas/Admiraliteit 109-112*; *Staten-Generaal 12561*; and *Staten-Generaal, LIAS-Frankrijk 6769, 1645-46*. His letters to the Directors of the Commerce with the Levant are kept in the volume *Levantsche Handel 282, 1631-1700*. The following documents have been consulted:
The ratification of Reynst's election and the instructions from the States General (abbreviated

According to the instructions drawn up by the States General on August 2, 1645, Reynst's mission was to be financed by the directors of the Levant Company. His letters to these directors as well as to the States General are most detailed and provide a careful account of his negotiations. Jan arrived in The Hague on July 28, 1645, collected his missives and letters of introduction to the French court and left the Netherlands on August 26, arriving in Paris on September 6. Van Lier and Reynst were unable to meet with French officials until the end of October (letter to directors of 30 October 1645), while the first discussions with the young king Louis XIV, Mazarin and the commissaries were held early in December. The slow pace of the negotiations made Reynst impatient and he asked several times in his letters to be allowed to return to Italy, most insistently on February 10, 1646, after he had received word that his 'confrater' had died. This 'confrater' must have been Abraham Heijrmans, Jan's close associate in Venice, who had died there in 1645.[95] Jan nevertheless stayed on and helped re-negotiate a naval treaty between France and the United Provinces which was drawn up and sent to the States General on February 3, 1646. The emphatic insistence by the States General that Dutch ships could not be searched by other nations was finally modified. It was agreed upon that inspectors were to be allowed aboard when encountering warships to verify the bill of

s.g.) to Reynst of 2, 16 August 1645, informing him about his mission to Paris. Reynst's letters to the s.g. of 9, 16, 23 September 1645; 11 November 1645; 9 December 1645; 13 January 1646; 3, 10, 17, 24 February 1646; 3, 31 March 1646; 7, 14, 22 April and 2 May 1646. Reynst's letters to the Directors of the Commerce with the Levant in Hoorn of 10 August, 21 September, 7, 21 October, 30 December 1645; 3, 24 February, 21 April and 2 May 1646; Jan Reynst's detailed, forty-seven page long report to the s.g. about his mission: *Rapport ende verbael, van 't geene bij Gio. Reijnst is gebesoigneert op de Commissie die de Hooghe Mogende Heeren Staten Generael vande Vereenighde Nederlanden hem hebben gelieft te geven, aen de Con : Ma^t van Vranckrijck, om ter vergaderinge van hare Ho : Mo : E^d gelevert te worden, door Sig. Johan Copes, derwijle vermits sijn vertreck van hier naer Italien geen mondeling rapport aen hare Ho : Mo : E^d can coomen doen.* (Staten-Generaal 8407: exhibitum of 14 April, 1646).

[94] Quoted at end of preceding footnote, '*Rapport ende verbael*'... . The only reference to a person named Copes is found in Brulez, *op.cit.*, p. 304, no. 916, where a Rizzardo and Henrico Copes were mentioned in an act of 30 April 1599, in Venice, apparently working for Jan and Jacques Nicquet. Johan Copes, therefore, possibly was a descendant of theirs.

[95] The final settlement of affairs between Gerard Reynst as heir of his brother Jan, deceased in Venice, and Abraham Heijrmans, heir of Abraham Heijrmans, merchant in Venice, is found under Not. Publ. J. v/d Ven, N.A.A. no. 1110, p. 26, dated October 7, 1654. I owe this information to S.A.C. Dudok van Heel.

[96] The original treaty is preserved in the Algemeen Rijksarchief, The Hague, *Secretekas Frankrijk, Staten-Generaal 12587 116*. A copy of it is in the Archives diplomatiques du Ministère des Affaires Etrangères, Quai d'Orsay, Paris, *Correspondance politique Hollande*, 36, fol. 156–163 v. Why Jan Reynst used an unidentified seal next to his signature instead of the coat of arms of the Reynst family remains unknown. I would like to thank W. Wijnaendts van Resandt at the Centraal Bureau voor Genealogie, The Hague, for his efforts in trying to identify this seal (letter of October 18, 1974).

[97] See following footnote.

goods issued to each merchant vessel. The official treaty for a duration of four years was signed on April 18, 1646 (text fig. 6).[96] Louis XIV, furthermore, conferred the knighthood of St. Michael upon Jan Reynst in appreciation for his services.[97] In his last letter to the States General from Paris, on May 2, 1646, Reynst writes about the various gifts he presented to members of the French court on behalf of the United Provinces and tells about the one that he had received from the king, valued at 5,000 guilders. The following day, May 3, Reynst left for Venice by way of Lyons and the Provence. Ridolfi's dedication of the first part of the *Maraviglie dell'Arte*, dated June 25, 1646, must have been written shortly upon Jan's return, because it already reflects his recent mission and adds that Jan was knighted by Louis XIV for his services. The chain represented around the Reynst coat of arms in Ridolfi's title-page to the *Maraviglie* probably alludes to this honor Jan had received at the French court (text fig. 7).

Jan did not live to see Ridolfi's *Maraviglie* published, however, (it appeared only during the summer of 1648), for he died in Venice on July 26, 1646.
We learn from a receipt of 1651 that Jan had a testament drawn up in Antwerp on March 23, 1645, which he handed over to the notary public J. Warnaertsz in Amsterdam on June 18, 1645.

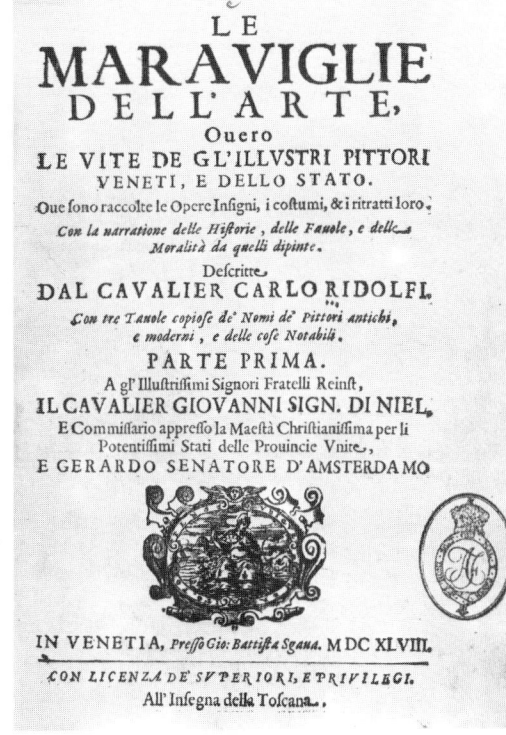

7

7a, b

ILLVSTRISSIMI
Signori miei Colendissimi.

ON è difficile il persuadere, che la Pittura traesse generosi natali, hauendo origine dalle humane menti à fine solo di dilettare gli animi gentili, ed essendo alleuata nella Reggia di Filippo Macedone, del Magno Alessandro, di Tolomeo di Egitto, e d'altri famosi Rè, che le diedero titolo non di serua, ma d'amica.

E' noto ancora quanto la Pittura fosse stimata da Candaule, e da Attalo Regi, e da Cesare dittatore, e come di quella si adornassero i Tempij della Grecia, e di Roma, le Curie, le habitationi de gli Augusti, e le loggie famose; come quelle di Pompeo, e di Ottauia.

Ma che bado à tempi andati? Ecco le Galerie de' maggiori Prencipi arricchite di pretiose pitture, come in Roma, Venetia, Fiorenza, Vienna, Parigi, Inghilterra, & Olanda. Tralascio il numero degli studij

studij priuati, & à quello delle VV. SS. Illustrissime solo mi appiglio, hauendo elleno con generoso dispendio erette in Venetia, & in Amsterdamo due famose Galerie, oue si ammirano opere molte di Raffaello, di Gio. Bellino, del Correggio, del Parmegiano, di Titiano, del Tintoretto, di Paolo, e di qual si voglia altro insigne Pittore: onde vengono frequentate da Personaggi, e da begli ingegni nella guisa a punto, che soleuano i popoli della Grecia visitare il Tempio di Venere in Gnido per la di lei statua da Prasitele scolpita.

Non m'inoltro à commendare questa eccellente disciplina: ma aggiungerò solo, che non meno dell'Historia ci incamina con singolari esempij all'acquisto d'ogni Virtù, mouendo ella più efficacemente gli animi con l'oggetto, che ci rappresenta: onde disse Quintiliano, che la Pittura: Sic internos penetrat affectus, vt ipsam vim dicendi nonnunquam superare videatur.

Nè meno vagliono le Imagini de' Maggiori ad eccitarci ad opere degne, & ad intraprendere magnanime imprese, rammemorandosi con quelle i loro gloriosi gesti: Quindi Filostrato: Quicunque picturam minimè amplectitur, non modò veritatem, verum & eam, quæ ad Poetas pertinet, iniuria afficit sapientiam. Eadem enim est vtriusque ad Heroum tam species, quàm gesta intentio.

Si

7c, d

Si che offerendole vn'Historia mista di Pitture, e di racconti, sarà proportionato il dono al gusto delle VV. SS. Illustrissime, riferendosi in questa fatti egregij d'Heroi, attioni pie, morali componimenti, diletteuoli racconti, riferiti dagli Historici, cantati da' Poeti, e coloriti da' Pittori.

Riceuino dunque con la solita humanità questa picciola mia fatica; picciola in riguardo dello scrittore: mà laboriosa per la raccolta delle cose molte (sparse in varij luoghi, e poco meno, che smarrite nella memoria degli huomini, e degna ancora per la serie degli illustri Pittori mentouati, che non hanno da inuidiare i più celebri degli antichi tempi, de' quali hauendo elleno le Pitture raccolte, era ben di douere, che comparendo gli stessi autori di nuouo in vita per mezzo delle stampe, fossero ancora raccomandati alla loro protettione. Et à chi più doueuasi la tutela de' Pittori, che alle VV. SS. Illustrissime? dalle cui liberali mani riceuono i Professori viuenti continue gratie, e fauori? Lo attestino i medesimi; ed il Signor Nicolò Renieri loro particolare amico, soggetto di tale stima, che molti Prencipi hanno ambito condurlo alle Corti, per vedersi effigiati dall'industre suo pennello, e mercè del suo valore, destinato dal Christianissimo Rè per suo Pittore in Italia.

Non entro negli honori sublimi della nobilissima sua

sua Casa, non essendo luogo proportionato il breue spatio di questo foglio, lasciando, che la Fama stessa per mille bocche n'adempisca l'vfficio: ma dirò bene, che sì come l'Eccellentissimo, e generoso loro Padre resse con tanta sua gloria col titolo di Generale nell'espeditione dell'Indie l'Armata Olandese, con accrescimento di stato à quella potentissima, e nobilissima Natione; così l'vna delle VV. SS. Illustrissime nel carico di Commissario per la medesima appo la Corona di Francia ha dato saggio ne' più importanti maneggi d'impareggiabile destrezza, e valore; per attestatione di che hà conseguito, trà gli altri segni di honore, l'ordine Regio di Caualiere, e la nobiltà della Francia; l'altra, come Zelante, e prudente Senatore, si è adoperata altresì nel gouerno publico, e nelle deliberationi più graui della Patria con vniuersale sodisfattione, & infinita sua lode.

Gradiscano per tanto il voto della mia diuotione, che al nome tutelare delle loro SS. Illustrissime consacro; poiche insignito di questo carattere, non pauenterò gl'incontri de' Critici, e dell'Inuidia. Le riuerisco per fine, e le desidero ogni prosperità.

Di Venetia, il dì 25. Giugno 1646.
Delle VV. SS. Illustrissime

Diuotissimo Seruitore
Il Caualier Ridolfi.

This *besloten*, i.e. handwritten testament is not known, but the same receipt states that Gerard Reynst was the universal heir of Jan Reynst.[98] Once a person, as in this instance Gerard, was designated overall heir, there was little necessity to draw up an inventory. This might explain why no such inventory has been found for Jan. Jan's possessions probably were transferred to Amsterdam soon after his death.[99] Six or seven paintings from Jan's collection remained in Venice where they reappeared in the collection of Nicolò Renieri.[100] Since Renieri was a close friend of the Reynst brothers, as mentioned by Ridolfi, we may suppose that Jan bequeathed them to him (see cat.nos. 44, 47, 48, 51, 52, 53, and 57).

It is of special interest that both Gerard and above all Jan Reynst were mentioned several times in the correspondence of Constantijn Huygens. Jan apparently was a very good friend (*amicissimus*) of David Le Leu de Wilhem,[101] who became counselor of Prince Frederick Henry in 1631. De Wilhem was a most learned man, a so-called *mercator sapiens*. While living in the Near East for two years he collected Egyptian art and may have interested Jan in it. De Wilhem referred to Jan in a letter to Huygens in 1645, as a former 'compagnon de negoce et de voyage'.[102] This does not surprise, because David Le Leu de Wilhem was a younger brother

---

[98] Not. Publ. J. v/d Ven, N.A.A. no. 1097, p. 109, dated May 26, 1651. Constantia Coymans acknowledged in this document the receipt of 9,250 guilders which her uncle Jan Reynst had bequeathed to her in his testament drafted in Antwerp on March 23, 1645 and confirmed by Warnaertsz on June 18, 1645. The passage about Jan and Gerard Reynst reads as follows: ... *comparanten van de heer Gerard Reijnst raat en oudt schepen deser stede hunne oom als universele erffgenaem van hun comparanten ander oom Joan Reijnst in sijn leven Ridder Van S. Michiel in Frankrijck overleden tot Venetien broeder was van de welgemelten h. Gerard Reijnst...* I am grateful to S.A.C. Dudok van Heel for the reference to this document and for his explanation about the implication of 'universal heir'.

[99] Mr. Silvano de Tuoni, Fondazione Cini, Venice, has been unable to find material on Jan Reynst in the Archives in Venice. Savini-Branca, *loc.cit.*, who used the archives extensively in preparation for her basic study on collections in that city during the 17th century has found no documents on Jan either.

[100] See Savini-Branca, *op.cit.*, pp. 264–68 for the paintings that passed from Jan Reynst into Renieri's possession. According to her, Renieri bought them from Jan (*op.cit.*, p. 269).

[101] Worp, *op.cit.*, I, p. 410, no. 802. See also Elias, *op.cit.*, II, p. 602, no. 236, Bijl. 1b, and C. C. van Valkenburg, 'Het regentengeslacht (Le Leu) de Wilhem,' *Jaarboek van het Centraal Bureau voor Genealogie*, part 25, 1971, pp. 132–80; 157–61. David Le Leu de Wilhem was born in Hamburg on May 15, 1588 and moved to Holland to study theology at Leiden. On January 26, 1633, he was married in The Hague to Constantia Huygens, the youngest sister of Constantijn. De Wilhem died in The Hague on January 27, 1658. I would like to thank S.A.C. Dudok van Heel for information on De Wilhem.

[102] *Ibidem*, IV, p. 189, no. 4065.

of Paulo de Wilhem, Jan's business partner. We also learn that Jan was living in Venice in 1633,[103] and that in 1645, he delivered a letter to Huygens on behalf of De Wilhem.[104] There is one letter by Huygens to Princess Amalia van Solms, in which he told of Jan's mission to the French court, assisting Van Lier in the negotiations.[105] Finally, two notes by Jan himself are preserved, written to Huygens from Paris, informing him about his mission.[106] Jan, furthermore, is mentioned in the correspondence between the humanist Nicolaas Heinsius (1620–1681) and the latinist Johannes Fredericus Gronovius (1611–1671). Heinsius wrote to Gronovius from Paris in April, 1646, about his impending trip to Venice in the company of Jan Reynst.[107] Isaac Vossius (1618–1689) informed Heinsius in another note that he had written letters to him which were delivered to Jan Reynst's house in Venice.[108]

Jan's death is mentioned in a letter Caspar van Baerle wrote to Huygens on September 27, 1646. Van Baerle added that Gerard Reynst had asked him to write a eulogy for Jan,[109] a very special honor for the Reynst family, if one realizes what an important role he played in the humanistic circles of Amsterdam during those years. If we recall further that Jan travelled to Italy in the company of the humanist Heinsius and that he was also on friendly terms with David Le Leu de Wilhem and knew Huygens, one becomes aware that Gerard and Jan Reynst were acquainted with some of the outstanding people of their time.

---

[103] *Ibidem*, p. 410.

[104] *Ibidem*, p. 189.

[105] *Ibidem*, p. 191, no. 4071.

[106] *Ibidem*, p. 218, no. 4136 and p. 297, no. 4318.

[107] Petrus Burmanus, ed., *Sylloges epistolarum*, III, Leiden, 1727, pp. 164–66, 168–71. I am grateful to Dr. I.I.E. van Gelder-Jost for this reference.

[108] *Ibidem*, pp. 563–65.

[109] Worp, *op.cit.*, IV, pp. 353–54, no. 4461.

Gerard Reynst	Wijntje Reynst	Jan Reynst	Constantia Reynst
1599–1658	born 1600	1601–1646	1603–1671
m. 1631 Anna Schuyt	m. Isaac Coymans		m. 1621 Joan Carlo Smissaert, Heer van Niel
1605–1671			1590–1644
			m. 1650 Willem de Raet

Joan Reynst	Constancia Reynst
1636–1695	1638–1674
m. 1667 Eva Hooftman	m. 1666 Pieter Schaep
	1635–1685

Anna Reynst	Anna Schaep
1671–1691	1667–1727
m. 1685 David de Wildt	m. 1689 Franco Pauw

III. THE COLLECTIONS OF GERARD AND JAN REYNST

A. THEIR CONTENTS

Due to the lack of any testament, inventory or other documentary evidence which would help establish the contents of the collections of Gerard and Jan Reynst, special importance is attached to the two sets of prints after a selection of paintings and antique sculptures in the possession of Gerard Reynst, the so-called CAELATURAE and the so-called ICONES. These prints, one hundred and forty-three in total, are here discussed as the primary sources. The secondary sources are a number of contemporary descriptions by visitors to the house on the Keizergracht in Amsterdam on the one hand, and as far as Jan Reynst is concerned, by Ridolfi on the other.

To these notes, two additional sets of inventories should be added that furnish further insight : the seventeen manuscript catalogues of the collection of Andrea Vendramin in Venice, completed by 1627, and the inventory of paintings in the collection of Charles II of England. Since much or all of the Vendramin collection later came into the possession of Gerard Reynst, these illustrated inventories become another important source to draw from in the reconstruction of the Reynst collection. They preserve at least visually many objects that since have perished, in particular the entire section comprised under *naturalia*. The inventory of Charles' II paintings at Whitehall and Hampton Court identifies eight as 'Dutch Present' that were not engraved for the CAELATURAE but

8

VARIARVM
IMAGINVM
A
CELEBERRIMIS ARTIFICIBVS
PICTARVM
CÆLATVRÆ
ELEGANTISSIMIS TABVLIS
REPRÆSENTATÆ.

Ipsæ Picturæ partim extant apud viduam Gerardi Reynst. quondam huius urbis Senatoris ac Scabini, partim CAROLO II. Britanniarum Regi a Potentissimis Hollandiæ West-Frisiæque Ordinibus dono missæ sunt.

AMSTELODAMI.

that presumably came from the Reynst collection as part of the 'Dutch Gift' in 1660. None of them had previously been associated with Reynst. This inventory, furthermore, is of particular interest with regard to the painting of a *Reclining Venus* by Cariani (cat. no. 9; P9), a painting that was also illustrated in the Vendramin catalogue DE PICTURIS and therefore establishes that the Reynst brothers did acquire pictures from Vendramin in Venice.

I. CAELATURAE

The principal source for the reconstruction of the collection of paintings formerly owned by Gerard Reynst is the volume of thirty-four prints after thirty-three paintings that appeared with the following title page:

VARIARUM/IMAGINUM/A/CELEBERRIMIS ARTIFICIBVS/PICTARVM/CAELATURAE/ELEGANTISSIMIS TABVLIS/REPRAESENTATAE. / Ipsae Picturae partim extant apud/ viduam Gerardi Reynst, quondam huius/urbis Senatoris ac Scabini, partim/CAROLO II. Britanniarum Regi a/ Potentissimis Hollandiae West-Friesiaeque/Ordinibus dono missae sunt. / AMSTELODAMI. (text fig. 8).[1]

These prints were commissioned from a number of engravers. The project probably was initiated around 1655, the year Jeremias Falck arrived in Amsterdam. With the death of Gerard Reynst,

---

[1] 'Engravings of various subjects painted by most famous artists, presented in most elegant plates. These pictures are partly with the widow of Gerard Reynst, who during his life was a senator and magistrate of his city, and partly they were sent to Charles II, King of England as a gift from the most powerful states of Holland and West Friesland. In Amsterdam'. The average size of the plates is 370/80 × 278/88 mm. The prints, primarily engravings, at times worked up in etching in the background, have blank margins below the image except for CAELATURAE II which is after letters. Some of the prints were issued after letters in later states.
The date *1653*, introduced into the literature by Law (*A Historical Catalogue of the Pictures in the Royal Collection at Hampton Court*, London [1881], pp. 46 and 97, nos. 148, 149 and 306) for Visscher's print after Lotto's *Odoni*, Van Dalen's print after Titian's so-called *Sannazaro*, and Holsteyn's print after Giulio's *Isabella d'Este*, is unfounded. Although Law's date has been taken over in subsequent literature, there appears no reason to accept it. The first to correctly repudiate it was Mahon (*op.tit.*, 1949, p. 304, note 13). Law, furthermore, believed that all three engravings were by Visscher, an opinion no longer accepted either (only Lotto's *Odoni* was engraved by Visscher; see cat. nos. 12, 16, 34).

[2] Grief about this abrupt ending was expressed in a letter by Falck sent from Hamburg on December 10/20, 1658, to the astronomer Hevelius:
'ich bin sehr geylet worden mit dem Werke von dem Amsterdamschen Ratsherrn, und könnte es auch nicht hintansetzen, weil etwas dabei zu thun war, und ich by 2600 Gulden von seiner Arbyt genossen, und war Hoffnung noch wol 1000 Gulden by ihm zu verdienen gewäst, wenn der gute Herr nicht so ellendig um gekommen wär, welches zu beklagen ist, denn bey dieser Zeit wenig solche Liebhabers zu finden seindt" (published in J. C. Block, *Jeremias Falck. Sein Leben und seine Werke*, Danzig/Leipzig/Vienna, 1890, p. 11). The

however, three years later, the work seems to have come to a halt.[2] The individual prints were gathered in a volume which was published at a somewhat later date (here referred to as CAELATURAE). Since the title-page mentions the 'Dutch Gift' of 1660, as well as the widow of Reynst, who died in 1671, we may suppose that the folio appeared in the latter part of the 1660's. The publisher may have been Clement de Jonghe (died June, 1677), because his inventory of 1679 listed fifteen *Cunst Plaaten van Reijnst*.[3]

The thirty-three paintings known through these engravings undoubtedly were the best and most esteemed ones owned by Gerard and Jan Reynst. (None of them came from Andrea Vendramin).

Ten of these paintings are now in the English royal collections, eight are in museums or private collections in Europe, Russia and the United States, while fifteen are lost or can no longer be securely identified. (For a discussion of the individual paintings see the CATALOGUE).

The standard edition of the CAELATURAE comprises the following thirty-four engravings:[4]

---

letter was first published by W.Seidel, in *Neue Preussische Provinzialblätter*, Königsberg, IV, no. 1, 1847, p.6; see Jacobs, *op.cit.*, p.27, note 3).

[3] See D. de Hoop Scheffer and K.G. Boon, 'De Inventaris-Lijst van Clement de Jonghe en Rembrandts etsplaten,' *de Kroniek van het Rembrandthuis*, XXV, 1971.1, pp.1-2. The inventory was drawn up by notary J. Backer, who was also engaged by the Reynst family. De Jonghe may possibly have bought the plates from the Reynst family at the sale of the collection in 1670, or, as the publisher, he may have simply kept them after the various editions were pulled. The following plates were listed, with the corresponding number of the CAELATURAE added in parenthesis:
Cunst Plaaten van Reijnst
Een vrouwstroni Titiaan (1)
van d^{to} een manstroni met een boeck ynde hant (2)
daniel omtrent der Engelen schaar (13)
de Geboorte Cristi (20)
Oosterse boccio bandinello (26)
Susanna (28)
Sacharias met het kindeken (21 ?)
Strooperij de Laar inde Grotten (32)
De Luijsenberge bamboots (31)
de Hoovardij van bourdon (8)
de groote strooperij lantschap (19)
Ars pictoria (7)
Cruijsdraaginge J : Falck (12)
Maria met het kindeken (10 ; 18 ?)
petrus met sijn swart opsiende (6)
(Gemeentelijke Archiefdienst, Amsterdam, Inv. no. N.A. 4528, J. Backer, 11 February 1679).

[4] The numbering and sequence of the prints in the text and the catalogue section are based on the CAELATURAE in the Rijksprentenkabinet, Amsterdam. The following copies of the CAELATURAE have been consulted: Amsterdam, Rijksprentenkabinet (Inv. no. 325-A-10). Cincinnati (Ohio), The Public Library (Inv. no. R 769/f v 29). Dresden, Staatliche Kunstsammlungen, Kupferstichkabinett (Inv. no. B 92, 4).
Göttingen, Niedersächsische Staats- und Universitätsbibliothek (Inv.no. 2 p. Art. plast. VII, 3600). London, The Surveyor's Office; copy from the collection of Horace Walpole. (I would like to thank Sir Oliver Millar for referring me to this volume).

1. *Athena* by Parmigianino (Hampton Court), engraved by Cornelis Visscher (cat. no. 20 ; Pls. p 20, p 20a).

2. *Male Portrait* (so-called *Sannazaro*) by Titian (Hampton Court), engraved by Cornelis van Dalen II (cat. no. 34; Pls. p 34, p 34a).

3. *Male Portrait* by Lorenzo Lotto (Hampton Court), engraved by Cornelis van Dalen II (cat. no. 17 ; Pls. p 17, p 17a).

4. *Portrait of a Dominican Friar* by Tintoretto (Hampton Court), engraved by Cornelis van Dalen II (cat. no. 32; Pls. p 32, p 32a).

5. *Male Portrait* (so-called *Aretino*) by Titian (lost), engraved by Cornelis van Dalen II (cat. no. 33 ; Pl. p 33).

6. *St. Bartholomew* by Guido Reni (attributed to) ?, (lost), engraved by Theodor Matham (cat. no. 24 ; Pl. p 24).

7. *Allegory of Painting* attributed to Guido Reni (lost), engraved by Theodor Matham (cat. no. 26 ; Pl. p 26).

8. *The Old Courtesan* by Bernardo Strozzi (Moscow), engraved by Jeremias Falck (cat. no. 30 ; Pls. p 30, p 30a).

9. *Portrait of Isabella d'Este* by Giulio Romano (Hampton Court), engraved by Pieter Holsteyn II (cat. no. 12; Pls. p 12, p 12a).

10. *The Virgin and Child and St. Anne* by the School of Raphael (lost), engraved by Jeremias Falck (cat. no. 21; Pl. p 21A).

11. *The Virgin and Child and the Infant St. John* by Jacopo Bassano (lost), engraved by Theodor Matham (cat. no. 6; Pl. p 6).

12. *Christ Carrying the Cross* by Jacopo Bassano (Earl of Bradford, Weston Park), engraved by Jeremias Falck (cat. no. 4 ; Pls. p 4, p 4a).

13. *The Resurrection of Christ* by Veronese (lost), engraved by Cornelis Visscher (cat. no. 37 ; Pl. p 37).

14. *The Entombment of Christ* by Jacopo Bassano (lost), engraved by Jeremias Falck, substituted for Tintoretto's *Pietà* (lost), etched by Cornelis Visscher (cat. nos. 5, 31 ; Pls. p 5, p 31).

15. *The Vision of St. Peter* by Domenico Fetti (lost), engraved by Jeremias Falck (cat. no. 10; Pl. p 10).

16. *St. Paul in Ecstasy* by Jan Liss (Berlin), engraved by Jeremias Falck (cat. no. 14; Pls. p 14, p 14a).

17. *St. John the Evangelist* by Guido Reni (attributed to) ?, (Leipzig), stipple engraving by Jan Lutma II (cat. no. 25; Pls. p 25, p 25a).

18. *The Virgin and Child and St. Anne* by the School of Raphael (lost), second version of CAELATURAE 10, by an inferior engraver (cat. no. 21; Pl. p 21B).

19. *The Ambush* by Pieter van Laer (Naples), etching and engraving by Cornelis Visscher (cat. no. 40; Pls. p 40, p 40a).

20. *The Adoration of the Shepherds* attributed to Lorenzo Lotto (lost), engraved by Jeremias Falck (cat. no. 18; Pl. p 18).

21. *The Presentation in the Temple* by Andrea Schiavone (lost), engraved by Jeremias Falck (cat. no. 27 ; Pl. p 27).

22. *The Virgin and Child with Tobias and Saints* by Bonifazio de' Pitati, called Veronese (Hampton Court), engraved by Jeremias Falck (cat. no. 7; Pls. p 7, p 7a).

23. *The Virgin and Child with Tobias and the Angel* by the School of Titian (Hampton Court), engraved and etched by Cornelis Visscher (cat. no. 36; Pls. p 36, p 36a).

24. *The Marriage of St. Catherine* by Veronese (Hampton Court), engraved by Theodor Matham (cat. no. 38; Pls. p 38, p 38a).

25. *The Holy Family with St. John and Elizabeth* attributed to Titian (lost), engraved and etched by Cornelis Visscher (cat. no. 35 ; Pl. p 35).

26. *Portrait of Andrea Odoni* by Lorenzo Lotto (Hampton Court), etched and engraved by Cornelis Visscher (cat. no. 16; Pls. p 16, p 16a).

27. *The Concert* attributed to Giorgione (Hampton Court), engraved by Jeremias Falck (cat. no. 11; Pls. p 11, p 11a).

28. *Susannah and the Elders* by Guido Reni (not securely identifiable), engraved and etched by Cornelis Visscher (cat. no. 23 ; Pls. p 23, p 23a).

29. *Queen Semiramis Receiving News of the Revolt of Babylon* by Guercino (Boston), engraved by Jeremias Falck (cat. no. 13; Pls. p 13, p 13a).

30. *The Brothel* by Jan Liss (Cassel), engraved by Jeremias Falck (cat. no. 15; Pls. p 15, p 15a).

31. *The Large Limekiln* by Pieter van Laer (formerly Prince of Liechtenstein), etched and engraved by Cornelis Visscher (cat. no. 41; Pl. p 41).

32. *The Shot with the Pistol* by Pieter van Laer (Leningrad), etched and engraved by Cornelis Visscher (cat. no. 42; Pls. p 42, p 42a).

33. *The Annunciation to the Shepherds* by Jacopo Bassano (lost), etched and engraved by Cornelis Visscher (cat. no. 3 ; Pl. p 3).

34. *Abraham Leaving Haran* by Jacopo Bassano (lost) ; etched and engraved by Cornelis Visscher (cat. no. 2; Pl. p 2).

No text accompanies these engravings and the plates are unnumbered which accounts for the different sequences found in the various copies of the CAELATURAE. Furthermore, all the impressions are before letters safe for Bassano's *Virgin and Child and the Infant St. John* (CAELATURAE 11) which bears the inscription at bottom left: *Jac. de Ponte Bassan pinxit./ Theod. Matham Effigiavit*. This is the reason for the many discrepancies and uncertainties with regard to names of painters and engravers found in the respective literature. Hecquet (1751) was one of the first to give a description of the CAELATURAE in the introduction of his catalogue of prints by Cornelis Visscher, in which he listed the twelve he believed Visscher had executed after paintings formerly with Gerard Reynst. According to him Reynst had commissioned the prints in order to present them as a gift to his friends.[5] The second statement about the 'Cabinet de Reynst' is included in Heinecken's *Idée générale* of 1771.[6] Heinecken infers that the prints originally were made individually and gathered only later by Reynst's widow, who published them in form of an album.

The first full description listing every print was furnished in 1810 by I.G. Stimmel in the sales catalogue of the collection of the German banker Winckler.[7] Stimmel's identifications of painters and engravers have proven rather reliable and most of his attributions are sill valid. His list of the first 34 prints became the

London, British Museum, Department of Prints and Drawings (Inv. no. case 157★x, b 27).
London, Victoria and Albert Museum (Inv. no. 100/8).
New York, Metropolitan Museum of Art, Department of Prints (Inv. no. 30.19).
Paris, Bibliothèque Nationale (Inv. no. AC. 11 fol.).
The Hague, Royal Library (Inv. no. G. 6 fol. 1927, B. 7).
The Hague, Rijksbureau voor kunsthistorische Dokumentatie (Gift of F. Lugt).

[5] R. Hecquet, *Catalogue des estampes gravées d'après Rubens, auquel on a joint l'oeuvre de Jordaens et celle de Visscher*, Paris, 1751, p. 20:

'Ce qu'on appelle le cabinet de Reynst, est une suite d'Estampes qu'un Curieux d'Hollande a fait graver à ses dépens, pour faire présent à ses amis, & pour illustrer la collection considérable qu'il avoit faite d'après les plus beaux Tableaux qu'il avoit de différens Maîtres, dont Visscher a gravé une partie.'

A copy of the CAELATURAE and the ICONES was listed in the sale catalogue of Jan Six (April 6, 1702, 'Papierkonst', p. 18, no. 28 : *De Schilderyen van de Heer Renst. 34 stuks, en de Statuën 47 stuks, by een gebonden* ; no. 29 : *De Schilderyen noch apart ingebonden*).

[6] K.H. von Heinecken, *Idée générale d'une collection complette d'Estampes*, Leipzig/Vienna, 1771, pp. 82-85.

'Gerard Reynst, Senateur & Echevin de la ville d'Amsterdam, avoit recueilli, avec grand soin, plusieurs tableaux, statuës, bustes & autres curiosités, en sorte, que son Cabinet étoit en très grande réputation. Plusieurs graveurs de ce tems avoient entrepris de graver ces ouvrages, par complaisance pour leur possesseur qui chérissoit les arts & les artistes. Après sa mort les Etats-Géneraux de la Republique choisirent les morceaux les plus renomés de cette Succession, pour faire un présent à Charles II. alors Roi de la Grande-Bretagne. Cependant, pour perpétuer la Memoire de Gerard Reynst, comme d'un illustre amateur, la veuve rassembla les planches &

principal source for future publications on the CAELATURAE. Wussin[8] took it over in its entirety in his catalogue raisonné of the prints by Cornelis Visscher (1865), but not without editing it to a certain extent. Block[9] followed Stimmel for his catalogue raisonné of the prints by Jeremias Falck (1890), while Wurzbach (1905–10)[10] classified the prints individually under the various artists, based on Wussin's list. Both Heinecken and Stimmel referred to five additional engravings (four of them identical) that are usually not included in the standard edition of the CAELATURAE. According to Heinecken,[11] who based his opinion on a copy in Dresden, the following prints were also made after paintings formerly with Reynst:

a. *Four Cyclops in a Forge*, attributed to Caravaggio (lost), engraved by Jeremias Falck (Appendix 11 ; Pl. A 11).

b. *Esau Selling His Birthright to Jacob* by Mathias Stomer (Leningrad), engraved by Jeremias Falck and dated 1663 (Appendix 9 ; Pls. A9, A9a)

c. *Flora, Silenus, and Zephyrus* by Jordaens (not securely identifiable), engraved by Schelte A. Bolswert (Appendix 5 ; Pl. A 5).

d. *St. John the Baptist Preaching* by Abraham Bloemaert (not identifiable), engraved by Jeremias Falck, dated 1661 (Appendix 1 ; Pl. A 1).

e. *The Duet* by Cornelis van Haarlem (Göttingen), engraved by Jeremias Falck (Appendix 2 ; Pls. A 2, A 2a).

forma le Recueil, donton vient de rapporter le titre. Il consiste en trente trois estampes, d'après trente deux tableaux, parceque la Sainte Vierge d'après Raphael a été gravée deux fois, la première planche n'ayant pas reüssi. Dabord ces estampes parûrent avant la lettre. Les noms de Peintres & de Graveurs y fûrent ajoutés ensuite & quelquefois les premiers différemment. Il faut donc rechercher les premieres épreuves, pour avoir les plus beaux exemplaires.'

[7] I.G. Stimmel, *Catalogue raisonné du Cabinet d'estampes de feu Monsieur (Gottfried) Winckler*, v, Leipzig, 1810, v, *Ouvrages reliés*, pp. 309–17, no. 29.

[8] Johann Wussin, *Cornel Visscher.*

Verzeichnis seiner Kupferstiche, Leipzig, 1865, pp. 270–79, footnote 11.
Based on the opinion of Rudolf Weigel, Wussin included one additional print by Visscher, *The Stable*, after Pieter van Laer (*op.cit.*, p. 274, no. 40). This impression is not included in the standard editions of the CAELATURAE and Weigel's attribution is unconvincing (Appendix 8).

[9] *Op.cit*. A. Hagen, 'Ueber den Kupferstecher Jeremias Falck,' *Kunstblatt*, no. 16, 30 March 1848, pp. 61–64, lists the following engravings by Falck which he believes to have belonged to the CAELATURAE in addition to the ones enumerated by Wussin :

Christ Carrying the Cross after Van Dyck (no. 85) ; *The Virgin and Child with St. John Feeding the Lamb* after Stella (no. 86) ; *The Virgin Mary* after Egmont (no. 87). Hagen also thinks that the print of *Flora, Silenus and Zephyrus* after Jordaens was made by Falck rather than by S.A. Bolswert (no. 83). On the other hand, Hagen does not include CAELATURAE 14b, 15, 16, 20, and 21 which are standard. His improvements therefore are most questionable. (see Appendix A 3, 4, 10).

[10] A. von Wurzbach, *Niederländisches Künstler-Lexikon*, Vienna/Leipzig 1906–11 (reprint 1963), II, p. 457.

[11] *Op.cit.*, p. 84.

Stimmel's list concurs with Heinecken's with respect to the prints described under *a* through *c* and under *e*. As for *d*, Stimmel mentions Falck's engraving of an *Entombment*[12] instead. This print is found repeatedly in other editions of the CAELATURAE (Pl. P 5). Often it is substituted for Visscher's etching after Tintoretto's *Pietà* (cat. no. 31; Pl. P 31). This latter print figures only in the CAELATURAE in Paris and Cincinnati. Hecquet was the first to refer to it. Wussin included it in the catalogue raisonné of Visscher's prints with a reference to the Reynst collection, but omitted it in his list of prints found in the CAELATURAE. The lack of an inventory makes it impossible to establish whether both paintings once were owned by Gerard Reynst.

The five remaining prints, listed by Heinecken and Stimmel, on the other hand do not seem to have been made after paintings that formerly were with Reynst, despite Heinecken's statement. First of all, two of the engravings are dated 1661 and 1663 respectively. As we know, Falck left Amsterdam soon after Reynst's death in 1658, to work in Hamburg.[13] Thus, it is most unlikely that he would still have been employed by Reynst's widow, especially at such a late date. Since all five prints are by Falck, or at least once were attributed to him, it seems more probable that they were inserted at a later date by individual collectors. Stimmel mentioned that this was the case for the CAELATURAE formerly owned by Winckler. (The same must be also true of the copy in Cincinnati, where at a later date two prints by Visscher after Van Laer were added). We may therefore conclude that the paintings listed by Heinecken and Stimmel as a supplement to the standard edition of the CAELATURAE probably never were in Reynst's possession.

Gerard Reynst was able to attract a group of very fine engravers. More than half of the prints were executed by two artists only, Cornelis Visscher (Haarlem, ca. 1629 – Amsterdam 1658) and Jeremias Falck (Gdansk, ca. 1609/10–1677). Both seem to have arrived in Amsterdam in or about 1655, possibly upon learning that Reynst was looking for skillful engravers to reproduce paintings and sculptures from his collection into print. If Falck's assertion is true that he earned 2,600 guilders during his employ-

---

[12] Stimmel, *op.cit.*, p. 316, no. 39.

[13] See above, footnote 2.

ment, he would have been paid a little more than 200 guilders a print (presumably including the copper), an exceedingly high salary.[14] Falck, therefore, may also have made some of the prints in the ICONES.

Both Visscher and Falck engraved twelve paintings each, Cornelis van Dalen II (Amsterdam, 1638–1665) and Theodor Matham (Haarlem, 1605/6–Amsterdam, 1676) four each, and Pieter Holsteyn II (Haarlem, ca. 1614–Amsterdam, 1673) and Jan Lutma II (Amsterdam, 1624–1685) one each.[15]

The finest engravings were executed by Visscher, Van Dalen, Falck and Holsteyn. No pattern is apparent in the division of labor among the six engravers. No one was working after one painter only and no sequence could be established that would indicate that the task was handed out according to certain guidelines. It seems to have been a group effort.

Despite these engravings, it cannot be established beyond doubt whether Falck's print is based on Strozzi's *Old Courtesan* (cat. no. 30; Pls. P 30, P 30a) in Moscow or Bologna, nor which of the various versions of Reni's *Susannah* (cat. no. 23; Pls. P 23, P 23a) was engraved by Visscher.

In the course of this study a number of erroneous attributions of paintings and prints have been corrected and brought up-to-date.

---

[14] In comparison, Cornelis Galle received in the 1630's only about 80–100 guilders for his prints, including copper (Max Rooses, *L'Oeuvre de P.P. Rubens*, V, Antwerp, 1892, p. 73, note 3, or p. 74, note 1). Based on Falck's salary, therefore, Reynst must have paid some 7500 guilders to have the best works of art from his collection reproduced in print, a rather considerable amount of money.

[15] The second version of the engraving after the Raphael school painting of the *Virgin and Child and St. Anne* (CAELATURAE 18) is by such an inexperienced hand that it is here attributed to an unidentified engraver. It is surprising that this print was not omitted from the final edition.

The list of engravers in alphabetical order is as follows:
Cornelis van Dalen II: CAELATURAE 2 (Titian); CAELATURAE 3 (Lotto); CAELATURAE 4 (Tintoretto); CAELATURAE 5 (Titian).
Jeremias Falck: CAELATURAE 8 (Strozzi); CAELATURAE 10 (Raphael school); CAELATURAE 12 (Bassano); CAELATURAE 14 (Bassano); CAELATURAE 15 (Fetti); CAELATURAE 16 (Liss); CAELATURAE 20 (Lotto, attributed to); CAELATURAE 21 (Schiavone); CAELATURAE 22 (Bonifazio); CAELATURAE 27 (Giorgione, attributed to); CAELATURAE 29 (Guercino); CAELATURAE 30 (Liss).
Pieter Holsteyn II: CAELATURAE 9 (Giulio Romano).

Jan Lutma II: CAELATURAE 17 (Reni, attributed to?).
Theodor Matham: CAELATURAE 6 (Reni, attributed to?); CAELATURAE 7 (Reni, attributed to); CAELATURAE 11 (Bassano); CAELATURAE 24 (Veronese).
Cornelis Visscher: CAELATURAE 1 (Parmigianino); CAELATURAE 13 (Veronese); CAELATURAE 19 (Tintoretto); CAELATURAE 19 (Van Laer); CAELATURAE 23 (Titian school); CAELATURAE 25 (Titian, attributed to); CAELATURAE 26 (Lotto); CAELATURAE 28 (Reni); CAELATURAE 31 (Van Laer); CAELATURAE 32 (Van Laer); CAELATURAE 33 (Bassano); CAELATURAE 34 (Bassano).
Unidentified engraver: CAELATURAE 18 (Raphael school).

Many of the early attributions have proven to be overly optimistic and some of the paintings therefore had to be relegated to 'attributed to' or 'school of' (CAELATURAE 10, 23, 27), while others turned out to be by different artists than originally stipulated (CAELATURAE 3, 8, 30) and some of the painters can no longer be identified with certainty (CAELATURAE 6, 17).

2. ICONES

Gerard Reynst also had a selection of his antique statues and busts reproduced in prints. They again were published at a somewhat later date, probably in ca. 1670, by Nicolaes Visscher (here referred to as ICONES).[16] The additional title-page by Gerard de Lairesse is inscribed as follows:

SIGNORUM VETERUM ICONES/Per D. GERARDUM REYNST/
Urbis Amstelaedami Senatorem ac Scabinum/
dum viveret Dignissimum/COLLECTAE/
AFBEELDINGEN DER OUDE BEELDEN/
Bij een vergadert door/De Heer Gerard Reijnst/
in syn Leven Hoogwaardig Raad en Schepen/
der Stadt Amsteldam./
(G. Lairesse inv. et fecit/AMSTELODAMI EX OFFICINA
NICOLAI VISSCHER/Cum Privilegio Ordinum Hollandiae
et Westfrisiae.) (text. fig. 9)[17].

---

[16] W. Goeree, *Inleyding tot de Praktyk der algemeene Schilderkonst*, Middelburg, 1670, p. 55, refers to the ICONES which indicates that they were published. (*Van gelyk gaan ook verscheyde Statuën uyt, welke na eenige Beelden en Tronien geteekend zyn, die gevonden worden tot Amsterdam, in het kabinet van den Heer Reynst; doo't syn meest Tronien en Borstbeelden, byna op de zelve wyze als P.P. Rubens syn twaelf Antyken, na de oude marbere steenen geteekend en uyt laeten gaen heeft*). I would like to thank Professor J.G. van Gelder for this reference.

[17] 'Representations of old sculptures collected by the former Gerard Reynst, who was a most distinguished councillor and magistrate of the city of Amsterdam'.

The average size of the plates is 320/30 × 190/200 mm. They are mostly engravings, at times worked up in etching in the background. The following editions of the ICONES are known:
Brussels, Bibliothèque royale de Belgique, Cabinet des Estampes. (The title-page differs, for it omits the 'dum viveret'; probably a first state).
Amsterdam, Rijksprentenkabinet (Inv.no. 325/A 11).
Göttingen, Niedersächsische Staats- und Universitätsbibliothek (Inv. no. 2⁰ Archaeol. I, 1841, fol.).
Paris, Bibliothèque Nationale (Inv. no. AC 12-pt. fol.).
One additional edition in The Hague, Museum Meermanno-Westreenianum (Inv. no. 6A 21 fol.) has the same title-page but was published by Louis Renard (Renouard), a French refugee, who established himself in Amsterdam, in 1703. He collaborated closely with Visscher and may have become a publisher in his own right in 1714-18. The ICONES in the library of the Rijksmuseum van Oudheden, Leiden and in the Stadt- und Universitätsbibliothek, Frankfurt, are restrikes.

Lairesse's title-page shows Father Time who is about to destroy antique sculptures but is prevented by the personification of Prudence. Prudence, according to Ripa,[18] has the gift to judge the past, bring order into the present, and foresee the future. Lairesse, therefore, indirectly praised Reynst for his preserving works of art from the wrath of time.[19]

The ICONES render a basic idea of the antique sculptures owned by Gerard Reynst. Again, only a choice selection was put into print, because a number of additional pieces are known that also seem to have belonged to Reynst but were not engraved for the ICONES. One such case in point is the statuette of a *Youth* in East Berlin (cat. no. 111; Pl. S 111) that was reproduced in the Vendramin catalogue DE SCULPTURIS as *Bacchus*. Later, it was part of the collection of the Great Elector and undoubtedly came from Reynst together with a number of other sculptures also in East Berlin and Dresden.[20]

The engravings in the ICONES reproduce the sculptures rather faithfully. The artists responsible for these prints remain unidentified. The prints themselves have not been noted for their artistic merits either. The first one to refer to them was Clarac (1850), but he merely published a selection of the plates.[21] He was also aware that the identifications on the prints were often most arbitrary and that restorations on the sculptures were not indicated.

Since Knorr von Rosenroth's descriptoin of the antique sculptures seen in the house of Gerard Reynst in 1663 follows the sequence in the ICONES almost *verbatim* we have to assume that all the prints were finished by that year. This excludes Lairesse as the engraver, because he did not move to Amsterdam until 1664, (J.J.M. Timmers, *Gérard Lairesse*, Amsterdam, 1942, pp. 83, 102 and nos. 113, 128-231, attributed all the prints in the ICONES to this artist).

[18] *Op.cit.*, pp. 416-18.

[19] Karl Arndt, 'Chronos als Feind der Kunst...,' *Nederlands Kunsthistorisch Jaarboek*, 23, 1972, pp. 337-38, with a full interpretation of Lairesse's allegorical design. Lairesse must have known François Perrier's title-page for the *Segmenta nobilium signorum et statuarum quae temporis dentem invidium evasere...*, Rome, 1638, where we already find the concept of Father Time gnawing away at antique sculpture.

[20] Cat. nos. 10, 21, 25, 34, 43, 53, 65, 71, 94 and 111. See footonte 28.

[21] F. de Clarac, *Musée de sculpture antique et moderne*, III, Paris, 1850, pp. CCLXV-VI. The sculptures mentioned by him are ICONES 6, 9, 11, B, 15, 1, 12, D, C, 93, 14, 18, 16, 4, 10, 7, 13, 8, A, 2, 5, 3, 17 (enumerated in the order given by him).

[22] Ch. Le Blanc, *Manuel de l'amateur d'estampes*, II, Paris, 1856, p. 484, s.v. Lairesse, no. 71-180; Recueil des statues antiques du cabinet Reynst. 110 p. (without any reference to the title-page by Lairesse). The prints are not listed by Wurzbach under Lairesse. M.D. Henkel, in the entry on Lairesse in *Thieme-Becker*, XXII, Leipzig, 1928, p. 236, attributes them to the artist, but considers them to be of little value.

Clarac, furthermore, indicated that only the title-page was by Lairesse, while the rest of the prints were by different artists. Le Blanc (1856)[22] attributed all the engravings to Lairesse, however, as did Timmers, who enumerated them individually for the first time.[23] The prints are numbered in the plate and their sequence within the individual copies of the ICONES remains more or less constant. The sculptures numbered in letters *A* through *M* were bought by the States of Holland and West Friesland and donated to king Charles II in 1660, as pointed out by Knorr in 1663. One of these busts, the so-called '*Faustina*' (ICONES K), was recently located at Hampton Court.

About half of the prints in the ICONES bear inscriptions identifying the portrayed. As will be shown later, many of these identifications were copied from the Vendramin catalogue DE SCULPTURIS that must have been on hand at the time the engravings were made.[24] Numerous captions are untenable, however, and probably reflect a traditional interpretation. Only those identifications were taken over in the CATALOGUE section that are still acceptable by today's scholarship.[25] Most of the busts represent private portraits executed primarily during the first to third century AD, with a number of later imitations. In many instances descriptive titles were therefore preferred. The prints without captions (ICONES 14–18, 57, 67–98) could not be identified except for ICONES 89.

[23] See under footnote 17.

[24] The Vendramin catalogue DE SCULPTURIS is lost since World War II. I would like to thank Dr. H.-E. Teitge, Deutsche Staatsbibliothek, East Berlin, for allowing me to consult the *Nachlass Jacobs* and use Jacobs' concordance. The present locations for the sculptures are indicated in parentheses: *L* for *Leiden*, *B* for *East Berlin*, *D* for *Dresden*, *H* for *Hampton Court* (see pages 244–247).

[25] I should like to thank Sheldon A. Nodleman for assisting me in identifying the various statues and busts and for reference to the most important relevant literature.

[26] Listed according to Papenbroek numbers (*Pb.*) with the corresponding ICONES numbers added in parentheses: Pb. 87 (7); Pb. 90 (4); Pb. 91 (1); Pb. 92 (2); Pb. 93–94 (6); Pb. 96–97 (3); Pb. 104 (96); Pb. 105 (71); Pb. 109 (86); Pb. 111 (5); Pb. 112 (63); Pb. 123 (49); Pb. 130 (31); Pb. 132 (32); Pb. 137 (98). In addition, the following sculptures must also have come to Leiden via Reynst: Pb. 101; Pb. 114; Pb. 118 (this is the *Bacchus-Indicus* that Oudendorp erroneously identified with the so-called *Plato* in ICONES 95); Pb. 125, and Pb. 144. For all of them, Oudendorp gave a provenance from the Vendramin collection. Finally, the following four monuments reproduced in the Vendramin catalogue DE ANTIQUORUM TUMULIS formerly also belonged to the Reynst collection: Pb. 32; Pb. 46; Pb. 53; Pb. 79. (I owe this identification to Professor H. Brunsting).

The earliest source for the sculptures formerly with Reynst is the catalogue by Franciscus Oudendorp, *Brevis veterum monumentorum ab amplissimo viro Gerard Papenbroekio Academiae Lugduno-Batavae legatorum descriptio*, Lugduno Batavae (Leiden), 1746.

For a brief survey of the Papenbroek bequest, see J.Q. van Regteren Altena and P.J.J. van Thiel, *De portret-galerij van de Universiteit van Amsterdam en haar stichter Gerard van Papenbroek 1673–1743*, Amsterdam, 1964, pp. 34–35, 51–56.

Twenty-six of the sculptures represented in the ICONES 1–98 have been preserved. By far the largest group of seven statues, nine busts and one head, reproduced in ICONES 1–7, 31–32, 49, 63, 71, 86, 89, 96–97 (top) and 98 is in the Rijksmuseum van Oudheden in Leiden. The sculptures were part of the bequest by Gerard van Papenbroek to the University of Leiden in 1738, and later formed the core of the collection of antiquities in the newly founded Rijksmuseum van Oudheden.[26] Their initial installation in the Orangerie of the Botanical Garden in Leiden is known from four drawings by Jacob van Werven (ca. 1745) (text fig. 11).[27] These statues and busts together with five tomb monuments, Palma Giovane's painting in the Six collection in Amsterdam, as well as a scarab in Leiden are the only items from the collection of Gerard Reynst that are still preserved in Holland today. One statue and four busts, reproduced in ICONES 10, 21, 25, 43, and 53 are in the Staatliche Museen, East Berlin. Originally, this group also included an additional statue and three busts which were given in exchange to Dresden, in 1724/26 (ICONES 34, 65/6, 72, 95).[28]

Many of the surviving sculptures are badly worn and are hard to recognize if compared to the corresponding prints in the ICONES. Some of the statues were already heavily restored during the 16th and early 17th century. In a few cases these modern restorations

[27] Gemeente Archief, Leiden; Top. Atlas no. 16002.

[28] Jacobs, *op.cit.*, p. 35, was the first to point out that sculptures once owned by Reynst later appeared in the collection of the Elector of Brandenburg in Berlin. Jacobs identified four (ICONES 10, 34, 65) as well as the statuette of a *Youth* (*Bacchus*), illustrated only in the catalogue DE SCULPTURIS. Neugebauer added ICONES 21, 25, 43, and 53. One further bust, ICONES 95, should be included in this group. The first inventory of antique sculptures in Berlin was made by Chr. von Heimbach, in 1672. (Dr. G. Heres, Staatliche Museen, East Berlin, kindly sent me a transcript). The following twenty-two sculptures were listed (with references to illustrations in Beger's *Thesaurus Brandenburgicus* and present locations, if extant):

1. CAESARIS IMAGO (ThB 352); Dresden.
2. IMAGO POMPEIAE (ThB 352); Dresden.
3. IMAGO LIVIAE.
4. M. AURELII IMAGO.
5. ANTONINI COMMODI IMAGO.
6. IMAGO DEAE LIBERAE.
7. IMAGO DIONYSII (ThB 243).
8. REGIS AEGYPTI IMAGO.
9. REGINAE AEGYPTI IMAGO.
10, 11. DUAE IGNOTA IMAGINES.
12. CAPUT CLEOPATRA.
13. CAPUT GALLAE (ThB 340); Dresden.
14. CAPUT IOVIS (ThB 322, as PLATO); Dresden (ICONES 95).
15. CAPUT PHILOSOPHI (ARISTOPHANES).
16, 17. DUAE CAPITA NOVI OPERIS.
18. MARTIS NUDI STATUA (ThB 341 as TRAJAN); Berlin (ICONES 10).
19. STATUA CERERIS (ThB 286 as TRIPTOLEMUS); Berlin (ex Reynst, but not engraved in ICONES; reproduced in Vendramin catalogue DE SCULPTURIS, I, 18 as BACCO).
20. CUPIDINIS STATUA.
21. STATUA PRIAPI (ThB 261); Dresden (ICONES 65, 66).
22. FIGURA SPHYNGIS.

At the time Beger became curator of the collection, there were still twenty-one sculptures extant. The pieces that he did not reproduce in his *Thesaurus Brandenburgicus*...(1696) may not have been exhibited in the 'Antiquitätenkabinett' proper but in one of the other rooms of the castle.

were removed to such an extent that the statues have survived as fragments only (ICONES 1, 2, 3, 5, 6).²⁹

The overall quality of the sculptures collected by Gerard and Jan Reynst is not very high. Many pieces are adaptations of well known statues: ICONES 14, for example, reflects the famous *Flora Farnese* in Naples (Reinach, I, p. 212, no. 795 D); ICONES 16 is reminiscent of the *Aelia Flaccilla* in the Bibliothèque Nationale, Paris (Reinach, II., p. 668, no. 4); ICONES 27 probably is a Renaissance adaptation of the so-called *Vitellius* in the Museo archeologico in Venice, while ICONES 42 shows the *Jupiter Verospi* type head, known from the statue in the Vatican (Reinach, I, p. 186, no. 666).

The collection as a whole nevertheless signified a rather important moment in the Netherlands as far as collecting of antiquities was concerned. It was by far the largest known collection of antique sculptures of that period. In 1650, the magistrates of the city of Amsterdam were interested in it and considered acquiring it for their new town hall, while Le Blon tried unsuccessfully to sell it to Queen Christina of Sweden.³⁰

The sculptures were also appreciated by Reynst's contemporaries, for a number of statues and busts turn up later in the collections of Jan Six (1618–1700),³¹ of Johan de Vries, burgomaster of Amsterdam,³² Gosuin Uilenbroek (died in 1741),³³ Nicolaes Witsen

---

Since the Elector of Brandenburg was the first to collect antique sculptures one should not exclude the possibility that all of the pieces listed in the Heimbach inventory may have come from Reynst (via Uylenburgh). The first record of a purchase of antique sculptures in Berlin dates from 1696, when the collection of Bellori was acquired. I owe this information to Dr. G. Heres.

²⁹ The restorations on the sculptures formerly with Reynst in Leiden apparently were removed already in the 19th century. See also J.G. van Gelder, 'Jan de Bisschop's Drawings after Antique Sculpture,' *Studies in Western Art, Acts of the Twentieth International Congress of the History of Art*, III, Princeton, N.J., 1963, p. 53 and footnote 11.

³⁰ Mentoined by Le Blon in his letter of 18/28 January 1650, preserved in the Riksarkivet, Stockholm, volume Hollandica 23 (which contains primarily letters from Harald Appelboom, the Swedish commissioner in the Netherlands). The letter was mentioned by K.E. Steneberg, 'Le Blon, Quellinus, Millich and The Swedish Court Parnassus',' *Analecta Reginensia* I, Stockholm, 1966, p. 339. I would like to thank Professor Van Gelder for this reference and Folke Ludwigs for his assistance in locating Le Blon's letter. The relevant passage reads as follows:

'Ma Dame ... Avec unne tres humble & particulliere advertençe du traité finalement conclu avec Monsʳ Reynst, touchant son Amas ou grand Cabinet, Consistant en Statues et pourtraits de Marbre Anticque, Medailles d'or, d'argen & Cuivre, urnes & estranges creations, Tableaux, Livres et presque toute sorte de rarritees, le tout accordé pour le prix de quarante Mille R/d : argent entier & comtant, ou trente Mille R.S. tout le reste sans les tableaux. Mais d'autant qu'il montre par ses quitances que cest amas a cousté plus qu'unne foisa utant, et qu'il s'imagine que v : M : ne l'ayant veu que par ouuie dire, ne le pouant cognoistre, targe a agreer cest Accort, et que le Magistrat de ceste ville comançe a le Carresser pour en optenir les Antiquitees, afin d'en orner leur nouveau Batisment ou Maison de ville, il m'a suplié de

(1641–1717),³⁴ Jeronimo de Bosch I (1677–1767),³⁵ Laurens van Campen³⁶ and Gerard van Papenbroek (1673–1743),³⁷ whose bequest to Leiden in 1738 has preserved the largest and most representative group of sculptures formerly with Gerard Reynst. Yet another Dutch collector who owned sculptures and paintings that once belonged to Reynst was Nicolaas Antoni Flinck (1646–1723).³⁸

Gerard Reynst must have thought very highly of his collection and must have treasured it, to have it reproduced so lavishly. Only a few such series documenting private collections are known in the seventeenth century and Reynst's CAELATURAE and ICONES figure prominently among them.³⁹

The largest of these collections reproduced in print belonged to the Marchese Vincenzo Giustiniani in Rome. Between 1633 and 1637, he employed, among others, Cornelis Bloemaert, Reijnier a Persijn and Theodor Matham (who also worked on the CAELATURAE some twenty years later in Amsterdam) to engrave his collection of antique sculptures. The album with the title *Galleria Giustiniani del Marchese Vincenzo Giustiniani (in Roma)*, was published ca. 1640. It consisted of two volumes, reproducing one hundred and forty-eight statues, one hundred and ten busts, one hundred and seventy-nine reliefs and twelve altars.⁴⁰

prier v:M: que puis qu'elle fait peut estre difigulté d'agreer l'accord susdit, Il plaise a v:M: de n'en divulger le prix a persone du Monde, pour ne lui retrancher l'espoir et l'apparence qu'il a d'en optenir bien davantage et a grande comodité. J'ey optenu finalement qu'il ne peult n'y ne doit penser a rien faire avec persone du Monde que n'ayons une benigne resolution ou reponse de v:M: ou que ie ne le pourei asseurer que le prix susdit &conclud, ne s'evente. etc : Si doncques v:M: a le dezir de s'en faire Maitresse, il sera necessaire de resoudre et m'honnorer de ses bonnes vollontees le plus promptement qu'il sera possible, affin que cest honnest home ne soit reduit a commetre quelque faute. ..."

31 As Professor H. Brunsting established ('Twee gouden eeuwen,' *Archeologie en Historie*, Bussum, 1973, p. 189), Six owned as many as twenty, possibly twenty-three sculptures and at least two reliefs that belonged to Reynst: ICONES 1 (Six catalogue 35); 2 (36); 3 (9); 4 (19); 5 (20); 6 (18); 7 (?) (22); 11 (23); 13 (21); 16 (28); 18 (29); 20 (?) (2); 22 (8); 23 (?) (14); 31 (11); 33 (5); 39 (7); 40 (4); 49 (1?); 55 (10); 63 (6); 89 (15?); 93 (?) (25). (All were listed in the sale catalogue of the Six collection of April 6, 1702 and could be identified through corresponding identifications such as 'consul', 'Tullius Hostilius' etc). Two reliefs should be added to this list; both were reproduced in the Vendramin catalogue DE ANTIQUORUM TUMULIS (fols. 22r, 23r). They are now in the Rijksmuseum van Oudheden, Leiden, where they were identified by Professor H. Brunsting. (Inv.no. Pb. 46; Pls. s 121, s 121a and Pb. 53, Pls. s 120, s 120a).

32 *Catalogus van een Cabinet van uytmuntende Schilderyen, konstige en uytvoerige teekeningen en prenten, van voornaeme Italiaansche, Franse en Nederlandsche Meesters; antique en moderne beelden, en eenige antique penningen; Merendeels by een verzamelt door wylen den wel Ed. Heer Joan de Vries, In zyn wel E. leven Burgermeester en Raed der Stad Amsteldam...*, 13 October 1738, no. 1 : *Faustina minor*; no. 2 : *Augustus*; no. 3 : *Hadrianus*; no.4 :

10

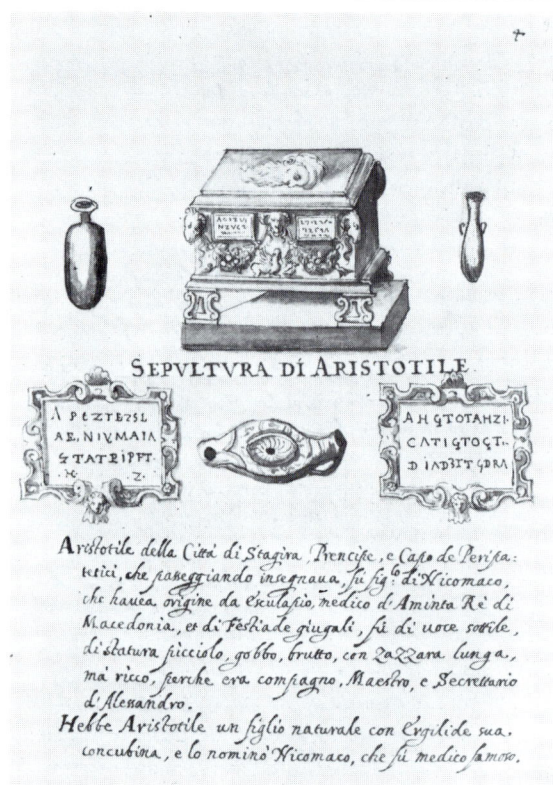

Julia Mammea; no. 5 : *Terentia*; no. 6 : *Een jonge Faunus*. A *Julia Mammea* (ICONES 40) and a *Terentia* (ICONES 39) were listed earlier in the Six collection and we may assume that they are identical with the ones listed with De Vries. Other sculptures in this same collection such as the ones listed under nos. 1, 2, 3 and 6 possibly also came from Jan Six, and ultimately from Reynst. (Kindly pointed out by Professor J. G. van Gelder).

33 Uilenbroek wrote a catalogue of his collection of coins, gems and sculptures which was published posthumously by Sigebert Havercamp, *Museum Uilenbroekianum* (Amsterdam, 1741). The following sculptures, listed on pp. 287–89, were associated with the Reynst collection :

2 *Cleopatrae..insigne Caput velatum* (ICONES 32), Leiden ;
4 *Bacchi sive Bacchae..caput* (ICONES 92), lost ;
5 *Julii Caesaris nudum caput* (ICONES 89), Leiden ;
7 *Hadriani Imp. caput nudum* (ICONES 85), lost ;
8 *Lucillae Imperatricis caput* (ICONES 44), lost
15. *Matronae stolatae Statua* (ICONES 16), lost ;
16 *Imperatoris juvenis palliati, Statua* (ICONES ?), lost ;
17 *Statua gladiatoris* (ICONES 15), lost.

34 The following sculptures from Reynst passed into this collection : the *Bust of Apollo*, today in Leiden (Inv.no. Pb. 104) and the *Male Portrait Head*, also in Leiden (Inv.no. Pb. 137) ; reproduced in ICONES 96 and 98 respectively. Furthermore, a funerary chest, which is reproduced in the Vendramin catalogue DE ANTIQUORUM TUMULIS (Inv.no. Pb. 32). Oudendorp no longer knew that this piece had passed from Vendramin into the collection of Gerard Reynst ; according to him it had been brought from the Orient. The identification of this chest with the illustration in the Vendramin catalogue is due to Professor H. Brunsting.
See cat. no. 117, Pls. s 117, s 117a.

35 The one bust purchased by De Bosch at the Papenbroek sale and donated later to the Academy is reproduced in ICONES 41. It can no

longer be traced in the Rijksmuseum van Oudheden, Leiden. For data on the De Bosch family, see E. Pelinck, 'Een Amsterdams Familiestuk,' *Jaarboek...Amstelodamum*, 44, 1940, pp. 105–09. See also Van Regteren Altena and Van Thiel, *op.cit.*, p. 51.

36 *Catalogus van ... edele gesteentens, ... nagelaten ... Laurens van Campen*, 17 July 1724 ff., under *..Marmere en albaste Beelden, en Beeldwerk, Anticq en modern*, no. 19 : *Een gladiator van Grieks Marmer, voor dezen gekomen uit het Cabinet van de Heer Reinst*. I would like to thank Professor E. H. Begemann for this reference.

37 See footnote 26 above.

38 According to Houbraken (*op.cit.*, II, p. 21), Flinck '..*heeft by een vergaderd een Zaal met de uitgelezenste Konststukken van oude Italiaansche Meesters, als Titiaan, den ouden Palma, Permens, Carats, Guido, N. Ponzyn, ... En tot meerder luister heeft hy daar tusschen nog geplaatst verscheide oprechte marmere Antique statuen ; wel eer gekomen uit de Konstkabinetten van den Hartog van Buckingham, de Heeren Reinst, en Borgermeester Six. In welke te beschouwen hy, (nu omtrent 70 Jaren oud geworden) zyn grootste vermaak vind*'. We may recall briefly that one of the portraits by Titian, engraved for the CAELATURAE by Van Dalen, the so-called *Pietro Aretino*, formerly in the collection of William III (I am grateful to Professor J. G. van Gelder for the reference to the 1713 sale), was last recorded in the list of paintings owned by Flinck, sold in Amsterdam in 1754 (see cat. no. 33). The only time Lugt mentions the Reynst collection in his *Les marques de collections, de dessins et d'estampes*, Amsterdam, 1921, is under no. 959 (N.A. Flinck), where he quotes this passage from Houbraken. The sculptures in Flinck's collection that formerly belonged to Reynst were dispersed sometime before 1754 (see J.G. van Gelder, 'Het kabinet van de Heer Jaques Meyers,' *Rotterdams Jaarboekje*, 1974, p. 168).

A somewhat more detailed description of the antique sculptures owned by Flinck was furnished by Uffenbach, who visited the collector in Rotterdam on No-

The other example that appeared just about simultaneously with the CAELATURAE and the ICONES was David Teniers' *Schilder-Thoneel... uyt de Schilder-Camer van den...Arts Hertog Leopold Wilhelm in't Hoff van Brussel...* (with texts in Latin, French and Spanish). On two hundred and twenty-nine folios, two hundred-and forty-four of the best paintings in the collection of Archduke Leopold Wilhelm were engraved by a number of printmakers under the supervision of Teniers. These two publications of prints after sculptures and paintings in a particular collection may have inspired Reynst to commission one of his own. There were also at least two known precedents by private collectors. Between 1639 and 1641, Lopez had his three well known paintings, Titian's *Ariosto* and *Flora* and Raphael's *Castiglione* reproduced in print by Persijn and Sandrart.[41] Some ten years later, Hollar made engravings after nine Italian paintings from the Van Veerle collection in Antwerp,[42] and in 1669, Jan de Bisschop engraved Annibale Carracci's painting of *Christ and the Woman of Samaria at the Well* and dedicated the print to Jan Six.[43]

vember 27, 1710. The busts of *Geta* (ICONES 45) and *Hadrianus* (ICONES 37, 49 or 61) as well as the statues of a togate figure with a modern head (ICONES 1) and of a dancing matron (ICONES 6) indeed belonged to the Reynst collection. See Zacharias Conrad von Uffenbach, *Merkwürdige Reisen durch Niedersachsen, Holland und Engelland*, III, Ulm, 1754, p. 515. I am grateful to Professor J.G. van Gelder for this reference.

[39] Contemporary with the ICONES was Jan de Bisschop's *Signorum veterum Icones*, published in The Hague in 1668–69, which comprised one hundred etchings after antique sculptures from various sources. Although a few of the sculptures chosen by De Bisschop were in Amsterdam collections at the time (Scholten, Gerrit Uylenburgh), none from Reynst were included. See J. G. van Gelder, 'Jan de Bisschop,' *Oud Holland*, 86, 1971, pp. 1–88.

[40] Clarac, *op.cit.*, III, pp. CCXLVIII–CCLI. See also Luigi Salerno, 'The Picture Gallery of Vincenzo Giustiniani, I,' *Burlington Magazine*, CII, 1960, pp. 21–27. G.E. Rizzo, 'Sculture antiche del Palazzo Giustiniani,' *Bolletino della commissione archeologica comunale di Roma*, XXXII, 1904, pp. 3–66, and XXXIII, 1905, pp. 3–61, believes that the prints were published as early as 1631, because one of them is dated to that year. 1631, on the other hand may indicate the beginning of the project. (Information provided by Professor J.G. van Gelder).

[41] For the Lopez collection see E. Bonnaffé, *Dictionnaire ...*, Paris, 1884, pp. 191–93 and Lugt, *Oud Holland*, p. 114; Wurzbach, *op.cit.*, II, s.v. *Reijner a Persijn*, p. 318, nos. 1 and 5; *ibidem*, s.v. *Joachim von Sandrart*, p. 557, no. 1.

[42] Lugt, *op.cit.*, p. 131 and note 64; Gustav Parthey, *Wenzel Hollar*, Berlin, 1953, nos. 1339, 1345, 1348, 1359, 1367, 1379, 1408, 1455 and 1511.

[43] Formerly in the Six collection, now in the Szépmüveszeti Múzeum, Budapest (Inv.no. 3823; A. Pigler, *Katalog* [1967], pp. 126 f). See also Van Gelder, *op.cit.*, pp. 15, 25.

3. DESCRIPTIONS FROM VISITORS TO THE REYNST COLLECTIONS

AMSTERDAM: AERNOUT VAN BUCHELL (1639)

The earliest description of works of art assembled by Gerard Reynst in the house *De Hoop* in Amsterdam, dates from 1639. In the morning of September 4, Aernout van Buchell visited the collection and recorded his impressions in his *Notae Quotidianae*, preserved in the library of the university of Utrecht.[44] Van Buchell was especially interested in the antique sculptures which apparently were installed in two rooms on the upper floor. He also mentioned illustrated books, coins and three paintings.[45]

Thanks to these diary entries we are for the first time able to connect some of the objects seen in the Reynst collection in 1639 with illustrations in the manuscript catalogues of the collection assembled by Andrea Vendramin in Venice (1627) and with prints after works of art owned by Reynst, reproduced in the CAELATURAE and the ICONES. Instrumental as a connecting link is the *Funerary Chest of Aristotle*, illustrated in the Vendramin catalogue DE ANTIQUORUM TUMULIS (fol. 9r; text fig. 10) and singled out by Van Buchell during his visit. Knorr von Rosenroth, who saw the Reynst collection in 1663, mentioned it once more and transcribed its two Greek inscriptions on the front, thus positively identifying it with the chest reproduced in the Vendramin catalogue. The chest is no longer extant.

---

[44] Ms. 1827, pp. 113 and 145–49. Published by Hoogewerff and Van Regteren Altena, *op.cit.*, pp. 96–100 and Van Campen *op.cit.*, pp. 77–78, 94–96. Van Campen's transcript is preferable since it is more accurate and complete.

[45] Transcript of Van Buchell's entries in his *Notae Quotidianae*, relating to his visit to Reynst: Verstae van neef van Benthem dat eenen Reinst seer curieux sijnde tot Amsterdam getrout met sijn nichte aldaer veel antiquiteijten, cunstige wercken ende fraeyheden van Venetien becomen heeft, uuyt Turckien, Grieckenlant ende Italien versamelt, als sijn statuen, tomben, schilderien, medaillien, prenten ende diergelycke, welcke alsoo de princesse Amelia onbekent ver-socht hadde te sien, wanneer Haere Hoocheyt de Coninginne van Vran(c)ryck geselschapte, ende groot vermaeck hadde in een vroubeelt Cleopatram repraesenterende, werde tselve haer vereert.
(Van Campen, *op.cit.*, pp. 77–78).
3. Septembris met neef Benthem gereist tot Amsterdam, ...
4. 's Morgens besocht de heer Reyns, een treffelick coopman, soon van den generael Reyns, wiens broeder, een groot amateur van de Roomse ende Griexe antiquiteyten, met groote costen boven het faveur van de Republyck veele antique statuen van Venetien aen sijn broeder heeft gesonden, waervan een raer van Cleopatra in marbre, het niet willende vercopen, door begeerte van de stadt Amsterdam tselve de princesse heeft vereert met goede recompense; daervan de selve princesse seer amoureus sijnde ende ialours, sorgende dat de coninginne van Vrancryck daer sinne toe mochte crijgen, het daechs voor haer aencomste hadde laten verbrengen in haer lustpallais tot Rijswijck, dat den prince met grote costen daer heeft doen bouwen. Hebbe daer in den hof cassen gesien met heele antique marmore beelden, grooter als het leven, representeerende vrouwen ende manspersoonen, het een was maer tot de helfte habitu armato was grooter als dandere, credebatur esse Tiberius Imperator. In 2 cameren boven den anderen staen seer veele Deorum Dearumque, Imperatorum et Imperatricum aliorumque magno nu(me)ro vultus,

12

pectore tenus. Spectabantur et urnae antiquae plures, inter quas una ex marmore pulcra elaborata cum inscriptionibus Graecis. Coperculum continebat Aristotelis effigiem, credebanturque cineres in illa eiusdem conservatos. Lychnae plures antiquae diversae formae aliaque suppellectilis Romana, antiquae icunculae ex marmore, ex aere. Mirabar tantum antiquarum rerum thesaurum apud unum Batavum invenire cum vix in media Italia apud ullum reperirentur. Librorum aderat praeterea copia in quibus de Romanis antiquitatibus plura tractabantur, inter quos quinque Aeneai Vici, quorum duos habui, in quibus 12 imperatorum et imperatricum numismata ab illo expressa. Vidi et Petrarchae Amores ante plures annos ex ipso archetypo descriptos, cuius literae non valde absimiles illis, quibus porcaria poema tempore Nicolai V pontificis Romani scriptum erat inscriptum, quod apud me in membranis servatur, circa nempe 1450 videtur, 75 annis post ipsius Petrarchae obitum.

Vidi etiam librum ibidem oblongum ex purissimis membranis in quo recentiore manu ad vivum quam maxime fieri potuit suis coloribus artificiosissime depicti pisces omnis generis, marini, fluviatiles, conchae, item et alia maris frutices corallini etc. qui liber 1000 flor. pretio aestimabatur.

Vidi et varias res petrificatas, lapides quoque scissiles in quorum (intestinis) pisces spectabantur adeo integre ac si iam primum ex aqua depormerentur, quales in Saxonia iuxta Lutgeri patriam reperiri tradit Munsterus. Nec deerant numismata antiqua, Graeca, Romana, inaequali magnitudine et forma, aerea, argentea, aurea.

Spectabantur et apud eundem maximi pretii picturae a summis artificibus factae, inter quas una quam nuper in auctione Uffeliana ipse emerat 900 flor., quae in catalogo ita exprimitur: 'Het conterfeitsel van een man op syn antycqs. Halve figuer, seer raer, van Titiaen.' Eminebant inter omnes tabulae satis magnae, quibus nullo quantumvis magno pretio carere vellet, prior Titiani manu Diva Virgo cum filio et Josepho, posterior manu Corregii in qua vir admodum vivide expressus, quam tabulam vel quovis pretio ambivit Britanniae rex, sed frustra, cum vel multa millia offerentur. Adeo apud quosdam picturae ars in pretio est ut ad antiquorum aestimationem proxime accedere videatur.
(Van Campen, *op.cit.*, pp. 94–96).

The antique statues displayed in cases in the courtyard and the many busts of 'gods' and 'goddesses' installed in two rooms on an upper floor probably also came from Vendramin. None of the paintings Van Buchell pointed out can be traced to this Venetian collection, however. The *Portrait of a Man with his Antiques, Half-figure, very rare, by Titian*, most likely is identical with Lorenzo Lotto's *Portrait of Andrea Odoni* which was recorded in the Odoni family in Venice as far back as 1532. It was part of the 'Dutch Gift' in 1660, and is now at Hampton Court (cat. no. 16; Pl. P 16a).[46] More tentative is the identification of Correggio's *Portrait of a Man* with Lotto's *Portrait of a Gentleman*, also at Hampton Court (cat. no. 17; Pl. P 17a),[47] while Titian's *Holy Family* can no longer be traced. The Reynst brothers, therefore, also bought paintings on an individual basis, probably directly from other Venetian families. Judging from the descriptions by Van Buchell and Knorr, the objects were kept together and exhibited in groups (text fig. 12).[48] According to Van Buchell Gerard Reynst had received many art objects and beautiful things from Venice which his brother had shipped to him at great expense, and which came from Turkey, Greece and Italy. Among them were statues, tomb monuments, paintings, medals, prints and other items. We may suppose that the many costly sculptures sent to Holland by Jan were the ones purchased recently from Andrea Vendramin.

[46] Van Buchell's inference that the painting had passed through the Van Uffelen sale on April 9, 1639, in Amsterdam, led Hoogewerff, Van Regteren Altena (*op.cit.*, p. 99, note 3) and others to identify it with Titian's *Ariosto*, now in the National Gallery, London, a supposition correctly rejected by C. Gould, *National Gallery Catalogues, The Sixteenth-Century Venetian School*, London, 1959, p. 116, note 15.

[47] Also suggested by Hoogewerff and Van Regteren Altena, *op.cit.*, p. 100, note 1.

[48] Judging from this illustration in the Vendramin catalogue and from various representations on title-pages of some early printed catalogues of collections, the objects were exhibited within a rather confining area, crowding each other on shelves that completely filled the available wall space (see Julius von Schlosser, *Die Kunst-und Wunderkammern*, Leipzig, 1908, figs. 88–91).

Van Buchell informs us further that Princess Amalia van Solms asked to visit the collection at the time she was entertaining Queen Marie de Medicis, in September 1638. The Princess was most impressed by the sculpture of *Cleopatra* and succeeded in receiving it in form of an official gift. Reynst apparently did not like to part with it and had to be persuaded by the city of Amsterdam to present it to the Princess in exchange for a good remuneration.[49] This *Cleopatra* may have been a marble copy after the famous antique sculpture of the *Sleeping Ariadne* in the Vatican.[50] The inventories of Princess Amalia's possession in the Huis ter Nieuwburg list two sculptures representing a *Cleopatra*, however, a reclining one, based on the Vatican original: *Een figuir van witte marber representerende eene Cleopatra leggende, lang 2 voeten*,[51] and a larger one, representing her seated: *Een figuer van eene Cleopatra sittende, omtrent vier voeten hoog*.[52] Since neither sculpture seems to have been preserved in the Netherlands, it is no longer possible to decide which of the two came from Gerard Reynst's collection.

The Princess had the sculpture brought to her 'lustpallais' at Rijswijck, the Huis ter Nieuwburg, the day before Marie de' Medicis was to arrive, in order to impress the French Queen. The seated *Cleopatra* which was about twice the size of the reclining one, would have been better suited for this purpose.

[49] *Thesauriersrekening* of the city of Amsterdam, 1638, p. 110 v. See Hoogewerff and Van Regteren Altena, *op.cit.*, pp. 97–98, note 3, and S.W.A. Drossaers and Th.H. Lunsingh Scheurleer, *Inventarissen van de Inboedels in de verblijven van de Oranjes en daarmede gelijk te stellen stukken 1567–1795*, I, *Inventarissen Nassau-Oranje 1567–1712*, (*Rijks Geschiedkundige Publicatiën*, 147), 's-Gravenhage, 1974, p. 260, note.

[50] W. Amelung, *Die Sculpturen des vaticanischen Museums*, II, Berlin, 1908, pp. 636–43, no. 414, fig. 57.

Amelung mentions only two replicas that have been preserved, one in the Museo archeologico, Florence, another one at Wilton House, England. W. Müller, 'Zur schlafenden Ariadne des Vatikan,' *Römische Mitteilungen*, 53, 1938, pp. 164–74 accepts one more replica in Madrid. Müller identified one further variant in bronze in the Louvre which was cast in Fontainebleau ca. 1540 from molds taken from the sculpture in the Vatican. See also Hoogewerff and Van Regteren Altena, *op.cit.*, pp. 97–98, note 3.

[51] Drossaers and Lunsingh Scheurleer, *op.cit.*, p. 260, no. 637; with reference that Willem II mentions a *Cleopatra* in a letter to his father Frederick Henry in August, 1639. The passage reads: *…'ick sal U.H. oock seggen dat het boovenste van de nicke van Kleopatra is al bijcans ghemaeckt…'* (see S.I. van Nooten, *Prins Willem II*, 's-Gravenhage, 1915, p. 32). A reclining *Cleopatra* was in Dresden by 1733 (reproduced in Le Plat, pl. 16 top).

[52] Drossaers and Lunsingh Scheurleer, *op.cit.*, p. 260, no. 641. These authors suggest that the *Cleopatra* was placed either in the garden or

AMSTERDAM: BARTHOLD NEUHAUSER (ca. 1650)

About 1650, Barthold Neuhauser, the later bishop of Erfurt, paid a visit to the Reynst collection to see in particular its Egyptian art. Neuhauser probably came in order to make drawings after some of the more interesting objects for his friend, Athanasius Kircher (1601–80).[53] Kircher wanted to study original Egyptian hieroglyphs which he claimed (unjustly) to have deciphered. Nineteen of Neuhauser's drawings after eleven works of Egyptian art from the collection of Gerard Reynst were incorporated by Kircher and reproduced in the third part of his *Oedipus Aegyptiacus*, a history of the wonders and mysteries of ancient Egypt, published in Rome between 1652 and 1654 (pp. 435, 457, 458, 511, 523–25).[54] Thanks to these drawings we have proof that Reynst owned several pieces (text figs. 13–22).

Only the three scarabs (text figs. 19–21b) can be traced to later Dutch collectors. These pieces were reproduced in 1695, in the second part of the *Dactyliotheca* by Abraham van Goorle, republished by Jacob Gronau (nos. 557, 558, 559). The scarab with the knight on horseback was identified by Stricker in Leiden (text figs. 21, 21a, b). It had been acquired in 1841 from the Utrecht apothecary G.J. van Klinkenberg. Although Stricker mentioned in concluding that all or at least a large section of the Vendramin

in the gallery of the house ter Nieuwburg. Hoogewerff and Van Regteren Altena, *op.cit.*, pp. 97–98, note 3, thought that based on Jacob de Hennin's *De Zinrijke Gedachten...*, Amsterdam, 1681, p. 9, the sculpture probably was installed in the garden of Honselaarsdijk. A seated *Cleopatra* was listed once more in the inventory of 1702 of the Huis Honselaarsdijk; Drossaers and Lunsingh Scheurleer, *op.cit.*, p. 475, no. 580: *Een sittende Cleopatra van marmer op een houdte pedestal*. Everything included in this second inventory was ceded to Prussia.

[53] Kircher had brought together the largest encyclopaedic collection in Italy, exhibited in the Jesuit college in Rome. The first catalogue of the *Musaeum Kircherianum* was written by Georgius de Sepibus (Amsterdam, 1678).

[54] B.H. Stricker, 'De Verzameling Reynst, Egyptische Antiquiteiten,' *Vooraziatisch-egyptisch Gezelschap 'Ex Oriente Lux', Mededeelingen en Verhandelingen*, VII, 1947, pp. 261–64, with illustrations, was the first to associate Kircher's drawings with works from the Reynst collection.

1. *Funeral Vase*
Reproduced in Kircher recto and verso, with representations of a mummy and a hieroglyphic inscription. Reproduced identically in the Vendramin catalogue DE SACRIFICIORUM, ET TRIUMPHORUM VASCULIS.., fol. 34. (text figs. 13, 13a).[55]

2. *Herm*
With illegible inscription. (text fig. 14).

3. *Head of Ramses II*
With an inscription identifying the portrayed.
Possibly not from Reynst. (text figs. 15, 15a).

4. *Slab with two Human Figures*
With illegible inscriptions. (text fig. 16).

5., 6. *Two Hellenistic Amulets*
(text fig. 17).

7. *Fragment of a Sphinx*, with inscription referring to Ramses II. (text fig. 18).

8.–10. *Three old Egyptian Scarabs.*
Inscriptions and ornamentation added in hellenistic time, according to Stricker. All three scarabs were illustrated in the Vendramin catalogue DE ANNULIS, ET SIGILLIS, AEGYPTIORUM SCARABAEIS..[56], fol.12 (text figs. 19, 19a); fol 13 (text figs. 20, 20a) and fol 14 (text figs. 21, 21a, b; Rijksmuseum van Oudheden, Leiden ; 43 mm. long, 30 mm. wide, 15 mm. high).

11. *Old Egyptian Heart Scarab.*
Possibly reproduced in the same Vendramin catalogue, fol. 10 (text fig. 22).

---

[55] According to Stricker probably late with a base that does not belong to the original vase.
The Vendramin manuscript is in the Bodleian Library, Oxford (Ms. D'Orville, 539).

[56] The manuscript is in the British Museum, London (MS. Sloane 4005).

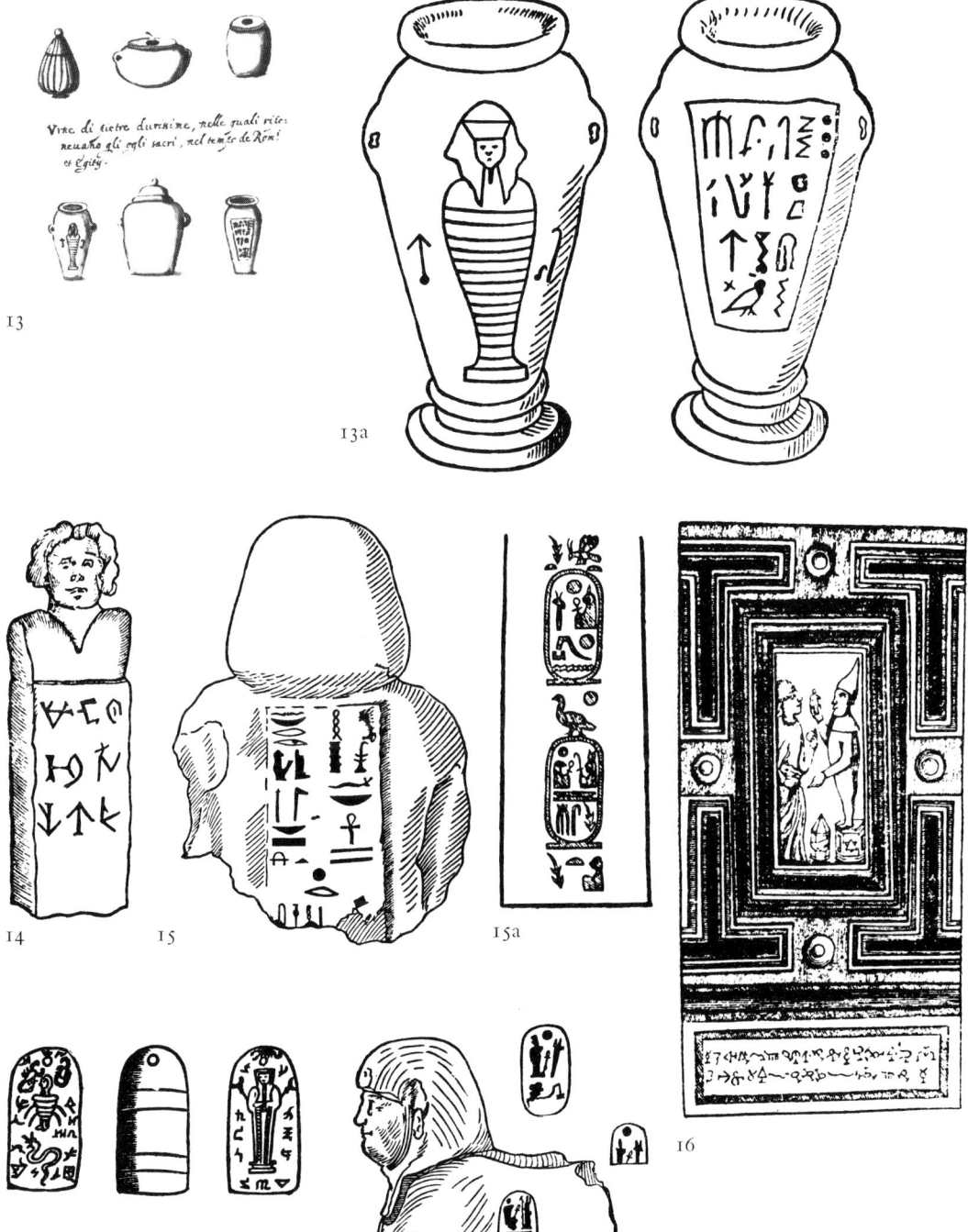

13-18

19-22

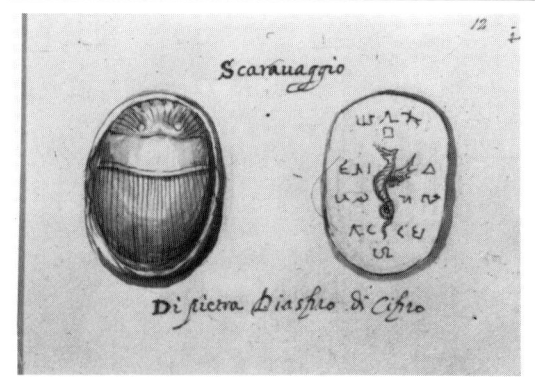

19

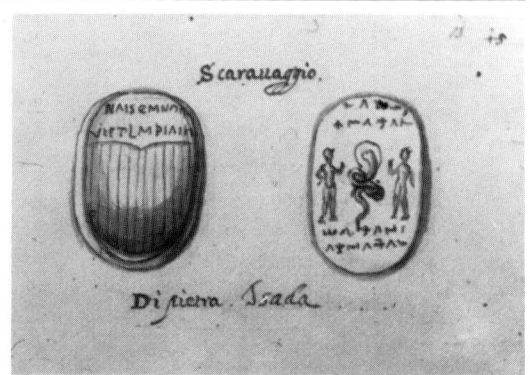

20

19a

20a

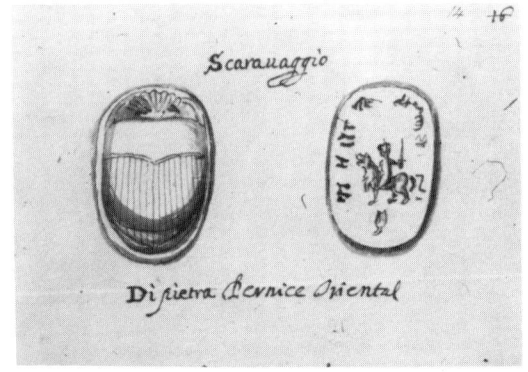

21

21a

22

21b

collection in Venice passed into the possession of the Reynst family, he did not establish that the objects known through these drawings by Neuhauser originally belonged to Andrea Vendramin, since they were reproduced in the fifth volume of his manuscript catalogue DE SACRIFICIORUM, ET TRIUMPHORUM VASCULIS (fol. 34) and in the eighth volume, DE ANNULIS ET SIGILLIS AEGYPTIORUM (fols. 10?, 12, 13, 14). Thanks to these illustrations in Kircher we have proof that Gerard Reynst had acquired works of Egyptian art from Vendramin.

AMSTERDAM: CHRISTIAN KNORR VON ROSENROTH (1663)

The fullest description of the collection owned by Gerard Reynst was given by Christian Knorr von Rosenroth, who visited the house on the Keizersgracht in the spring of 1663, or five years after the death of Reynst and three years after the best Italian paintings and twelve antique sculptures had been purchased from the collection by the States of Holland and West Friesland.
Knorr (1636–89),[57] who studied philosophy, theology and law at the university of Leipzig, undertook an extended trip through the Netherlands, France and England from 1663–66 to further his knowledge. He recorded his impressions in his *Itinerarium*, a manuscript in Latin, preserved at the library in Wolfenbüttel (West Germany).[58] The diary entries of his visit to Amsterdam were published separately by Fuchs.[59]
Knorr began his description of his visit to the house *De Hoop* with the paintings which hung in the first room. Some of them, according to him, were valued at 4,000 guilders and a number of them had been presented to king Charles II. Among the few he deemed worthwhile mentioning were a *Story of Susannah* (one of the versions of Reni's painting ; CAELATURAE 28 ; cat. no. 23 ; Pl. P23a); *The Head of St. Bartholomew* (lost; CAELATURAE 6; cat. no. 24; Pl. P24); a *Representation of Rustic Life* (lost; not engraved); a *Battle Scene* (lost; not engraved); *A Tavern* (Jan Liss, Cassel; CAELATURAE 30; cat. no. 15; Pl. P15a); *The Salvator Mundi* (lost; not engraved; possibly identical with Giovanni Bellini's painting reproduced in the Vendramin catalogue DE PICTURIS, fol 1)[60]; a full-length figure of *Mercury* (lost ; not engraved) and of *Mars* (lost ;

---

[57] Fuchs, 'Christian Knorr von Rosenroth," *Zeitschrift für Kirchengeschichte*, XXXV, 1914, pp. 548–83 ; K. Salecker, *Christian Knorr von Rosenroth*, Leipzig, 1931; J. N. Jacobsen Jensen, *Reizigers te Amsterdam, Suppl.*, Amsterdam, 1936, p. 19, S. 8; Friedhelm Kemp,

'Christian Knorr von Rosenroth,' *Neue Zürcher Zeitung*, 9 May 1971, pp. 51–52 (Fernausgabe, no. 125).

[58] *Anonymi cuiusdam Itinerarium.* 253. 1. Extr. 4⁰. Ao 1663 incoeptum. pp. 57–119 (see Fuchs, *op.cit.*, p.571, II, no. 9).

[59] J. C. Breen and Fuchs, 'Aus dem 'Itinerarium' des Christian Knorr von Rosenroth,' *Jaarboek van het Genootschap Amstelodamum*, 14, 1916, pp. 239–45.
VI. In Aedibus Dn. G. Reynst Mercatoris ditissimi in fossa

not engraved), as well as full-length figures of *Nudes*. Knorr continues with a *Journey of Merchants* (Pieter van Laer, Naples; CAELATURAE 19, cat. no. 40; Pl. P40a); *The Transfiguration of Christ* (lost; CAELATURAE 13; cat. no. 37; Pl. P37) and *The Virgin* (lost; not engraved) and ends with 'etc.', implying that there were more paintings.

The second room contained fifty antique sculptures with a few additional modern pieces. The enumeration which follows and which supposedly was assembled by Knorr, is so close to the ICONES not only in the identification of the statues and busts but also in the sequence they are listed in that one can hardly imagine that Knorr could have written it without consulting a copy of the ICONES. Not only do the identifications correspond to the respective names on the prints, but they also seem to stop whenever the engravings are before letters. Therefore, it seems safe to assume that the prints in the ICONES already existed by 1663, the year of Knorr's visit (see also p. 47, note 17). The sculptures installed in this second room numbered about fifty-five, because the identifying names correspond to the prints in the ICONES numbered 1–13, 19–41, 43–55, with 14–18 and 42 missing.[61]

The sculptures in the third room were described very briefly, starting with *Sylvius Posthumus* (ICONES 56) and continued with

Imperatoria habitantis sequentia videre licuit. In Conclavi primo asservabantur picturae nobilissimae et pretiosissimae, inter quas tabulae 4000 fl. aestimatae, e quibus aliquas coemtas Ordines Hollandiae Regi Angliae dono dederunt. E paucis, quas notare licuit, sunt: Historia Susannae: effigies S. Bartholomaei: Idea vitae rusticae: Praelium Taberna. Salvator mundi, Mercurius integer; Mars integer; itidem nudi; delineati saltem: Iter mercatorum: Transfiguratio Christi. S. Maria etc. etc.

[60] Reproduced in Tancred Borenius, *The Picture Gallery of Andrea Vendramin*, London, 1923, pl. 1.

[61] Knorr's enumeration matches the plates in the ICONES as follows (with the respective plate numbers added in parenthesis):
In Conclavi secundo Statuae antiquae marmoreae locum habebant, cum paucis recentibus ; omnium numerus 50 erat. Catalogum contexui sequentem : Consul Romanus togatus (1). *Tullus Hostilius, pectore tenus* (2). *Venus pudica nuda, manum inquinibus praetendens* (3). *Flora nuda* (4). *Venus alia* (5). *Abundantia cum cornu copiae* (6). *Hercules cum exuviis leonis pectore tenus ; artificio insigni elaboratus* (7) ; *Iulia pectore tenus* (8). *Adonis nudus* (9). *Gladiator nudus dextra pugionem, sinistra vaginam tenens* (10). *Apollo nudus cum arcu* (11). *Cupido nudus* (12). *Hermaphroditus nudus cum Salmace mutata, vestibus virgineis, pudendo masculino* (13). *Hercules alius, pector* (19). *Augustus pect.* (23). *Tiberius pect.* (24). *Caesar pectore tenus artificiosissime sculptus, Graecum opus* (20). *Agrippina maior pector. praestantissima, vetustate paululum exesa* (21). *Poppaea Sabina* (22). *Calphurnia* (25). *Octavia* (26). *Vitellius* (27). *Domitianus* (28). *Pallas cum sphynge in galea* (29). *Nervonus* (30). *Livia* (31). *Cleopatra* (32), *Traianus* (33). *Gordianus* (34). *Helena* (35). *Octavia aliter* (36). *Hadrianus* (37). *Caracalla* (38). *Terentia* (39). *Julia Mammaea* (40). *Agrippina minor* (41). *Aristaeus* (43). *Pallas cum triplici sphynge in galea* (44). *Geta* (45). *Flavia : pectore tenus omnes* (46). *Faunus cum pelle caprina ridens* (47). *Antonius* (48). *Hadrianus aliter* (49). *Cupido alius* (50). *Faunus alius* (51). *Cyrus tiara tectus* (52).

Papyrius (ICONES 59). The detailed enumeration ends with the *Priap* (ICONES 65/66), just as it does in the ICONES itself, where the plates following this herm are all before letters. Knorr expresses this indirectly by concluding 'with many other unknown sculptures'. He did mention a bust of *Cicero*, however, and a *Fauna*, possibly ICONES 92.

Finally, Knorr stated that eleven sculptures (he omitted the bust of *Domitian*, ICONES H) were bought together with the above mentioned paintings by the States of Holland and presented to king Charles II of England. As indicated earlier, one of these busts (ICONES K : *Faustina* ; Pl. S 108) was identified at Hampton Court and Knorr's statement, therefore, is trustworthy.[62]

Knorr continued enumerating other objects,[63] among them the chest containing the ashes of Aristotle, a foot and a half long, and a foot wide and high. This is the very funerary chest that was reproduced in the Vendramin catalogue DE ANTIQUORUM TUMULIS (text fig. 10) and discussed for the first time by Van Buchell in 1639. The foot of a large statue (*pes colosseae statuae*) is still extant in Leiden (cat. no. 116 ; Pl. S 116).

Other objects enumerated by Knorr such as horns of goats, corals, antique lamps, fish, shells, pieces of wood and so on, most likely came from Vendramin, because they seem to correspond to items illustrated in various catalogues of this Venetian collection.

Flavia alia (53). *Lucilla* (54). *Faustina* (55) ; *pectore tenus pleraeque.*

[62] Knorr's list reads as follows (with the corresponding ICONES numbers in parenthesis) :
In tertio conclavi adhuc aliae erant statuae : nempe, Sylvius, Pothumus (56).*Papyrius* (59). *Antonia maior* (60). *Hadrianus denuo* (61). *Claudia* (62). *Bacchus* (63). *Severus, pectore tenus* (64). *Priapus inguine tenus* (65/66) ; *cum incognitis aliis quam plurimis. Item Ciceronis caput integrum cuius facies alibi quoque e marmore saltem elevata conspiciebatur. Fauna itidem* (92) *etc. Atque hinc ablatae sunt statuae Sabinae* (A), *Caracallae* (B), *Aesculapii* (C), *Cupidinis* (D), *Scipionis Africani* (L), *M. Bruti* (M), *Commodi* (I), *Faustinae* (K), *Tiberii* (G), *Vestae* (E), *Cybeles* (F),

etc. ab Ordinibus Hollandiae cum supradictis picturis Angliae Regi oblatae.

[63] *Praeterea in hoc conclavi reposita erant sqq :*

Loculus cinerum Aristotelis marmoreus, cuius operculo insculpta erat facies Philosophi elevata e lapide : lateri anteriori inscripta erant.

   hinc
ΑΡΙϹΤΟΤΕΛΗϹ
Ο ΚΑΛΛΙϹΤΟϹ ΤΩΝ
ΦΙΛΟϹΟΦΩΝ

   illinc
ΑΡΙϹΤΟΤΕΛΗϹ
ΟΝΙΚΟΜΑΧΟΥ
ΦΙΛΟϹΟΦΟϹ

longitudo cistae erat sesquipedalis,

latitudo cum altitudine unius pedis ; figura huius cistae expressa extabat in libro quodam A. 1612 Patavii impresso sub titulo : Aenigma Aristotelicum. Similes loculi adhuc duo inibi servantur, alter cum inscriptione latina, alter sine inscriptione ;

Urnae quaedam marmoreae albae, angusto orificio, insigni magnitudine ; Urnae minores aliae similes ; Membra quaedam humana marmorea, inter caetera pes colosseae statuae, arte multa insignis ;

Cornu caprae Indicae ;
Cornu ibicis ;
Corallinum fruticosum longitudine quatuor pedum ;
Lampades sepulchrales antiquae viginti duae, variarum figurarum ;

The last foreigner to mention his visit to the house of Reynst was Prince Cosimo de' Medici, in January, 1668. Soon afterwards Reynst's widow, Anna Schuyt, went to live with her son, Joan Reynst and the collection eventually was sold in the spring of 1670.

VENICE: CARLO RIDOLFI (1646)

Ridolfi not only was acquainted with Jan Reynst personally, but he was also the only contemporary person who referred to him as a collector of paintings. Thanks to Ridolfi, who listed twenty-one paintings in his *Maraviglie dell'Arte*[64] of 1648 that supposedly were owned by Jan and exhibited either in his 'galleria' in Venice or in his house(s) in Amsterdam, we are able to obtain some idea of the type of paintings Jan was interested in. Only four of them are extant and are now at Hampton Court: the two paintings by Schiavone of *Christ before Pilate* and the *Judgment of Midas*, the Titian school piece of *Madonna and Child with Tobias and the Angel* as well as Veronese's *Marriage of St. Catherine* (cat. nos. 28, 29, 36, 38).[65]

We know nothing about the antique sculptures Jan was supposed to have collected, despite the fact that Van Buchell called him a 'groot amateur van de Roomsche ende Griexe antiquiteyten'.[66] No documents are known either that would tell of purchases or shipments of works of art by the Reynst brothers.

Figura fasciculi florum urnulae impositorum, cum papilionibus quibusdam, acu picta ;

Piscis Columbus, Orbis species, globosus ferme ;
Historia perpetua picturis Illustrium virorum omnia ab orbe condito, usque ad Ottonem M. expressa : opus antiquum ;

Conchylia varia ;
Petrefacta multa ; inter caetera Os Ilium, quod diffractum interius naturam osseam adhuc retinebat : Lingua vitulina : fungi, ligni, frusta, conchae, nautilus etc. ; praecipue tres pisces lapidibus inclusi, non adumbrati saltem per naturam ut Islebici, sed excavato mirum in modum lapide ita inserti, ut spinae parte una, parte altera illarum vestigia manifesto

appareant. In Occidentali enim littore Italiae cum continuatis per multum tempus Subsolani flatibus mare paulum a terra recedit, limus quandoque partim adurente sole partim exiccante vento ita indurescit, ut lapidis naturam assumat, atque tum si vel pisces in eo vel conchae vel ossa relicta haeserint, lapide haec deinceps quasi vestita reperiuntur, fissaque massa interius quod latuit mirifice oculis subicitur. Porro peculiari scrinio repositae erant antiquitates Aegyptiacae variae, aereae potissimum ; nempe Isidis Icunculae quam plurimae diversae magnitudinis ; Osiridis effigies pedis altitudine, Astacurum figurae, Apis simulacrum aeneum rauri figura, cum tubulo ori inserto, per quem oracula quondam edidisse creditur. Sphinx item et idola varia

alia ; Unguentaria et lachrymatoria marmorea diversae magnitudinis quam plurima ;

Cista cum capsulis multis, in quibus gemmarum asservabatur magna quantitas, antiquarum praesertim, cum sculpturis diversis ;
Duae cistulae numismatibus antiquis plenae, inter quae nummus Pertinacis IMP. CAES. P. HELV. PERTIN. AVG. Cap. Aug. rad. et barbatum. PROVIDENTIAE... COS. III. Providentia stolata stans utramque manum coelum versus protendens ;

Nummus Severi, cum tribus in aversa figuris stolatis cornua copiae tenentibus ;

Num. Pescennii Negri cum inscript. Graeca : aversa exhibet Dianam polymammiam Ephesiam ...ΕΦΕΣ... ΔΙΑΝ... ;

4. THE COLLECTION OF ANDREA VENDRAMIN

One collector mentioned repeatedly in this context who was of such significance with regard to the formation of the collections assembled by Gerard and Jan Reynst was Andrea Vendramin (1565?–1629) in Venice. Although no documents are known that would confirm that the Reynst brothers purchased much or all of the Vendramin collection, the connection was still known to Cosimo de' Medici, who mentioned in 1668, during his visit to the collection in Amsterdam that the Reynst family had acquired it from Vendramin in Venice.[67] Uilenbroek was also aware of it in 1729, when he wrote that Reynst had brought the antique sculptures from Italy.[68] This tradition finds further support in objects owned by Gerard Reynst that can be traced back to illustrations in the manuscript inventories of the so-called *Musaeo Andreae Vendrameno*. Thus it seems appropriate to dwell briefly on this Venetian collector.

Andrea Vendramin, a member of the well known Venetian family, had assembled a rather large collection of antiquities, *naturalia* and about one hundred forty paintings, referred to in 1615 by Vincenzo Scamozzi in his *Dell' idea dell' architettura universale*.[69] The entire holdings were catalogued in manuscript form, possibly with a sale in mind. Each object was reproduced in a simple drawing in pen and brown ink and brown wash, accompanied by titles and at times by explanatory texts. This inventory of the *Musaeo Andreae*

Num. Prusiae, Mithridatis, Herculis in cuius aversa Zodiacus. Ottonis, Magnentii, Graecarum civitatum etc. etc. ;

Sex Typi nummorum chalybei antiqui, quorum unus faciem Claudi exhibebat, reliqui aversarum saltem erant ; monumentum antiquitatis insigne ;

Nummus Amstelodamensis in memoriam depulsi ab Urbe Principis Auriaci Friderici Henrici ; inscr. :
Ex uno crimine omnia nosces. Equus ferociens.

Magnis tamen excidit ausis. Phaeton ; Nummus in memoriam Synodi Dordracenae. Inscr. : Asserta religione. Concilium Theologorum. Erunt ut mons Sion

MDCXIX Templum in monte, sole illustratum ;

Nummi in memoriam pacis Belgicae, cum Hispanis, item cum Anglis ;

Nummus Rudolphi II aureus ; Effigies Urbani VIII aenea in nummo ;

Cistulam alteram Laocoonornabat aeneus, in altera rosa Hierichuntina asservata erat.

64 *Op.cit.*, I, pp. 62, 145, 201, 226, 251, 267, 289, 340; II, pp. 55, 85; see also Savini-Branca, *op.cit.*, pp. 269–70.

65 Gerard, as 'universal heir', inherited Jan's paintings in 1646. Four of them were engraved and included in the CAELATURAE (18, 36–38). The seven (six ?)

pictures mentioned by Ridolfi as in Jan's collection that turned up in the forced sale of paintings owned by Nicolò Renieri in Veniceon December 4, 1666 (cat. nos. 44, 47, 48, 51–53, 57) are all lost. The remaining eight paintings, also mentioned by Ridolfi, are no longer identifiable.

66 Van Campen, *op.cit.*, p. 94.

67 G.J. Hoogewerff, *De twee reizen van Cosimo de' Medici, Prins van Toscane, door de Nederlanden (1667-69)*, Utrecht, 1919, pp. 77–78 : 'Martedì 3 (January, 1668)... di lì a casa dei Reims, parenti del borgomastro di tal nome, già nominato, che anno una buona raccolta di pitture antiche e de' migliori maestri di Lombardia, residuo di un gran gabbinetto

Vendrameno was finished by 1627. It consisted of seventeen small volumes, six of which are still extant today.[70] Thanks to these drawings a sizeable part of this former Vendramin collection can be reconstructed and many items are later found in the possession of Gerard Reynst. Most important with regard to the Reynst collection are the three manuscript catalogues of DE SCULPTURIS, DE ANTIQUORUM TUMULIS and DE PICTURIS since a number of objects illustrated in them are still known today.

Jacobs[71] was able to establish that a total of sixty-four of the over two hundred antique statues and busts reproduced in DE SCULPTURIS could be matched against a corresponding number of prints in the ICONES. This means that about two thirds of the best antique sculptures assembled by the Reynst brothers came from Andrea Vendramin. Fourteen are still known, nine of them are in Leiden, four in East Berlin and one is in Dresden. Other sculptures that were not engraved for the ICONES but were listed in old inventories with a provenance from the Vendramin collection, probably were also illustrated in DE SCULPTURIS.

The catalogue DE ANTIQUORUM TUMULIS contains the 'sepultura di Aristotile' which is of particular interest since it proves that by 1639, objects from the collection of Andrea Vendramin were in the house of Gerard Reynst in Amsterdam. In addition, Professor H. Brunsting was able to identify five tomb monuments in this

comprato da' loro antenati in Venezia del Vandramino, et oltre alle pitture si viddero alcune statue e marmi greci e latini con quantità di medaglie, tra le quali alcuni coni, delle più antiche, cosa in vero per la rarità stimabilissima.'

[68] In the sale catalogue of his books (1729) Uilenbroek added the following comment after the Vendramin catalogue DE SCULP-TURIS: *Marmora et statuae quae Ampl. D. Gerardus Reinst ex Italia in has oras attulit ac edidit in hoc Manuscripto inter alia reperiuntur delineata et ex hoc museum translata sunt.*

[69] Venice 1615, part I, p. 305: *Il Chiarissimo Sig. Andrea Vendramin a San Gregorio nella sua casa sopra Canal grande ha disposto due stanze, dove con triplicato ordine si ritrovano non poche statue, e cento quaranta petti di varie grandezze, e torsi, e bassirilievi, e vasi, e pietre nobili ed altre petrificate, e buon numero di medaglie antiche, e sette statue del Vittoria in un suo scrittoio d'olivo ed ebano e forse cento quaranta quadri grandi e piccoli di buone pitture.* See also Savini-Branca, *op.cit.*, p. 285. How quickly the memory of this Vendramin collection had faded in Venice can be judged from Ridolfi's *Maraviglie*. By 1646, he could only remember one painting that he believed to be still in the possession of Andrea Vendramin, namely Giorgione's *David with the Head of Goliath*, reproduced in the Vendramin catalogue DE PICTURIS, fol. 15 (illustrated in Borenius, *op.cit.*, pl. 3 and pp. 22–24). He apparently was unaware of the fact that Andrea had died in 1629 and that he had owned many more paintings. Ridolfi never alluded to the likely sale of this collection nor to its transfer to Amsterdam. (The text in Ridolfi, *op.cit.*, I, p. 101 reads as follows: '... vi si ritrasse [Giorgione]*in forma di Dauide con braccia ignude, e corsaletto in dosso, che teneua il testone di Golia: haueua da vna parte vn Caualiere con giuppa, e beretta all' antica, e dall' altra vn Soldato, qual Pittura cadè dopò molti giri in mano del Signor Andrea Vendramino.*').

70 A full description of the *Musaeum Vendramenum* was given by Jacobs, *op.cit.*, pp. 19–22, who followed the order given in the sale catalogue of the Bentes library, beginning with the paintings, sculptures and religious objects, etc. The following sixteen catalogues were listed:
I. De Picturis; II. De Sculpturis; III. De Deis Oraculis, Idolis et Antiquorum Sacerdotibus, addita explicatione eorundem rituum et habituum in sacrificiis usus; IV. Habitus diversarum Nationum; V. De Sacrificiorum et Triumphorum Vasculis, Lucernisque Antiquorum, Urnis a liquoribus, Lacrimis atque Vasculis vitreis; VI. De Antiquis Romanorum Numismatibus; VII. Illustrium Venetorum Numismata; VIII. De Annulis et Sigillis Aegyptiorum Scarabaeis, Emblematibus ornatis et aliis signis et figuris in Gemmis et Lapidibus a natura delineatis et incisis; IX. De Rebus naturalibus mixtis atque compositis et in omni genere petritis; X. De Buccinis, Cochleis et Conchis maritimis diversarum Mundi partium; XI. De Mineralibus; XII. De Rebus Indicis et ex aliis mundi regionibus, tam Orientalibus quam Occidentalibus, valde curiosis et visu dignis; XIII. De libris Chronologiarum universalium figuris et coloribus ornatis. De Iconibus aere et ligno incisis Alberti Aldegravii et aliorum Pictorum insignium. De Animantium, Piscium et Avium cujusvis generis forma et Historiis. Plantarum et florum nobiliorum viridario. Mirandis Romanae Urbis Vetustatibus et aliis rebus visu delectabilibus; XIV. De Variis Rebus, peculiarem locum no habentibus; XV. De Auctorum Insignium de Christo redemptore scriptis, consideratione dignis; XVI. De Manuscriptis.

Eight of them were recorded in 1729, in the sale of the library of Gosuin Uilenbroek: nos. I (Uilenbroek 779); II (1036); III (1060); V (922); VI (1017); VII (1156); IX (555); XI (641); after Uilenbroek's death, five were listed in the second auction of his books in Amsterdam, in 1741: II (823); III (842); V (703); VI (707); VII (991).
The following Vendramin catalogues are extant:

I. DE PICTURIS, London, British Museum, Ms. Sloane 4004 (Samuel Ayscough, *Catalogue of the Manuscripts preserved in the British Museum...*, I, London, 1782, p.380). The title-page reads in full: DE PICTVRIS/*in musaeis Dni Andreae*/*Vendrameno*/*positis.*/*Anno Domini.*/M.D.C.XXVII. The manuscript contains 86 fols., measuring ca. 23,2 × 16,5 cm. each. Fols. 3 r.–9 v. give a general history of painting during the period of classical art followed by an index (fols. 10–11) and an alleged letter from Lentulus, the chancellor of Herod, relating Christ's appearance. The catalogue includes 155 drawings after paintings in the collection considered to be worthwhile reproducing as well as a list of some 150 paintings and watercolors merely listed on fols. 85 r.–86 r. For a list of the individual paintings that were illustrated see CATALOGUE, pp. 171–74. Published in full by Borenius, *op.cit.*; reviewed by Oskar Fischel, *Zeitschrift für bildende Kunst*, LVIII, 1924, no. 3-4, pp. 28–29.

V. DE SACRIFICIORUM ET TRIUMPHORUM VASCULIS, Oxford, Bodleian Library, Ms. D'Orville 539. (F. Madan, *A Summary Catalogue of Western Manuscripts in the Bodleian Library at Oxford*, IV, Oxford, 1897,

p. 137, no. 17417, bought by Jacques d'Orville in 1741). The title-page reads in full: DE SACRIFICIORVM, ET TRIVM=/*phorum uasculis, Lucernisque. Antiquorum, Vrnis à liquoribus, lacrimis, atque*/*uasculis uitreis, in Andreae*/*Vendrameno musaeo repositis.*
Double-page drawing of a wall with niches, where the vases are exhibited (text fig. 12), followed by an index to other writings in the collection and by 44 illustrated pages, representing vases, cups, rhytons, lamps, plates, and glasses. Size of the page 232 × 169 mm.

VIII. DE ANNULIS ET SIGILLIS AEGYPTIORUM, London, British Museum, Ms. Sloane 4005 (*Catalogue*, II, p. 365).
The title-page reads in full: DE ANNVLIS, ET SIGILLIS, *AEgyptiorum scarabaeis, emblematibus ornatis, et aliis signis et figuris in gemmis et lapidibus à natura delineatis, et incisis, in musaeo Andreae, Vendrameno repositis. Anno Dni.* MDCXXVII. The manuscript has 37 pages, measuring 230 × 163 mm. The illustrations represent scarabs, cameos, vases, rings and gems. There is a discourse in the beginning on the veneration by the ancient Egyptians of scarabs (fol. 3–5) followed in the end by an essay on seals and carved precious stones (fol. 38–49).

IX. DE REBUS NATURALIBUS, London, British Museum Ms. Sloane 4006 (*Catalogue*, II, 656). The title-page reads in full: DE REBVS NATVRALIBVS/*Puris, mixtis, atq.ue compositis*/*et in omni genere pe :*/*tritis. in Musaeo*/*Dni Andreae Vendrameno*/*repositis.*/*Anno Domini.* M.D.C.XXVII. Fol. 2 contains an index of essays by various authors on minerals, snails and sea animals, kept separately; fols. 3r.–5v. contain general in-manuscript. Four drawings corresponded to funerary chests and reliefs in the Rijksmuseum van Oudheden, Leiden, and one to a relief in the Centraal Museum, Utrecht[72] (cat. nos. 117–21; Pls. S117–S121a). These tomb monuments came to Leiden through the Papenbroek bequest, in 1738. The same bequest also included a number of antique sculptures that not only were illustrated in DE SCULPTURIS but also in the ICONES and their provenance from Reynst remains beyond doubt. We may, therefore, safely assume that the tomb monuments again were with Reynst before they came to Papenbroek and Leiden.

Judging from the catalogue DE PICTURIS Andrea Vendramin had brought together some three hundred paintings and watercolors.[73] One hundred and fifty of them were reproduced in pen drawings while an identical number was considered of less interest and therefore merely listed.

The paintings were predominantly by Venetian artists, many with overly confident attributions. Thus we find no less than thirteen believed to be by Giorgione, nine by Giovanni Bellini, and five each by Titian and Palma Vecchio, four by Andrea Schiavone, and two by Tintoretto. Most of them were portraits (ca. eighty) with a few mythological and religious representations. Only six of them have been identified. Three of them, the *Reclining Venus* by Cariani (cat. no. 9; Pl. P9), the so-called *Ceres* by Giorgione and the

troductory remarks on the fossils illustrated in the catalogue with their various origins; fols. 6r.–7r. give the index of the manuscript. Fols. 8–91 illustrate crabs, shells, sponges, starfish, sea weeds, tongues of serpents, stones in various shapes, at times resembling human organs and forms; petrified snails, charcoal, fig or date pits; reeds; various pieces of wood, among them a piece from the Arch of Noah, with an explanatory text; corals, crystals and other objects.

XI. DE MINERALIBUS, London, British Museum, Ms. Sloane 4007 (*Catalogue*, II, p. 678). The title-page reads as follows:
DE MINERALIBVS/*omnis generis, tam mettallicis,/et puris lapideis, quam et/ gemmatis/In musaeo D. Andreae Vendrameno positis./Anno Domini* M.D.CXXVII. Fol. 2 contains an index of a number of separate essays on the diversity of minerals and metals and on medicinal waters. Fols. 3r.–4r. give an index of the cut stones, and 7r. of the seven basic metals. Fols. 8–69 illustrate stones containing metals with accompanying texts; fols. 66r.–67v. describe medicinal characteristics of some types of soil; fols. 70r.–87v. contain discussions on the virtues of many precious stones illustrated in the present catalogue, based on ancient manuscripts.

XVII. DE ANTIQUORUM TUMULIS, East Berlin, Deutsche Staatsbibliothek, Ms. Phill. 1893. (Formerly in the Bibliotheca Meermanniana…, IV, 1824, no. 815). The title-page reads in full: DE, ANTIQVORVM, TVMVLIS/ *vasculis Vrnis à cinere, atque/ mortuorum monumentis in/Andreae Vendrameno musaeo collocatis./Anno Domini.* MDCXXVII. The manuscript consists of an index and nineteen pages with line drawings, six with a commentary.

This volume was not in Bentes' possession. It apparently always stayed separate from the rest but it should not be excluded that is was also in the Reynst collection. This supposition seems to be confirmed through the traditional identification of the 'sepultura di Aristotile' as it is found in Andrea Vendramin's explanatory text (see text fig. 10).

Venetian Senator by Tintoretto can be traced to the collection of Gerard Reynst. Most or all of the paintings from the Vendramin collection probably were purchased by the Reynst brothers but the old inventories are too general to allow for positive identifications with illustrations in DE PICTURIS.[74] As mentioned earlier, none of the paintings reproduced in this Vendramin catalogue were later engraved for the CAELATURAE. Since we do have proof, however that at least some of the paintings formerly with Andrea Vendramin were later owned by Gerard Reynst, we may suppose that they were bought together with the sculptures and the *naturalia*. A list of these paintings illustrated in DE PICTURIS, therefore, was appended to the CATALOGUE (see pp. 171–74).

We have circumstantial evidence that at least sixteen Vendramin catalogues, and most likely also the seventeenth, were purchased by the Reynst brothers at the time the objects as such were acquired. Sixteen volumes figured in the sale of the library of Albert Bentes (1643–1701) in Amsterdam, on April 24, 1702.[75] The only person who could have been interested in them previously was Gerard Reynst who owned many of the objects illustrated in them. Since the sixteen volumes were kept in a special case and therefore were considered an entity in themselves they must have accompanied the Vendramin collection at the time it was transferred to Amsterdam. They most likely served as a

The manuscript II. DE SCULPTURIS was formerly in the library of the Staatliche Museen, Berlin, but has been lost since World War II. The best description of it is found in Jacobs (*op.cit.*, p. 21, note 3). According to him, volume one contained 48 pages with seventy drawings of statues and statuettes, partly with titles and about ten drawings of torsi (fols. 47, 48); the second volume had 71 pages with about eighty drawings of busts, almost all with titles; pages 72–91 showed sculptures of less value which were used for ornamental purposes in the courtyard, under stairs and elsewhere. This lot included about seventy busts, statuettes and reliefs, almost all without titles. The concordance of DE SCULPTURIS with the ICONES is based on Jacobs' notes in his *Nachlass* in East Berlin (see following footnote).

[71] *Op.cit.*, p. 26, note 1. For the full transcript from Jacobs' papers in Berlin, see pp. 244–247.

[72] Professor H. Brunsting, who intends to publish the Vendramin catalogue together with the surviving monuments elsewhere, has kindly given me permission to include them in the present study.

[73] See CATALOGUE, pp. 171-174.

[74] *The Venetian Lady* by Pordenone, DE PICTURIS, fol. 60, may be identical with a painting appraised by Dujardin and Dodijns, in 1672. Giorgione's representation of a *Man and a Woman*, fol. 19, is perhaps identical with a picture sold from the collection of Pieter Six (1704), no. 25 : *Een Man en Vrouw in een Stuk, van Giorgione* ; sold for 11 guilders. Variants of this painting are known in Edinburgh, Berlin, Bassano and Oxford (see Borenius, *op.cit.*, p. 25).

[75] On October 6, 1708, Albert's son Hillebrant Bentes (1677–1708) advertized a sale on the 16th that included a 'Cabinet van Andreas Vendrameno van Antiquiteyten'. The younger Bentes, therefore, owned antique sculptures that most likely once were with Reynst and that he may have inherited from his father. See Dudok van Heel, *op.cit.* p. 170, no. 122.

source of information on the various aspects of the collection which is only understandable since Reynst probably took the entire collection over *en bloc*. This is corroborated further if we recall that the catalogue DE SCULPTURIS was used for the captions on the prints for the ICONES.

The fact that these manuscript inventories were found in the Netherlands in the 17th century further strengthens our belief that Jan purchased the collection *tale quale* and shipped it North, accompanied by these explanatory, illustrated catalogues. From Knorr's *Itinerarium* we also learn that the collection was exhibited more or less according to the individual categories pre-established in these manuscripts. This is another indication that the two were interrelated and dependent on one another.

Unfortunately we cannot determine when these works of art and objects from the collection of Andrea Vendramin were purchased. One would expect, however that Jan who lived in Venice since 1625, made the acquisition either alone or jointly with his brother sometime after Vendramin's death, in 1629.[76] Much or all of it was then sent on to Amsterdam, where it was installed in the house on the Keizersgracht which was owned by the two brothers.

To buy collections *en bloc* was not that unusual during the early part of the seventeenth century. We know of at least three such transactions, all negotiated via Venice during the late 1620's and

---

[76] The assertion by Kenneth Clark (*Rembrandt and the Italian Renaissance*, New York, 1966; paperback edition, New York, 1968, p. 102) that the 'gallery of Andrea Vendramin was brought to Amsterdam in the 1640's, and the sale was handled by Saskia's cousin, Gerrit van Uylenburgh, needs to be repudiated. As we now the collection had been acquired by Gerard and Jan (?) Reynst almost in its entirety and brought to Amsterdam to be exhibited in the house of Gerard Reynst not to be sold. Clark's implication that only the paintings (*gallery*) were sold in Amsterdam is even less trustworthy since the Cariani, apparently one of the best paintings formerly in the Vendramin collection, was still in the Reynst collection in 1660 when it was selected for the 'Dutch Gift'. Clark's statement was repeated by Julius S. Held, in *Rembrandt's Aristotle and Other Rembrandt Studies*, Princeton, N.J., 1969, p. 97, note 36.

The date 1633, referred to by Savini-Branca (*op.cit.*, p. 71) as the year the remainder of the collection was partly in Amsterdam and partly in England, must be a misprint, since her information obviously is based on the title-page of the CAELATURAE which she dates, without reason, to 1663 (*loc.cit.*, note 45).

1630's: the sale of the Gonzaga collection to Charles I through the services of Daniel Nijs, in 1627;[77] the sale of the collection of Bartolomeo della Nave to the Marquess of Hamilton with the assistance of the English ambassador in Venice, in 1636/38;[78] and the sale of the collection of Daniel Nijs to Arundel, in 1638[79] (limited to his coins and gems).

Van Buchell's *Notae* tell that the antique sculptures filled two entire rooms and that numerous full length statues were displayed in cases in the courtyard. We may therefore surmise that by 1639 at the latest the majority of the sculptures had been received in Amsterdam, together probably with most of the collection purchased from Vendramin.

The seventeen manuscript catalogues are the only reminder of the former Vendramin collection. Even though they have survived only partially they nevertheless give a fairly accurate idea of the type and extent of the collection acquired by Gerard Reynst since most of the objects themselves have disappeared.

Judging from descriptions of works of art and antique sculptures seen by the *Anonimo Morelliano* (probably identical with the Venetian scholar Marcantonio Michiel who died in Venice in 1552), between 1525 and 1543 in Venetian houses it becomes evident that the Vendramin collection continues the tradition initiated in the sixteenth century by men like Andrea Odoni, the Venetian noble and cleric Pietro Bembo, Michele Contarini, Giovanni Ram and Gabriele Vendramin.[80] By the end of the century, Venice had emerged as one of the leading centers of private, non-princely art and antiquarian collections and their contents were included in guide books to the art treasures of Venice such as Francesco Sansovino's *Venetia città nobilissima et singolare descritta*, published for the first time in 1581. For the early seventeenth century Vincenzo Scamozzi's *Idea dell' Architettura*, published in 1615 is most revealing. It describes six collections that consisted of paintings and sculptures, among them the ones of

---

[77] Alessandro Luzio, *La Galleria dei Gonzaga venduta all' Inghilterra nel 1627–28*, Milan, 1913.

[78] E.K. Waterhouse, 'Paintings from Venice for seventeenth-century England,' *Italian Studies*, VII, 1952, pp. 1–23; Savini-Branca, *op.cit.*, pp. 61–68 and 251–54.

[79] Mary S. Hervey, *The Life, Correspondence and Collections of Thomas Howard, Earl of Arundel*, Cambridge, 1921, pp. 409–10; Savini-Branca, *op.cit.*, pp. 70 and 254.

[80] A brief survey of Venetian collectors in the sixteenth and seventeenth centuries is given by Oliver Logan, *Culture and Society in Venice 1470–1790*, New York, 1972, Appendix pp. 297–321.

[81] Logan, *op.cit.*, pp. 161–62 and Appendix.

Andrea Vendramin, Bartolommeo della Nave, Carlo Ruzzini and Daniele Nis. Vendramin's collection is the best inventoried among them and thus serves as a point of reference in evaluating other early seventeenth century collections assembled by private citizens.[81]

5. THE 'DUTCH GIFT'

The States of Holland and West Friesland promised a number of valuable presents to king Charles II of England at the time of the latter's departure from Scheveningen on June 2, 1660.[82] Accordingly, the council of the provinces of Holland and West Friesland agreed in the meeting of June 21, to try to obtain from the Princess Dowager a bedstead complete with all the accessories for 100,000 guilders to be presented to Charles II. The members of the council were also asked to think of further valuable gifts for the English monarch.[83] A letter of September 2, to Cornelis de Vlaeming van Outshoorn specified that the king preferred Italian paintings and antique sculptures to pictures by modern masters. Van Outshoorn, therefore, was asked to contact the widow of Reynst since that collection included such objects. He was specifically instructed to try to persuade her to sell to them at a reasonable price and if necessary to ask the assistance of burgomasters or members of the governing board of the city of Amsterdam.[84] The following September 17, Van Outshoorn reported that twenty-four paintings and twelve antique sculptures were to be purchased from the Reynst collection for which the widow would receive 80,000 guilders.[85] The works of art for this so-called 'Dutch Gift'[86] were selected by the sculptor Artus Quellinus (1609–88) and the art dealer Gerrit Uylenburgh (ca. 1626–ca. 1690). Uylenburgh was also asked to supervise the transport to England.

In addition to the paintings and sculptures purchased from the widow of Reynst the States of Holland and West Friesland donated four paintings and a stately yacht called Mary. (The latter was presented on behalf of the city of Amsterdam).[87] One of these additional paintings, bought from the Amsterdam burgomaster

---

[82] *Relation/en forme de journal,/du/ Voyage et sejour,/que/ le serenissime et tres-puissant Prince/Charles II/ Roy de la Grand' Bretagne, &c./A fait en Hollande, depuis le 25. May,/jusques au 2 Juin 1660./A La Haye./chez Adrian Vlacq./*M.DC.LX, Avec Previlege des Estats d'Hollande & West-Frise, p. 100 : 'Sur le soir, Mr de Wimmenum (Van der Broeckhorst), se servant de l'occasion que le Roy luy donna, en parlant des temoignages d'affection que les Estats d'Hollande luy avoient rendus, dit à sa Majesté, que l'intention de Messieurs les Etats d'Hollande estoit de faire quelque chose de plus, s'il se fust trouvé en leur Etat des raretés que l'on eust pû presenter à un si grand Prince. Toutefois qu'ils

24

DE KUNSTKROON
Voor den Koningk van groot Britanje &c.

Aen den E.E. Heer

SYMON van HOOREN,

Burgermeester van Amsterdam, staende met zijne medegezanten reisvaerdigh naer Engelant.

Téque adeò decus hoc ævi, te consule, inibit.

Oe zal men met een' braven trant
  Het staetgezantschap best geleiden,
    Nu gy, ten dienst van 't vaderlant,
    Ter Maze uit streeft naer 't Engelsch strant,
  Door Nereus groene waterweiden,
    Den Teems op, daer het juichend hof
Ten hemel rijst op STUARTS lof.

De Roos van Engelant verspreit
  Op 't rijzen van die zon haer geuren,
    En d'onderdruckte Majesteit,
    Zoo lang met hartewee verbeit,
  [Terwijl de vrede en wetten treuren,
    In eenen nacht van haet en twist,]
Gaet schooner op uit zulck een' mist.

Gy zult den grooten koning zien,
  Den helt, ter heerschappy geschapen,
    En gansch Britanje op zijne knien
    Hem eer en offergaven biên,
  Die, zonder zwaert en bloedigh wapen,
    Het rijck herwon, en, vol gedult,
Ontlaste van die zwaere schult.

Nu bloeien alle staeten weêr.
  De ridders draven, als voorheenen,
    Ten hove, en d'adel, in zijne eer,
    Begroet met vreught den jongen heer,
  Daer hy, van diamant bescheenen
    En gout, uit 's vaders hoogen troon,
Hen overstraelt met zijne kroon.

Met welck een gunst zult gy [de mont
Van zeven staeten] KAREL groeten,
    Daer Liefde en Trou het staetverbont
    Bezeglen, en, oprecht van gront,
  Eleckandren liefelijck gemoeten,
    En Eilanden, verknocht aen een,
De welvaert bouwen van 't gemeen!

Hy, die de kunsten, lange stom
  En balling, weder uit het duister
    Te voorschijn brengt, en geeft alom,
    In 't opgaen van zijn koningsdom,
  Haer' eersten glans, en vollen luister,
    Zal Hollants kunstgaef niet versmaên,
Maer zien het hart des offraers aen.

Dees goude tijt vergadert vast
  De meesterstucken met verlangen.
    De Batavier, door Pallas last,
    Als perlen, die te zamen past,
  Om tot cieraet op 't hof te hangen,
    Op dat de Koning zijn gezicht
Magh weiden in dit schilderlicht.

Hun schutsheer STUART hoort met vreught
Het stomme doeck en marmer spreecken,
    En kent elck werckstuck, en zijn deught,
    Die 's kenners oogh en hart verheught.
  Alle Italjaensche Apellesstreecken
    Ontfangen haeren prijs by hem,
En door zijn wijze orakelstem.

Op 't spoor van 's konings voorbeelt wort
  De renbaen van de kunst ontsloten,
    Een rijcke geest, die d'eélsten port
    En noopt, ten boezem ingestort.
  De graven, en de hofgenooten
    Om strijt naer puick van meestren staen,
Die d'ouden naertreên op hun baen.

De zangbergh gaet hierop ten dans,
  Niet met Apol van Delos eilant,
    Maer van Britanje, met zijn' krans
    Van roosmarijn en roozeglans,
  Gewelkomt, als der rijcken heilant.
    Hoe wedergalmt die gansche kust?
Hier bloeit de kunsteeu van Augusst.

                                        J. V. VONDEL.

t'Amsterdam, voor de Weduwe van ABRAHAM de WEES, op den Middeldam. 1660.

se donneroient la liberté de luy faire accomoder, & de luy envoyer à la premiere occasion, quelques presents, qu'ils supplieroient sa Majesté de considerer, comme des preuves de leur bonne volonté, plutost que comme des effets de leur pouvoir. Le Roy s'en voulut défendre, en disant, qu'il ne luy failloit point d'autres asseurances de l'affection de Messieurs les Etats d'Hollande, que celles qu'ils venoient de luy donner en l'occasion presente, qu'il en estoit satisfait, & qu'il les remercioit, non seulement des effets du passé, mais aussi de la bonne volonté, qu'ils luy tesmoignoient pour l'avenir.'

83 'Den XXI Junij 1660; Presenten de Heeren van Wimmenum, Druijvesteijn, Meerman, van der Graeff, van Hoorn, Groenendijck, Jonassen ende den Raedt Pentionaris;
... Is naer deliberatie goetgevonden dat onder anderen aenden Heere Coninck van Groot Brittannie sal werden geoffereert ende vereert seecker kostelijck geborduert Ledicant toekomende haere Coninckljcke hocheijt de princesse Douariere van Orangien, ende dat aen deselve daer vooren sal werden aengebooden ende vereert de somma van hondert duijsent Guldens, het welcke den heere van Wimmenum versoght werd hooghgemelte haere Hoocheijt inder beste ende bequaemste termen ende manieren ter eeren vanden Lande voor te draegen ende smaeckelijck te maecken. Ende sijn de vordere Heeren van haere Edele Mo: College versoght ende werden mede versocht mits desen bij gelegentheijt inden haeren te onderstaen hoedaenige kostellijcke ende sortabele rariteijten voorde hooge waerdicheijt van Sijnne Ma^t. souden mogen te vinden wesen, omme neffens het voorsz. Ledicant aen hoochstgemelten Coninck vereert te mogen werden, ende van haere ondervingen rapport te doen, omme alsdan verder te werden geresolveert.' (Algemeen Rijksarchief, The Hague, *Staten van Holland, 1572–1795*, no. 3010). See also Leupe, 'Schilderijen en Statuen voor Karel de Tweede, Koning van Engeland 1660,' *De Nederlandsche Spectator*, 1876, p. 184.

84 Letter sent by the council in committee published by Leupe, *loc.cit*. 'Aenden Heere van Outshoorn, gecom.^{e} raedt & ende oudt Borgerm(eester) der Stadt Amstelredam.
Eerentfeste &.
Wij werden in't seecker beright, selffs door middel van den geenen die't emploij heeft van op te schicken ofte in ordre te brengen het Cabinet vanden Coninck van Groot-Brittannien dat sijnne Maj^t. niet soo veel gesint is op schilderijen van moderne meesters, als wel deselve particuliere speculatie heeft op Antique stucken, ende van Italiaense Meesters. Ende dewijlle in het Cabinet vande wedue van Renst eenige soodaenige stucken berusten, soo versoecken wij dat UE. de moeyte gelijven te nemen van met de voorsz. weduwe naerder te spreecken ende de selve tot het overlaeten van dien voor een reghtmaetige prijs te disponeren, des noodigh wesende gebruijckende ten selven eijnde de persuasive entremise van Borgem(eeste)ren ende reg(eerder)s der stadt Amsterdam, Ende bij aldijen onder de weduwe van den gewesen Agent Le Blon ofte andere dergelijcke kunst, de speculatie van hooghgemelte sijnne Maj^t. waerdigh sijnde, soude mogen te vinden wesen, soo gelieve UE. wijders de moeijte te nemen van het selve te onderstaen ende daer over te handelen, soo als UE. best aghten sal, ons dijenthalven volcomentlijck gedraegen(de) tot UEs. oordeel ende dispositie. Ende onder des, Erntefeste & den IIe September 1660 Gecommitteerde Raeden.' (Algemeen Rijksarchief, The Hague, *Staten van Holland, 1572–1795*, no. 1396).

85 'Den XVII September 1660; presenten de Heeren van Wimmenum, Druijvesteijn, van der Graeff, Meerman, van Hoorn, Groenendijck ende Jonassen. rapport Inkoop Cabinet Juff. Reijnst.
Den Heere van Hoorn versocht sijnde omme met de Weduwe van Gerrit Reijnst tot Amsterdam te spreecken, ende te verdraegen overden koop van desselfs Cabinet ofte een goet gedeelte van dien, heeft gerapporteert dathij uijt het selve Cabinet in koop heeft aengestaen vieren twintig stucken schilderije ende twaelff statuen sijnde alle het beste ende curieuste naer het ordeel van die geenne die haer de kunst van schilderen met het geenne daer aen dependeert verstaen, ende dat voorde somma van tachtig duijsent Guldens te betaelen met rente Brieven overde respective comptoiren op het eerste consent tot negotiatie bij haere Edele Groot Mogende te dragen. Hebbende hem particuliere gelijck in ende omtrent de keure van de voorsz stucken ende statuen gedijent vande addresse ende advis van Gerrit Uijlenburg ende Culinus; waer op sijnde gedelibereert is den voornoemden Heere bedanckt ende werd bedanckt mits desen voorde moeijte bij hem genomen, ende is voorts het voorsz gehandelde

goetgevonden ende geapprobeert, gelijck het selve ende goetgevonden ende geapprobeert werd mits desen. Ende op dat de voorsz stucken, ende statuen behoorlijck gepackt ende wel versorght mogen werden is goetgevonden dat daertoe sal werden gebruijckt den dienst vanden voornoemden Uijlenburg die oock het transport van deselve naer Engelandt bij woonen ende so onderwege als aldaer besorgen soo veel mogelijck dat alle schaeden verhoed werden, daer voor hij in redelijckheijt sal werden beloont; Ende sal van deselve gegeven werden detwel (?) aenden meergemelten heere van Hoorn omme te dienen to sijnne naerrichtinge.'
(Algemeen Rijksarchief, The Hague, *Staten van Holland, 1572–1795*, no. 3010). See also Leupe, *loc.cit.*, pp. 184–85.
Knorr von Rosenroth intimated (see p. 63) that some of the Italian paintings presented to Charles II were appraised at 4,000 guilders, a very large sum if one compares for example the prices paid on August 31, 1660 for paintings by leading Dutch and Flemish artists: thus Rembrandt's painting of *Susannah* brought 560 guilders, his portrait of *Adriaen Banck* only 150 guilders, and a large *Hunt* by Rubens only 300 guilders (see C. Hofstede de Groot, *Die Urkunden über Rembrandt [1575–1721]*, [*Quellenstudien zur Holländischen Kunstgeschichte*, III], The Hague, 1906, pp. 275–76, no. 232).

86 The first so-called 'Dutch Gift' took place in late April, early May, 1610 when the States General presented Henry, Prince of Wales, the elder brother of Charles I, with several paintings among them a *Seastorm* by Porcellis and the *Battle at Gibraltar* by Hendrick Cornelisz. Vroom (Amsterdam, Rijksmuseum, Inv. no. 2606). Several members of the English court were presented with two sets of tapestries, a coach with four horses, four large bezoar-stones, an ivory fan and damask. See J. G. van Gelder, 'Notes on the Royal Collection –IV: The 'Dutch Gift' of 1610 to Henry, Prince of 'Whalis', and Some Other Presents,' *Burlington Magazine*, CV, 1963, pp. 541–44. In 1636, the States General presented Charles I of England with another 'Dutch Gift' consisting of seven white horses with richly trimmed red velvet harnesses, a watch, a small mother-of-pearl chest, a lump of amber, and four paintings: Geertgen tot Sint Jans' *Lamentation* and *The Legend of The Relics of St. John the Baptist*, both in the Kunsthistorisches Museum, Vienna (Inv. nos. 991; 993), Mabuse's *Adam and Eve* (probably the painting still at Hampton Court, Inv. no. 580), and a *St. Jerome*, attributed to Lucas van Leyden (perhaps identical with a painting in the Rijksmuseum, Amsterdam, attributed to Aertgen van Leyden, Inv. no. 14 334). Furthermore, a piece of embroidery was presented, along with damsk, China and cows. The purpose again was political, to improve relations with the English king who openly advocated his 'Right to the Dominion of the Seas'. See J. Bruyn and Oliver Millar, 'Notes on the Royal Collection – III: The 'Dutch Gift' to Charles I,' *Burlington Magazine*, CIV, 1962, pp. 291–94.
A detailed reconstruction of the paintings included in the 'Dutch Gift' of 1660 was made by D. Mahon, 'Notes on the 'Dutch Gift' to Charles II,' *Burlington Magazine*, XCI, 1949, pp. 303–05; 349–50; *Burlington Magazine*, XCII, 1950, pp. 12–18; 238. See also the introduction by P.J.J. van Thiel in the exhibition catalogue *Het Nederlandse Geschenk aan Koning Karel II van Engeland 1660*, Amsterdam (Rijksmuseum), 1965, pp. 3–7, with a reconstruction of the gift on p. 8. Furthermore, Michael Levey, *The Later Italian Pictures in the Collection of Her Majesty the Queen*, London, 1964, p. 19, 39; and E.K. Waterhouse, 'A Note on British Collecting of Italian Pictures in the Later Seventeenth Century,' *Burlington Magazine*, CII, 1960, p. 54.

87 C. G. 'T Hooft, 'Een geschenk van Amsterdam aan Karel II van Engeland in 1660,' *Jaarboek...Amstelodamum*, XIX, 1921, pp. 1–13. The ship perished in 1675, but a drawing by Willem van de Velde the younger in the British Museum, London still gives an idea of its splendor (reproduced in 'T Hooft, opposite p. 7).

88 'Den XXIII September 1660; presenten de Heeren van Wimmenum, Druijvesteijn, van der Graeff, Meerman, van Hoorn, Groenendijck, Jonassen ende de Raedt Pentionaris.
Arbitrage schilderij Borgermeester de Graeff.
Den Heere Andries de Graeff Borgermeester der Stadt Amsterdam ter begeerte van haere Edele mogende afgestaen hebbende seecker stuck schilderije omme onder anderen mede begrepen te werden onder de vereeringe te doen aenden Coninck van Groot Brittannien tot soodaenigen prijs als bij twee persoonen hinc inde te nomineren soude werden gearbitreert, is naer deliberatie goetgevonden daertoe van wegen haere

Andries de Graeff,[88] probably was Saenredam's *View of the Groote Kerk* at Haarlem, dated February 27, 1648,[89] today in the collection of the Marquess of Bute at Mount Stuart, Rothesay (text fig. 25).[90] The three other paintings were acquired from the painter Gerrit Dou.[91] According to a convincing reconstruction by Mahon,[92] one of them was the so-called *Young Mother* by Gerrit Dou, dated 1658, today in the Mauritshuis, The Hague (text fig. 23),[93] while the other one possibly was a version of Elsheimer's *Mocking of Ceres*.[94] The third painting remains to be identified.

The entire 'Dutch Gift' was shipped from Rotterdam shortly after October 18, 1660, for on this day the Dutch extraordinary ambassadors to England bid farewell to the States General. The event was commemorated by Vondel in a poem dedicated to one of them, Simon van Hoorn (text fig. 24).[95] This was the same Van Hoorn, who was a close friend of Gerard Reynst and who was included in Van der Helst's portrait of the *Governors of the Kloveniersdoelen* of 1655 (text fig. 4).

Van Nassau and Van Hoorn recounted in a letter of November 16/26, 1660, how favorably the presents were received by the king. Charles II with the entire court, all the important Englishmen as well as most of the foreign emissaries came to admire the presents. The king was most pleased by Titian's *Madonna and Child* (cat. no. 36; Pl. P 36a) and the paintings by Dou and Elsheimer.[96]

Edele mogende te versoecken ende te nomineren Gerrit Douw- Ende van wegen den Heere de Graeff volgende desselfs aenbiedinge te laeten assumeren Reijnier van der Wolff, ende sal van desen gegeven werden extract (?) aen meegemelten Heere de Graeff omme te dienen, tot sijnne naerrichtinge.' (Algemeen Rijksarchief, The Hague, *Staten van Holland, 1572–1795*, no. 3010). A letter sent to Van der Wolff and Dou is also preserved:
'Reijnier vander Wolff. woondende tot Rotterdam.
Eersamen etc. de heere Andries de Graeff Borgermeester der Stadt Amsterdam, heeft t' onser begeerte affgestaen seecker stuck schilderije, des dat het selve bij UE. ende Gerrit Douw, woonende tot Leijden geestimeert soude werden. Ende dijent derhalven desen ten eijnde UE. op morgen ter klocke thijen uijren wille hier wesen ende u laeten vinden voor onse vergaeder plaetse alwaer het voorsz: stuck schilderije UE. sal werden geexhibeert, ende ons daer toe verlaetende, Bevelen wij UE. onder des de bescherminge Godes. Geschreven inden Hage den XXVIII September 1660. Gecommitteerde Raeden.' (Algemeen Rijksarchief, The Hague, *Staten van Holland, 1572–1795*, no. 1396). See also Leupe, *op.cit.*, p. 185.

[89] See J.G. van Gelder, *op.cit.*, pp. 541–42, under note 1. Based on Mahon, *op.cit.*, p. 350 and p. 238, the painting was listed in the Royal inventories as follows:

Inventory of Charles II, p. 9, no. 137: *Peter Sanredam. Harlaem Church. A prospective. Dutch p^{r}sent. 5.8 (by) 4.9.* James II (Cat. ms., Whitehall, no. 71; Bathoe, no. 71). Anne Inv., Windsor, no. 17 in store with annotation 'removed to Somerset House'. Recorded at Somerset House on October 28, 1714 (British Museum, Add. ms. 19'933, no. 16), today in the collection of the Marquess of Bute (see following footnote).
Van Gelder (*loc.cit.*) pointed out that Saenredam previously had offered his painting of the *Grote Kerk* to stadholder Willem II (letter of May 21, 1648 to Constantijn Huygens).

[90] P.T.A. Swillens, *Pieter Janszoon Saenredam*, Amsterdam, 1935, p. 117, no. 171, fig. 71; and

25

The ceremony of the presentation of the gifts was described as follows in the *Mercurius Publicus* of November 8 – November 15, 1660, and in the *Parliamentary Intelligencer* of November 12 – November 19, 1660 :

'And yesterday the Lords Ambassadors from the States of the United Provinces, presented to his *Majesty* in the Banqueting House at Whitehall, an extraordinary crimson embroider'd velvet Bed, Cloth of State, Chaires and Stooles suitable, worth very many Thousands of pounds, and also an excellent Collection of Pictures of the most Famous, Auncient and Moderne Masters, with a great number of Statues of white Marble of excellent Sculpture : and to day these Lords Ambassadors had Audience, wherein His *Majesty* heartily thank'd them for so worthy a Present, and express'd his willingness to enter into a neerer Alliance with them.'

No list of the works of art purchased from the widow of Reynst has survived nor a list of the presents included in the 'Dutch Gift'. Its content was established thanks to Mahon, who reconstructed it from the royal inventories, in particular the inventory of Charles' II paintings at Whitehall and Hampton Court, drawn up ca. 1666-67,[97] where twenty pictures were designated as 'dutch present' : the three paintings by Dou, Saenredam and ? Elsheimer, and seventeen paintings by Italian artists from Reynst.

Catalogue Raisonné of the Works by Pieter Jansz. Saenredam (exh. cat.), Utrecht (Centraal Museum), 1961, cat. no. 58, fig. 60. The inscription on the base of the left choir-column reads : *Dit is de Cathedrale grote kerck van Haerlem in Hollandt. Pieter Saenredam, dese met schilderen voleijnt, den 27 februarij 1648* (This is the Cathedral Great Church of Haarlem in Holland, Pieter Saenredam finished painting this, the 27th of February, 1648).

[91] Not identified in the known documents.

'Aen Gerrit Douw, woonende over de Brouwerije vanden Hamer tot Leijden. Eersaemen vroomen discreten. Het sal noodigh wesen dat de drije stucken schilderije die in Onsen naeme van UE. gekocht zijn, woonsdach ofte donderdagh toecomende uijterlijck wrden getransporteert tot Rotterdam : Ende dient derhalven desen ten eynde UE. deselve stucken wel ende seeckerlijck inpacken, ende doen brengen sal binnen gemelte Stadt, die addresserende aen Pieter Puert, Coopman aldaer, Waertoe ons verlaetende. Bevelen ... XVIII October 1660. Gecommitteerde Raeden.'

'Aen Gerrit Douw woonende over de Brouwerije van Haemer tot Leijden. Eersamen etc. Opden uwen aen Ons geschreven in antwoorde vanden Onsen sub dato 18 deser loopende maendt, vinden wij goet dat UE. de bewuste stucken schilderye behoorlyck ingepackt sal laeten volgen met den brenger deses genaemt Gerrit Uijlenburgh die daermede sal handelen achtervolgende Onse ordre ende Ons daertoe verlaetende. Bevelen... Geschreven inden Hage den negenthijenden October XVIe ende sestigh. Gecommitteerde Raeden.' (Algemeen Rijksarchief, The Hague, *Staten van Holland, 1572-1795*, no. 1396). See also Leupe, *op.cit.*, pp. 185-86.

[92] *Op. cit.*, p. 304, note 21 and p. 350, no. A. Based on Mahon, the painting was described as follows : Inventory of Charles II, p. 23, no. 389 : *Dow. A Dutch woman at worke her childe in ye cradle, her maid by with fowle & severall other things. Dutch Present.* 2.5 (by) 1.10. James II Cat. (ms., Whitehall, no. 501 ; Bathoe, no. 500).

These seventeen were as follows (listed according to artists):

Barocci, *Woman with Dog* (lost; no. 14; cat. no. 1).

Bassano, *Christ Carrying the Cross* (Weston Park; no. 161; cat. no. 4).

Bordone, *Portrait of a Man* (Hampton Court; no. 167; cat. no. 8)

Cariani, *Reclining Venus* (Hampton Court; no. 544; cat. no. 9).

Giorgione, attr. to, *The Concert* (Hampton Court; no. 534; cat. no. 11).

Lotto, *Portrait of a Gentleman* (Hampton Court; no. 116; cat. no. 17).

Marco d'Oggiono, *Christ Child and St. John* (Hampton Court?; not securely identifiable; no. 335; cat. no. 19).

Parmigianino, *Athena* (Windsor Castle; no. 315; cat. no. 20).

Raphael, school, *Christ on a Lamb, the Virgin and St. Joseph* (lost; no. 390; cat. no. 22).

Schiavone, *Christ Before Pilate* (Hampton Court; no. 54; cat. no. 28).

Schiavone, *The Judgment of Midas* (Hampton Court; no. 169; cat. no. 29).

Tintoretto, *Portrait of a Dominican Friar* (Hampton Court; no. 103; cat. no. 32).

Titian, *Portrait of a Man* (Hampton Court; no. 21; cat. no. 34).

Titian, attr. to, *Holy Family with St. John* (lost; no. 166; cat. no. 35).

Titian, school, *The Virgin and Child with Tobias and the Angel* (Hampton Court; no. 532; cat. no. 36).

Veronese, *The Marriage of St. Catherine* (Hampton Court; no. 165; cat. no. 38).

Unidentified artist, *Christ and the Virgin* (lost; no. between 546/7; cat. no. 39).

Only one of these paintings, Bassano's *Christ Carrying the Cross*, has left the English royal collections.

---

[93] W. Martin, *Gerard Dou*, London, 1902, pp. 135–36, no. 164; idem, *Gerard Dou*, Stuttgart/Berlin, 1913, reproduced on p. 90; *Beknopte Catalogus van de schilderijen beeldhouwwerken en miniaturen*, Mauritshuis, 's-Gravenhage, 1971, p. 43, no. 32: panel, arched at the top, 73,5 × 55,5 cm. Signed: *G Dou 1658*. This is the most 'contemporary' painting included in the 'Dutch Gift'. Houbraken (op.cit., II, p. 4) stated that it was returned to the Netherlands by William III. He continued that it may originally have been purchased by the East India Company or by the States General from De Bie's collection (possibly Johan de Bye, Leiden) for 4,000 guilders to be presented to Charles II, an assertion that would need further proof. The painting was included in the exhibition *Het Nederlandse Geschenk aan Koning Karel II ..*, Amsterdam, 1965, no. 3.

[94] Mahon, op.cit., p. 350, no. B. Listed in the Inventory of Charles II, p. 20, no. 334: *Elschamor. An Olde woman holding a Candle & a woman drinckinge, a night peice. Dutch p*[r]*sent*. 0.11 (by) 0.9. James II Cat. (ms., Whitehall, no. 519; Bathoe, no. 518). Anne Inv., Kensington, no. 14. According to Mahon, the painting was still at Kensington during the reign of George II (1727–60), but has since left the royal collections. Christian Ludwig von Hagedorn (1713–80), in his *Lettre à un Amateur de la Peinture*, Dresden, 1755, p. 179, writes that 'Dou did not feel ashamed to copy the Painting of Ceres when the original had to go to England, where unfortunately it perished in the Whitehall fire' (published by Mahon, loc.cit.). One may therefore suppose that Elsheimer's painting, *The Mocking of Ceres*, was in Dou's own collection, and Dou copied it before selling the original to the States General. For Elsheimer's *Mocking of Ceres* and its versions see Heinrich Weiszäcker, *Adam Elsheimer, der Maler von Frankfurt*, Berlin, 1936, pp. 183–89, figs. 105–07; and *Adam Elsheimer* (exh. cat.), Frankfurt (Städelsches Kunstinstitut), 1966/67, cat. no. 32, fig. 30. A Dutch copy made in the second half of the 17the century has been recorded in the collection of Prussia since the end of the 17th century. This may possibly be

The transfer of five more Italian paintings into the English royal collections remains unclear. All of them were engraved for the CAELATURAE and therefore certainly belonged to the Reynst collection. Three of them were also included in this same inventory of Charles II but without the specification *Dutch Present*, namely Bonifazio Veronese's *Virgin and Child* (Hampton Court; no. 158; cat. no. 7); Lotto's *Andrea Odoni* (Hampton Court; no. 264; cat. no. 16) and Giulio Romano's *Isabella d'Este* (Hampton Court; no. 4; cat. no. 12). The fourth picture, Reni's *Allegory of painting* (cat. no. 26) is traceable only to the inventory of James II.[98]

The fifth one, Guercino's *Semiramis*, was in the possession of the Dukes of Grafton, who were descendants of Charles II through his liaison with Barbara Villiers.[99] The painting, therefore, probably also belonged to the English royal collections but was presented by the king to Barbara Villiers. Today it is in the Museum of Fine Arts, Boston (cat. no. 13).

Since at least three of these paintings were included in the same royal inventory of Charles II as the paintings clearly marked *Dutch Present* one must assume that they entered the royal collections at the same time. Whether they were purchased as part of the 'Dutch Gift' but not recorded as such in the inventories or whether they perhaps were acquired directly through emissaries of Charles II has to remain open. One fact is beyond doubt, however:

Dou's copy (?). (For a description of the painting see *Die Gemälde im Jagdschloss Grunewald*, Berlin, 1964 p. 62, no. 78, ill.; inv. no. GK I 10013). I would like to thank Professor J. G. van Gelder for this reference. Another version of Elsheimer's painting that is now in the Prado formerly belonged to Rubens.

95 Albert Verwey, *Vondel*, Amsterdam, 1937, p. 816. See also Jan Heringa, *De Eer en Hoogheid van de Staat*, Diss. Groningen, 1961, pp. 358–60 and p. 594. I would like to thank Professor J. G. van Gelder for these references.

96 Algemeen Rijksarchief, The Hague, *Staten van Holland, 1572–1795, 2810/3, Engeland aan Staten,*

1660–61, (Ambassadeurs L. van Nassau van Beverweerd and Simon van Hoorn). Also published by Leupe, *loc.cit.*

'Mijn Heere!
Wij hebben op maendagh laestleden aen Sijne Majesteit bekent gemaekt dat de Heeren Staten van Holland ende Westvriesland ons hadden gelast eenige stucken schilderijen ende andere fraeijicheden mede te nemen, ende te versoeken dat het sijne Maj. niet qualick geliefde te nemen, dat wij deselve aen hem uyt den naem van voorgemelde Staten offereerden, ende lieten brengen daer Sijne Maj. geliefde te ordonneren; waarop bij Sijne Maj. geantwoord sijnde, dat deselve hem seer aengenaem souden sijn, ende men due aenden Bewaerder van sijn Cabinet soude addresseren, hebben wij ordre gestelt dat de voornoemde presenten op dinsdagh aen 't Hof van den Conink sijn gebracht, ende op het Banquethuys, sijnde de grootste sael van Whitehall uijt de kassen gedaen, ende in ordre gestelt; het ledekant door den Concierge Boer ende de schilderijen en de statuen door den Schilder Uylenborgh, als sijnde alle beyde Lieden haer op die respective saken wel verstaende, alwaer deselve door den Conink ende het gansche Hof, alle de Grooten van Engeland, ende meest alle de vreemde Ministers sijn gesien, ende van allen gepresen, ende geestimeert voor een van de beste presenten die oyt aen eenen Prins konden geschieden. De

all five of them formerly belonged to Gerard Reynst.
Of the twenty-four paintings, therefore, that supposedly were purchased from the widow of Reynst on behalf of the States of Holland and West Friesland, no less than twenty-two can be traced to the English royal collections. Twenty are listed in Charles' II inventory yet only seventeen among them are specifically marked 'Dutch Present'. The inventory of Charles II thus proves not to be reliable enough to allow for a full reconstruction of the paintings included in the 'Dutch Gift' but it nevertheless furnishes us with a sound approximation.

Fourteen of these twenty-four paintings were engraved for the CAELATURAE and therefore have a firm provenance from the Reynst collection, namely the Bassano, Bonifazio Veronese, Giorgione (attributed to), Guercino, the two Lotto, the Parmigianino, Reni (attributed to), Giulio Romano, Tintoretto, three Titian or Titian school pieces, and the Veronese. The painting of a *Reclining Venus* by Cariani was illustrated in the Vendramin catalogue DE PICTURIS[100] and its provenance from the Reynst collection also appears secure (as we saw earlier, the brothers probably acquired all or at least the larger part of the Vendramin collection). The two Schiavone paintings are identical with the pictures referred to by Ridolfi in his *Maraviglie dell'Arte* as in the collection of Jan Reynst.[101] Since the remaining five paintings which are lost

Conink selve hoeft daer in genomen een groot werlgevallen, ende alles curieuselijck ende met opmerkinge doorsien, ende daechs daraen, naer dat wij onse propositie in de particuliere audientie hadden gedaen, ende Sijn' Maj. antwoord becomen, gelijk inde nevensgaende missive in't lange is verhaelt, ӕi lu filı ʿijue Maj. muu ons in verdere particuliere discoursen te treden, ende de presenten van haer Ed : Gr : Mo : seer te prijsen, ende te seggen, dat hij d'Heeren Staten voor denselve ten hoochsten bedanckte, sich voorts extenderende op verscheyde particuliere schilderijen, die S. Maj. wel de meeste scheenen te behagen, als dat van Titiaen, sijnde een marienbeeld met een kind, die van Douw ende Elshamer, alshoewel de Conink toonde, dat hijse in't generael altemael hooch achte.
Hiermede blijven wij.

Mijn Heere,

UEds. ootmoedige Dienaren,

(signed)
T. ılı Nɩ ıʋ ıu
Simon van Hoorn.

London, 16/26 November 1660.'

[97] Levey, *op.cit.*, p. 39, no. 16.

[98] Two copies of this composition have survived in the English royal collection (Windsor Castle Inv. nos. 1018 and 1102), but it can no longer be established, whether either one came from Reynst or whether an original version existed that is now lost. See Levey, *op.cit.*, p. 93, cat. nos. 582–83.

[99] D. Mahon, 'Guercino's Paintings of Semiramis,' *Art Bulletin*, XXXI, 1949, pp. 217–23, especially p. 219, figs. 3, 5.

[100] Reproduced in Borenius, *op.cit.*, plate 11

[101] Ridolfi, *op.cit.*, I, pp. 250–51.

are all by Italian artists we may assume that they too were purchased from the widow of Reynst.

These paintings undoubtedly represented the very best the Reynst collection had to offer, especially if we think of Lotto's *Portrait of Andrea Odoni*, Giulio Romano's *Portrait of Isabella d'Este* or Titian's *Portrait of a Man*, the so-called *Sannazaro*.[102]

Charles II may have purposefully specified a predilection for Italian art in order to replenish the royal collection since the Italian pictures collected under Charles I had been dispersed through the Commonwealth sale in 1649.[103] These twenty-four paintings from the collection of Gerard Reynst presented to Charles II at the Restoration were the only major additions to the royal collection of Italian pictures during the latter part of the seventeenth century.[104]

Together with these paintings the States of Holland and West Friesland also purchased twelve antique sculptures. Until very recently, none of them could be traced and it was feared that they had perished in the Whitehall fire of 1698. This, however, is no longer the case. Based on the prints in the ICONES, at least one of these sculptures could be identified, namely the *Bust of a Woman* at Hampton Court (Pl. s 108b)[105] which is identical with the bust reproduced as '*Faustina*' (ICONES K; cat. 108; Pl. s 108). This corre-

---

[102] Eight of these paintings were reassembled in an exhibition entitled *Het Nederlandse Geschenk aan Koning Karel II van Engeland 1660*, celebrating the 'British Week' from May 12 until June 20, 1965. They were Dou's *Young Mother*, The Hague (see note 92 above) and cat.nos. 9, 11, 17, 29, 32, 36 and 38.

[103] The inference that Gerard Reynst was one of the important purchasers at the sale of the collection of Charles I in 1649 was justly repudiated by Mahon (*op.cit.*, p. 303 and especially note 10). Mahon was able to trace its origin to the note-books of George Vertue, published in *The Walpole Society*, XVIII, 1929-30, p. 46.

After an entry of 1717, Vertue inserted: *Part of the Rentz Gallery presented to King Charles 2d after the Restoration. by the Widdow of ... of Rentz*. Later, in an entry in connection with three paintings in the possession of the Queen Dowager, Catherine of Braganza, the Reynst provenance was reworded. One of these three paintings was specified as *Dutch Present*, and Vertue continued that the *Dutch Presents were pictures bought at the sale of K Charles I collection and were got into the hands of ... Rentz in Holland. he dying the States b*^{ot}* them of his Widow. and in order to facilitate the peace with K Charles the 2d they sent these over to the King.* (Published in *The Walpole Society*, XXIV, 1935-36, p. 93).

Vertue's comment was perpetuated by every writer who treated the 'Dutch Gift' or the Reynst collection (Waagen, *Treasures*, I, pp. 13, 15; C. Blanc, *Le Trésor de la Curiosité*, Paris, 1857, p. XXV; F. Lugt, *Mit Rembrandt in Amsterdam*, Berlin, 1920, p. 66 to mention a few often quoted sources) until Mahon corrected it (*loc.cit.*). If we recall that Gerard Reynst apparently did not travel and that his brother Jan, who seems to have been the primary source for any acquisitions, had been dead for three years Vertue's implication sounds even less trustworthy.

[104] Waterhouse, *loc.cit*, 1960.

[105] Sir Geoffrey de Bellaigne, Surveyor of the Queens Works of

spondence, furthermore, lends proof to Knorr von Rosenroth's statement that the sculptures numbered in the ICONES A through M were part of the 'Dutch Gift' (see p. 65). The earliest record of the '*Faustina*' in the English royal collections is found in an illustrated manuscript preserved at the Royal Library at Windsor Castle, entitled *Busts & Statues in Whitehall Gardens* (Pl. S 108a).[106] This same manuscript illustrates at least one, possibly two other busts also known from prints in the ICONES that reproduced sculptures presented to Charles II: the *Cijbele* (ICONES F; Pls. S 104, S 104a) and the *Domitianus* (ICONES H; Pls. S 106, S 106a).
This implies that some of the sculptures from the 'Dutch Gift' were placed in the Whitehall Gardens rather than in the Palace and thus escaped the fire. We should not exclude the possibility that further busts might be identified in the English royal collections.

Despite the fact that 'His Majesty heartily thanked them (i.e. the States) for so worthy a Present, and express'd his willingness to enter into a neerer Alliance with them', the political as well as commercial friction between the two countries persisted. The new English Navigation Act signed on September 13, 1660, clearly was to benefit the English and discriminate against those vessels, notably the Dutch, which shipped into English ports. By 1664, the two countries were at war again.

Art, associated this bust for the first time with the illustration in the manuscript *Busts & Statues in Whitehall Gardens*, Windsor Castle, vol. A 49, Inv. no. 8923, bottom left (Pl. S 108a), but did not connect it with the print in the ICONES or the 'Dutch Gift'. (Letter of 29 March 1974 to Miss Jenifer Sherwood.) Three of these statues (A–C) and six of the busts (E–I, M) came from the Vendramin collection, where they were illustrated in the manuscript DE SCULPTURIS in 1627. For the concordance see pp. 244–247.

[106] For a discussion of the manuscript see A.H. Scott-Elliot, 'The Statues from Mantua in the Collection of King Charles I,' *Burlington Magazine*, CI, 1959, pp. 218–27. These illustrations are found among the third group of drawings (Inv.nos. 8917–27), measuring ca. 13 × 8¼ in. They are on thicker paper and drawn with pencil and brown wash, partly gone over with pen and brown ink. There are four busts to a page. Some of the illustrations are inscribed with pen and brown ink *B*, *br*, or *bras* (for brass or bronze) and *S* or *ST* (for stone ?). Miss Scott-Elliot's suggestion that the busts represented in this third (later) group of drawings were in the Royal collections after the Restoration seems confirmed by the fact that two, possibly three of them were included in the 'Dutch Gift' of 1660.

B. THEIR NATURE

1. FORMATION

Relatively little certain can be said about the formation of the collection due to the lack of documents. The emphasis on Venetian art and Van Buchell's statement that Jan sent objects acquired in Italy, Greece and Turkey to Holland suggest that Jan was the one who formed both collections, his brother's as well as his own. The nucleus of the Reynst collection were the works of art and objects taken over from Andrea Vendramin in Venice and transferred to Amsterdam by the 1630's. This nucleus was greatly enlarged, however, by paintings and sculptures purchased from other sources. As pointed out earlier, none of the works engraved for the CAELATURAE were illustrated in the Vendramin catalogue DE PICTURIS. The best paintings, therefore, came from other collections and most likely were bought individually. Forty-five of the one hundred and ten pieces of sculpture engraved for the ICONES also were bought in addition to the ones purchased from Vendramin. Little could be established about earlier provenances of these works. One, possibly two of the paintings in the Reynst collection were mentioned in 1532 by Marcantonio Michiel as hanging in the house of Andrea Odoni in Venice, namely Lotto's *Portrait of Odoni* (cat. no. 16; Pl. P 16a) and Titian's *Holy Family with St. John and St. Elizabeth* (cat. no. 35; Pl. P 35), and we may suppose that most of the other works with Gerard Reynst came from Venetian or North Italian collectors as well. The most contemporary painting in the Reynst collection was Guercino's *Semiramis*, painted in 1624, for Daniele Ricci in Bologna (cat. no. 13; Pl. P 13a). According to Malvasia the Reynst brothers also owned a large number of Carracci drawings (cat. no. 59).

At the time of Gerard Reynst's death in 1658, there were at least sixty pictures in his collection. This figure is derived from Van Buchell's description in 1639, the paintings included in the CAELATURAE, Knorr von Rosenroth's enumeration in his *Itinerarium* of 1663, and Dujardin's et al. appraisal of Italian pictures in 1672. This does not include the paintings inherited from his brother Jan in 1646, who owned about twenty-one. At least six of them can be traced to the collection of Gerard in Amsterdam, seven to Nicolò Renieri in Venice, with the remaining eight probably also ending up in Holland. In addition, Reynst most likely also owned some 150 paintings that previously belonged to Vendramin.

These various sources show that the Reynst brothers collected almost exclusively paintings by sixteenth century Venetian artists. The only known pictures by a Dutch artist were the three by

Pieter van Laer, who spent many years in Rome and supposedly was acquainted with one of the Reynst brothers. Unfortunately none of the paintings formerly in the possession of Gerard Reynst show any kind of a collector's mark that would facilitate retracing them.[1] The sculptures probably were purchased in Venice, where collections of antique statuary are recorded as early as 1523, the year Cardinal Domenico Grimani (1461–1523) bequeathed several statues and busts to the city. We learn from Michiel's visits to a number of collections that many included antique sculptures. Thanks to the donation by Giovanni Grimani, the Patriarch of Aquileia (ca. 1500–1593) to the city of Venice, two hundred and seventeen antique sculptures were on exhibition in the Antisala of the Bibliotheca Marciana in Venice by the summer of 1596.[2] They are proof of the lively interest for antique statuary in Venice already during the sixteenth century.

The collection of Gerard Reynst was even more extensive. It included about two hundred and thirty statues and busts acquired from Vendramin in addition to some forty-five works illustrated in the ICONES that did not figure in the catalogue DE SCULPTURIS. Thus, in total Reynst owned nearly three hundred pieces of antique statuary, not counting the antique tomb monuments and funerary chests. In all likelihood Jan was the one who made the purchases and acted as his own agent. Since he lived in Venice and was involved with the trade as a factor, he may have dealt occasionally not only in salt and grain but in art as well. This apparently was not unusual at that time for traders with international connections. De la Fontaine Verwey's comments with regard to Daniel Nijs,[3] a Flemish merchant in Venice also dealing in art, seems to be just as appropriate for the brothers Gerard and Jan Reynst. In his words there were 'many such internationally oriented Netherlandish merchants in the Golden Age. Residing abroad they were acting as merchants, ship-owners and bankers, supervising the loading of the ships and the dismanteling of armies. They were lending money to princes and adventureres alike and at times were

---

[1] Based on an examination of the paintings at Hampton Court that came from the Reynst collection. Information kindly provided by Sir Oliver Millar. A Wyatt's identification of G R as Gerard Reynst's mark is unfounded (see *Gazette des Beaux-Arts*, 1, 1859, pp. 176–77).

[2] See Marilyn Perry, 'The Statuario Publico of the Venetian Republic,' *Saggi e memorie di storia dell'arte*, 8, 1972, pp. 75–150, especially pp. 78–82. I should like to thank Dr. I.I.E. van Gelder-Jost for this reference.

[3] H. de la Fontaine Verwey, 'De Zaken van Daniel Nijs,' *Maandblad...Amstelodamum*, 56, 1969, pp. 79–82. See also I.H. van Eeghen, 'Het geslacht Nijs (Nederlandse cosmopolieten in de 17de eeuw),' *Jaarboek...Amstelodamum*, 59, 1967, pp. 74–102, especially pp. 76–77.

actively engaged in diplomatic missions. Once they had attained some wealth, they started collecting art which they often resold at a profit.' This is very plausible if one realizes that a number of rich merchants also were collectors such as the Reynsts, Van Uffelen or Van Veerle. They not only collected primarily Italian art but also spent some time in Italy.

If Jan Reynst needed advice he probably received it from Nicolò Renieri,[4] a close friend of his, who had arrived in Venice about the same time. Renieri was not only a painter but a dealer and collector as well who owned paintings by contemporary artists such as Guercino, Liss and Fetti, i.e. artists also represented in the collection of Gerard Reynst in Amsterdam. Another person who could have acted as an adviser or an intermediary was Daniel Nijs (1572–1647), referred to earlier, who again dealt in art and collected. Jan may have become acquainted with both Renieri and Nijs through his association with the large Flemish community living in Venice in the early 17th century among them his two uncles, Jan and Jacques Nicquet.

We know from correspondence that Gerard Reynst was acquainted with at least one gentleman dealer in art, namely Michel Le Blon (1587–1658),[5] who was acting as an agent besides being an artist and engraver. In the later 1630's, he introduced Samuel Bloemaert, a brother-in-law of Gerard and Jan Reynst, at the Swedish court.[6] Reynst[7] certainly was in contact with Le Blon by 1650, for the latter wrote to Queen Christina of Sweden[8] that Reynst was being courted by the magistrates of the city of Amsterdam, who hoped to purchase the antique sculptures for their new town hall. At the same time Le Blon was offering these very pieces to the Queen at a special price of 30,000 Swedish thalers. We have no further documentation that would tell just how seriously the city of Amsterdam was interested in these sculptures. In the end, neither the city nor Queen Christina purchased them.

---

[4] Born in Maubeuge (then part of the Southern Netherlands), ca. 1590; in Rome by 1621, in Venice ca. 1626, where he died in 1667. For a list of paintings in his collection see Martinioni (1663), pp. 377–78; Campori (1870), pp. 442–47; E.K. Waterhouse, 'Paintings from Venice for Seventeenth-Century England: Some Records of a Forgotten Transaction,' *Italian Studies*, VII, 1952, pp. 1–23; Savini-Branca, *op.cit.*, pp. 264–68. The recent article on Renieri by P. L. Fantelli, 'Nicolò Renieri 'Pittor Fiamengo',' *Saggi e Memorie di storia dell'arte*, 9, Florence, 1974, pp. 79–195 states only that the artist was in close contact with the Flemish community in Venice. Fantelli refers to the Reynst brothers in a brief quote from Ridolfi's introduction to the *Maraviglie dell'Arte*.

[5] See H. de la Fontaine Verwey, 'Michel Le Blon, graveur, kunsthandelaar, diplomaat,' *Jaarboek...*

Le Blon, furthermore, made an engraving of the coat of arms of the Reynst family (illustrated on the title-page).[9] There are no records, however that would indicate that Reynst bought from Le Blon. Gerard, for example, was not among the twenty-six Amsterdam merchants who acquired antique sculptures in 1646 that had been shipped to Holland from Antwerp through the services of Michel Le Blon,[10] nor did works of art from Van Dyck's collection enter into Reynst's possession as far as can be established.[11]

One art dealer, who is often mentioned in connection with Reynst, is Gerrit Uylenburgh. He was involved in the dispersal of the collection, however, rather than in its formation. Uylenburgh's first known contact with the Reynst collection dated from 1660, when he assisted Van Outshoorn and Artus Quellinus in the selection of the twenty-four Italian paintings and twelve antique sculptures for the 'Dutch Gift' (see p. 75). As was pointed out briefly before, the collection of Gerard Reynst was dispersed at a sale that took place the end of May, 1670, in Amsterdam. We learn this from a short reference in a letter by Constantijn Huygens the younger to his brother Lodewijk, dated May 29, 1670, where he wrote: '...nostre Bisschop[12]... est à Amsterdam ou se vend presentement le Cabinet tant renommé de Reinst ....'[13]

Amstelodamum, 61, 1969, pp. 103–25.

[6] *Ibidem*, p. 116.

[7] The 'heer Ranst', mentioned in Le Blon's letter to Musson of 1645, with regard to Van Dyck's estate, refers to another well-known Amsterdam family. (Jan Denucé, *Bronnen voor de Geschiedenis van de vlaamsche Kunst*, v, *Na Peter Pauwel Rubens*, Antwerp, 1949, p. 34, no. 49).

[8] Letter of 18/28 January 1650. See p. 50, note 30, above.

[9] F.W.H. Hollstein, *Dutch and Flemish Etchings, Engravings and Woodcuts, ca. 1450–1700*, II, Amsterdam, 1949 ff., p. 143, no. 105, ill.; 77 × 54 mm.

[10] Denucé, *op.cit.*, pp. 55–56, no. 79: bill of Jacques Brel of 13 September 1646; J.G. van Gelder, *Studies in Western Art*, pp. 51–52, and J.G. van Gelder and Ingrid Jost, 'Two Marble Statuettes from Seventeenth-Century Amsterdam Collections,' *Scripta Archaeologica Groningana* 6, Groningen/Bussum, 1976, pp. 297–304.

[11] Jenny Müller-Rostock, 'Ein Verzeichnis von Bildern aus dem Besitze des Van Dyck,' *Zeitschrift für bildende Kunst*, 57, (N.F. XXXIII), 1922, pp. 22–24.

[12] According to Malvasia, the 'signori Reinst' owned many Carracci drawings (cat. no. 59) and one wonders, whether the Carracci drawings recorded in De Bisschop's possession perhaps were bought at the dispersal of the Reynst 'cabinet'. (This idea was first expressed verbally by Professor J.G. van Gelder).

[13] *Oeuvres complètes de Christiaan Huygens*, VII, *Correspondance 1670–1675*, The Hague, 1897, no. 1808. The passage is also published in J.G. van Gelder, 'Jan de Bisschop,' *Oud Holland*, LXXXVI, 1971, p. 59, no. 15.

This 'cabinet de Reinst' must have attracted many artists, collectors and dealers. It probably was sold publicly rather than through an auction and no records of the sale have come to light.[14] A few of the paintings and sculptures can be traced to Jan Six in Amsterdam and to Gerrit Uylenburgh. Since the latter was the most renowned dealer in Italian art at that time, according to Houbraken,[15] he may possibly have been in charge of the sale, or may have taken the works on consignment.

In the summer of 1671, Uylenburgh offered thirteen Italian paintings and a number of antique sculptures from the Reynst collection to the Elector of Brandenburg for the sum of 30,000 guilders. Twelve of these paintings brought about a controversy in Holland, involving a large group of painters who testified for or against their authenticity. The first to repudiate them was Hendrik van Fromantiou, a Dutch artist, who had worked for Uylenburgh and who, therefore, was familiar with the latter's workshop practice of having Italian paintings copied by his assistants. At the time, Fromantiou was artistic counselor to Frederic William. According to Houbraken,[16] Fromantiou declared the paintings to be imitations and assured the Great Elector that he could show him the originals in Holland and elsewhere. This of course, made Frederic William decide against the purchase and Uylenburgh was asked to take the paintings back. Since he refused,

---

[14] The earliest issue of the *Amsterdamsche Courant* still extant dates from March 12, 1672 and we, therefore, can no longer check, whether the sale was advertised officially. See also S.A.C. Dudok van Heel, 'Honderdvijftig advertenties van kunstverkopingen uit veertig jaargangen van de Amsterdamsche courant 1672-1711,' *Jaarboek ... Amstelodamum*, 67, 1975, pp. 149-73.

[15] *Op.cit.*, III, Maastricht, 1953, p. 170. D. Mahon, in *Art Bulletin*, XXXI, 1949, p. 222, note 44, already suggested that paintings from the Reynst collection were disposed through Uylenburgh.

[16] *Idem, op.cit.*, II, pp. 231-33. For other discussions of this incident see Floerke, *op.cit.*, pp. 105-09; R. Dohme, 'Die Ausstellung von Gemälden älterer Meister im Berliner Privatbesitz,' *Jahrbuch der königlich preussischen Kunstsammlungen*, IV, 1883, pp. 126-27; Georg Galland, *Hohenzollern und Oranien* (Studien zur deutschen Kunstgeschichte, 144), Strassburg, 1911, pp. 95-96; Jan Six, 'La famosa Accademia di Eeulenborg,' *Jaarboek der koninklijke Akademie van Wetenschappen te Amsterdam*, 1925-26, 1926, pp. 238-40; Horst Gerson, *Ausbreitung und Nachwirkung der holländischen Malerei des 17. Jahrhunderts* (*Verhandelingen uitgegeven door Teyler's tweede Genootschap*, N.R., XII), Haarlem, 1942, pp. 228-29. Brochhagen, *op.cit.*, p. 7.

Fromantiou himself accompanied the pictures to Amsterdam, where they were to be judged publicly.[17] Among the painters who declared some of the paintings did have some merit and that all of them were good enough to hang in some collection of Italian art were Gerbrand van den Eeckhout, Johan Lingelbach and Philips Koninck.[18]

On May 12, 1672, a number of different artists, among them Adam Pijnacker, Jan Wijnants, Gerard de Lairesse, Willem Kalf, Roelandt Roghman and Lambert Doomer gathered in the 'Keijserkroon', an inn on the Kalverstraat, to judge the twelve Italian paintings that Fromiantou had contested. They concluded that these were rather poor Italian paintings worth at the most one tenth of the estimated prices. Some of the pictures were even declared worthless (*vodden*).[19] Uylenburgh, however, was not convinced and unrelentingly selected still different artists living in Amsterdam. On May 14, Philips de Momper, Abraham Begeyn, Pieter Codde and others, again testified to the contrary.[20] After the testimony of Carel Dujardin and Willem Dodijns on May 16,[21] the paintings were sent to The Hague, where they were exhibited in the 'Confrerie-Kamer' on Saturday, May 21, for further inspection. Among the artists living in The Hague who testified about the paintings were Theodor Verschuyr, Jacques Vaillant, Caspar Netscher and Dirck Dalens.[22] Vermeer and Johannes

[17] Bredius, *Oud Holland* 1886; the following transcript of the appraisers' list was published on pp. 279–80 (Prot. Not. F. Tixerandet, Amsterdam):

Op huyden den 16en May 1672 compareerden Sr. CAREL DU JARDIN, oudt omtrent 50 [jaren.....] Sr. WILLEM DODIJNS, oudt omtrent 41 jaren, beyde Mrs. Schilders alhier ter stede woonaghtigh, dewelcke ten versoecke van Sr. HENRY DE FROMANTIOU, Mr. schilder van sijn Cheurvorstelycke Hoocheydt van Brandenburg, sigh althans bevindende alhier ter steede, verclaerden hoe waer is, dat zij heden seeckere 12 distincte stucken hieronder gespecificeert *nacuwkeurigh besigtigt* en naer derselver kennisse en oordeel bevonden hebben als volght:

No. 1. Een Venus en Cupido, beelden grooter als 't leven, geseyt van MICHEL ANGELO BONAROTTI, maer dewijl dit stuck in teykeningh en gratie met het minste van de voorsz. MICHEL ANGELO niet te compareeren is, hebben sij attestanten haer verplicht gevonden naer waerheyt te oordeel, o tselve niet *van* maer wel *naer* de voorsz. Mr. te comen, waerom sij het oordeelen in geenen deele van sodanigen prijs te sijn als het vercocht is.

No. 2 en 3, die sij attestanten verclaerden haer onbekent te sijn.

No. 4. Een Conterfeytsel van GIORGION, geseyt door TITIAEN geschildert, bij overlegh geoordeelt in couleur en pinseel, veel min in teykeningh met de uytstekende wercken van TITIAEN te accorderen.

No. 5. Een herder en herderin van TITIAN, waerin (sij) eenige qualiteit van de voorsz. TITIAN bespeuren, soo in een goede armonie van dagingh als couleur, maer int geheel en alles niet geoordeelt soo uytstekend als voorgegeven wert.

No. 6. Een weergae, sijnde kinderticus en achter en heremitien, geseyt van denselven TITIAN, maer in geen deel te vergelijcken bij het bovenstaende, *hebbende geen kraght noch helderheyt van Couleur, daerin de voorsz. Titian genoegsaem boven alle meesters geëxcelleert heeft.*

Item. Een dans van naeckte kinderntiens, geseyt van JACOMO PALMA, heel naeckt en ontbloot van die qualiteyten, die in de voorsz. meester geresideert hebben, soo in couleur, teyckening, actie, ordonnantie, als dagingh, waarom

Jordaen even came from nearby Delft to testify.²³

It was to no avail, however. The Elector of Brandenburg chose to return all the Italian paintings and kept only Spagnoletto's *Head of St. John* as well as the sculptures in exchange for the 4,000 guilders he had made as a down payment at the time the negotiations with Uylenburgh were initiated.²⁴ Fromantiou left the paintings in the care of Jan Wils, who was most reluctant to hand them back to Uylenburgh. Uylenburgh finally succeeded in obtaining them on February 8, 1673,²⁵ and sold them publicly in Amsterdam the following February 23.²⁶

Among the people who sided with Uylenburgh was no other than Constantijn Huygens the elder (1596–1686) who pointed out that the paintings could not have been all that bad since for many years they were admired in the 'cabinet' of Reynst. He expressed these feelings in a letter to Gerardt Bernard von Pölnitz, an officier in the army of Brandenburg. The letter was written from The Hague on May 23, 1672, at a time this controversy between Uylenburgh and Fromantiou had reached its climax :

'Je ne sçaurois m'empescher de soustenir encor ceste fois l'innocence et la probité du pauvre Sieur Uijlenburg. Je l'ay trouvé si rudement persecuté par le peintre Fermenteau, qui remue icy toute pierre pour amasser des voix qui veuillent decrier pour copies

niet waerdigh geacht met de naem van soo braven meester verciert te werden.

Een Venetiaense dame van PARIS PORDENON (more likely BORDONE) wel versien met eenige gracelijckheyt, doch niet uytmuntende, veel min van soo hoogen prijs.

Een Conterfeytsel van een prelaet, geseyt van HANS HOLBEEN, twelck sij in geenen deele bevinden 't accorderen met de pinseel van HOLBEEN, veel min een origineel van de voorsz. HOLBEEN.

Een Ceres [met overvloet met veele naeckte Kindertjes] in 't kleyn, geseyt van GIORGION DE CASTELFRANCO, maer al te plat en ge [...], geplackt op een swarte grondt, en dat terwijl contrarie voorsz. M_r. in couleur soo uytmuntende is geweest, waerom sij oordeelen 't selve soo verre aft te (dwalen)? van de opregte originalen van de voorsz. GIORGION.

Een oudt man, geseyt van RAPHAEL, die sij oordeelen al te gering, om op den schouderen van soo divien meester te leggen en daarmede sijn reputatie te declineeren.

Een St. Paulus, half beelt, levensgroot, geseyt van JACOMO PALMA DEN OUDEN, doch al te mistekent, en insonderheyt een hant onbequaem om het swaert te houden, en de rest niet veel beter.

Een schoone Venetiaense Dame, met de handt int hayr, wel van de waerdigste in dese Collectie, maer niet onder de uytmuntende van Titian.

Een lantschap [met een Satier die de nimphe Caresseert], veel meer gelijckende naar ANDREAS SCHIAVONI als naer TITIAN, doch door tijt en ongeval soo bedorv n en op verscheyde plaetsen verschildert, dat daerinne niet veel deugt te bespeuren is.

Alle twelck sij verclaeren haer gemoedt en volgens een oprecht oordeel te hebben bevonden in soodanigen staet en conditie als hierboven gespecificeert staet. etc. (Tixerandet's acts for these years are damaged by fire and inaccessible).

¹⁸ Bredius, *op.cit.* 1886, p. 42. The others present were Dirck Santvoort, Jan Blom, Wallerant Vaillant, Anthony de Grebber, Abraham van den Tempel and Adriaen Backer.

¹⁹ Gemeentearchief, Amsterdam ;

Notariael Archief no. 4074. The original document was found by S.A.C. Dudok van Heel and published in parts by Lucius Grisebach, *Willem Kalf 1619–1693*, Berlin, 1974, p. 23 and Appendix I, document 20. Also present were Willem van Aelst, Jan André Lievens, Otto Marseus, Mattheus van Pellecum, Melchior de Hondecoeter, Bartholomeus Appelman, Hendrick van Someren, Barent Graat, Daniel Wolfraedt, Jacob Vennecool and Jan Wils. See Bredius, *loc.cit.*.

[20] Bredius, *loc.cit.*. The others present were Willem Strijcker, Dirck Ferreris, Pieter Pz. Niedeck, Lodewijk van Ludick, Harman Collenius, Christiaen Striep and David Eversdijk.

[21] For a transcript of the testimony see footnote 17. Brochhagen, *op.cit.*, p. 7 states that some of the Italian paintings Dujardin had to judge came from the dispersed collection of Gerard Reynst.

[22] Bredius, *op.cit.* 1886, pp. 42–43. Also present were Johanna van Aerssen van Wernhout, Pieter Monincx, Jan van Sandrart, Johan Monincks, Johannes van Haensbergen, Martinus Mijtens, Francois van Santwijck, Daniel Haringh. Johan de Baen, Johan Le Ducq and Jeronimus van Diest gave a separate testimony (A. Bredius, 'Italiaensche Schilderijen in 1672 door Haagsche en Delftsche Schilders beoordeeld,' *Oud Holland*, 34, 1916, pp. 88–93).

[23] Bredius, *loc.cit.*, 1916. I would like to thank Professor M. Montias for this reference. The Vermeer document is also published in Albert Blankert, *Johannes Vermeer van Delft 1632–1675*, Utrecht/Antwerp, 1975, p. 127, document 30. The document (Prot. Not. P. van Swieten, The Hague) reads as follows:

		Ryxdaelders
een	Venus en Cupido, beelden grooter als het leven van MICHIEL ANGELO BONAROTTI, Hollants gelt	350 : 320
een	Conterfeytsel van GIORGION DEL CASTELFRANCKO van TITIAEN, naer het leven geschildert	250 : 240
een	harder ende harderinnetje van TITIAEN	160 : 150
	Weergadingh van deselve groote van TITIAEN	120 : 110
een	dans van naeckte Kindertjens levensgroote van IACOMO PALMA	250 : 240
een	Venetiaense Dame van PARIS PORDINON (N.B. PARIS BORDONE)	160 : 150
een	Conterfeytsel van een Prelaet van HANS HOLBEEN	120 : 110
een	Ceres met overvloet, met veele naeckte Kindertjes van GIORGION DEL CASTEL FRANCO	120 : 110
een	out mans confeytsel van RAPHAEL URBIN	150 : 140
een	St. Paulus, halfbeelt, levensgroote, van de Oude JACOMO PALMA	80 : 70
een	Schoone Venetiaense vrouw van TITIAEN	200 : 185
een	lantschap van TITIAEN met een Satier die de nimphe Caresseert	240 : 230

sijnde alle de voorsz. Schilderijen gecachetteert met het signet van syne Ceurvorstelycke Doorluchticheyt VAN BRANDENBURCH, welcke voorsz. Schilderijen niet alleen niet en sijn uytmuntende Italiaensche Stucken ofte schilderyen, maer ter contrarie niet weerdich te sijn te draegen den naem van een goet meester, veel min den naem van soodanige uytmuntende meesters daer deselve voor uytgegeven werden gedaen te syn en vervolgens dselve by haer deposanten tot een prijs van merite niet connen werden getaxeert.

Eindigende, sy hiermede... etc.

[24] Dohme, *op.cit.*, p. 127. Since the painting by Ribera that the Elector of Brandenburg retained was estimated only at 180 thalers and since Uylenburgh never returned the rest of the money given to him as a down payment we may suppose that the Elector kept the antique statues in return. (see Emil Jacobs, 'Ein Bilderkauf des Grossen Kurfürsten,' *National-Zeitung*, December 28, 1904, no. 787, Feuilleton).

[25] Bredius, *op.cit.* 1916, pp. 91–93. Some of the paintings apparently were damaged during these transactions.

[26] Bredius, *op.cit.*, 1886, pp. 44–45.

une douzaine de pieces que ledit Uylenburg a vendues à Son Alt^e Electorale, contre ce que plusieurs de nos meilleurs peintres en ont declaré, que moy mesme j'ay voulu m'appliquer à en prendre connoissance, assité de plusieurs personnes d'honneur et de condition, qui par beaucoup d'usage et de commerce en matiere de tableaux italiens s'en sont acquis une experience plus assurée que la miene quoyque j'ose presumer d'y entendre quelque peu pour ma part, et puis vous declarer, Monsieur, qu'apres avoir visité le tout par le menu, nous avons bien trouvé que d'aucunes de ces pieces en surmontent d'autres en valeur, mais non pas qu'aucune du nombre puisse estre reprochée pour copie, comme en effet toutes ont esté avouées originales, par longues années, dans le fameux cabinet de feu le Sieur Reinst à Amsterdam. Nous avons donc bien jugé par la verité du faict, et mesme par des discours aigres et effrontez du persecuteur, que toute sa visée ne tend qu'à ruiner la reputation du persecuté, qu'on dit n'avoir pas merité ceste recompense pour des effects de sincere et ancienne amitié. Je me tiens si asseuré, Monsieur, de vostre generosité et de la peine que les plus matois auroyent à vous destourner du chemin de la justice, que je n'ose pas seulement m'avancer à vous prier de soustenir celle de cest'honnest' homme. Het doet mij genoegen, dat gij u in deze zaak op mij hebt beroepen.'[27]

One of the Italian paintings rejected by the Great Elector that was described as *Een ceres met overvloot, met veele naeckte kindertjens van Giorgion del Castel Francko* at the time the painters Karel Dujardin and others appraised it on May 12, 1672,[28] subsequently figured in the inventory of Uylenburgh's possessions, drawn up in 1674, under no. 47: *Een cherus met kinderties van Schorschon*.[29] Jacobs[30] was the first to point out that this must be the painting reproduced in the Vendramin catalogue DE PICTURIS, fol 11, as *La Diuitia di Zorzon*.[31] Thus a second painting from the Reynst collection can be traced back to Vendramin.

---

[27] The correspondence of Huygens is now preserved in the Royal Library at The Hague. Published by Worp, *op.cit.*, VI, pp. 303–04, mo. 6841. Von Pölnitz was married to Leonora, a natural daughter of Prince Maurits, and died in 1679. (Worp, *op.cit.*, V, p. 332, no. 5642). See also A.W.E. Dek, *Genealogie van het vorstenhuis Nassau*, Zaltbommel, 1970, p. 148. (I owe this reference to S.A.C. Dudok van Heel).

[28] See footnote 17.

[29] Bredius, *op.cit.*, XI, p. 1669.

[30] *Op.cit.*, pp. 35–37.

[31] Reproduced in Borenius, *op.cit.*, pl. 11.

Another painting included in this list of appraisals [32] as *Een Venetiaense dame van* PARIS PORDENON (read Bordone) *wel versien met eenige gracelijckheyt ...* perhaps is identical with the picture illustrated in DE PICTURIS, fol. 60.[33]

So far at least one of these Italian paintings that Uylenburgh offered to the Elector of Brandenburg has been preserved. The *Dance of Naked Children* (*Een dans van naeckte kindertkens, geseyt van Jacomo Palma..*) (Pl. P 58), figured in the sale of the collection of Jan Six in April, 1702, where it was acquired by his son, Jan.[34] Six may have purchased it from Uylenburgh after it was rejected by the Elector of Brandenburg. The painting remained in the Six family and is presently hanging in the house Amstel 218 in Amsterdam.[35] It is the only painting formerly with Gerard Reynst that has remained in Holland. Its provenance from Uylenburgh was first established by J. Six in 1925,[36] but the fact that it previously was part of the 'cabinet' of Reynst was no longer known. Actually, Jan Six must have purchased two additional paintings that came from Reynst, namely the *Venetian Senator* by Tintoretto,[37] and a *Brothel Scene* by Liss.[38]

No documents are known that would establish that Gerrit Uylenburgh's father, Hendrick (born in Cracow, ca. 1587; died 1661), who was active as a dealer during Gerard Reynst's lifetime, knew this Amsterdam collector or had sold works of art to him.[39]

---

[32] See footnote 17.

[33] Reproduced in Borenius, *op.cit.*, pl. 48.

[34] *Catalogus Van uitmuntende Konstige, meest Italiaansche Schilderyen, Deer voornaame Italiaansche, Fransche, Hoog en Nederlantsche Meesters geschilderd; Mitsgaders verscheyde schoone Antique Marmere Statuën, Borstbeelden, Koppen, Bazerelieve,..., Jan Six*, Amsterdam, 6 April, 1702, no. 7: *Tien dansende kinderen van Palma Giovene, levensgrotte.* (not listed in F. Lugt, *Répertoire des Catalogues de Ventes*, I, The Hague, 1938. Copies are in the Six collection, Amsterdam and in the Rijksbureau voor kunsthistorische Documentatie, The Hague; a photocopy is in the Frick Art Reference Library, New York). See also I.H. van Eeghen, 'De Familie Six en Rembrandts Portretten,' *Maandblaad...Amstelodamum*, 58, 1971, pp. 112–16 and S.A.C. Dudok van Heel, in *Jaarboek...Amstelodamum*, 67, 1975, pp. 145–46 for the Six family as collectors.

[35] I would like to thank Jhr. Six van Hillegom for allowing me to reproduce the painting.

[36] 'La famosa Accademia di Eeulenborg,' *Jaarboeck der koninklijke Akademie van Wetenschappen te Amsterdam*, 1925–26, 1926, pp. 238–40, ill. opp. p. 240.

[37] Sale Amsterdam, 6 April 1702, no. 23: *Venetiaansch Senaat, van de zelve (Jacomo Tintoret)*; sold for 22.5 guilders, probably to Pieter Six (sale Amsterdam, 2 September 1704, no. 12); Hoet, I, p. 72.

M.G. de Boer, 'Vergeten Leden van een bekend Geslacht,' *Jaarboek...Amstelodamum*, 42, 1948, pp. 20–27 and p. 33, no. 12. Probably identical with the illustration in DE PICTURIS, fol. 78, reproduced in Borenius, *op.cit.*, pp. 39–40 and pl. 66. Possibly today in the collection Oskar Reinhart, Winterthur, as the *Portrait of Senator Girolamo Grimani*; see *Sammlung Oskar Reinhart am Römerholz*, Winterthur (s.a.), p. 40, no. 122; canvas, 112×94 cm.

[38] Sale Amsterdam, 6 April 1702, no. 95: *Een Venetiaansch Lusthuis, zeer sierlyk en ongemeen curieus*; sale Amsterdam, 2 September 1704, no. 16; Hoet, I, p. 72; De Boer, *op.cit.*, p. 33, no. 16. The painting is now in Cassel (see cat. no. 15; Pl. P 15a).

Lugt's assertion that Reynst was in close contact with Uylenburgh cannot be substantiated. We have no documents either that would ascertain Lugt's supposition that Rembrandt had access to the Reynst collection through Hendrik Uylenburgh, a cousin of Rembrandt's first wife Saskia.[40]

We may therefore conclude that it was primarily Jan who formed the collection in agreement with his brother Gerard who must have supported it financially. The death of Jan in 1646, most likely put an end to new acquisitions. Gerard even considered selling part of it in 1650, if he were to receive a good price. Nothing came of it, however, and he had a selection of his best paintings and sculptures published in prints instead. Because of these very prints we are still able to realize how many fine paintings and sculptures the two brothers collected over the years.

---

[39] Bredius, *Künstler-Inventare*, XI, pp. 1684–90.

[40] Lugt, *Oud Holland*, pp. 117–118.

2. CHARACTER OF
THE COLLECTIONS

Thanks to the descriptions by various visitors we were able to establish that the Reynst collection not only included paintings, sculptures, tomb monuments, coins and printed books, but also minerals, metals, sea animals and sea plants as well as rare and curious objects found in nature and exotic items from the Orient. Thus, the collection contained *artificialia* (paintings, sculptures, coins) and *naturalia* (animals, vegetables, minerals or fossils) that is objects created either by man or by nature. The Reynst collection, therefore, was not just an art collection but rather a 'Kunstkammer', reflecting the manifold aspects of the universe.[41]

The initial attempt at a more methodical grouping of the Reynst collection goes back to the Vendramin catalogues which were not mere inventories but had the material organized according to the individual species. The various divisions in those manuscripts were not unlike the ones found in Quichelberg's innovative catalogue of an imaginary, ideal collection, the *Theatrum sapientiae* of 1565 (probably inspired by the collection of Duke Albrecht V of Bavaria) which differentiated between the following five groups (*Inscriptiones*): objects pertaining to religion, views and historical data relating to the founder of the collection; *artificialia* (sculpture, coins, medals, goldsmith's works); *naturalia*; *artes mechanicae* (scientific and musical instruments), and works of art (paintings and prints).[42] Through this systematization the collections became

[41] The best and most comprehensive surveys of the various aspects of collecting are found in : Julius von Schlosser, *Die Kunst- und Wunderkammern der Spätrenaissance*, Leipzig, 1908 (basic study on early collections); Barbara Jeanne Balsiger, *The Kunst- und Wunderkammern : A Catalogue Raisonné of Collecting in Germany, France and England, 1565–1750*, Diss. Pittsburgh, 1970 (most detailed treatment of collections with valuable insight into collecting and the motivations of the respective collectors); Luigi Salerno, 'Arte, scienza e collezioni nel manierismo,' *Scritti di Storia dell'Arte in onore di Mario Salmi*, III, Rome, 1963, pp. 193–214 (best evaluation of Italian collections); Christian Theuerkauff, 'Zum Bild der 'Kunst- und Wunderkammer' des Barock, '*Alte und moderne Kunst*, XI, Sept./Oct. 1966, pp. 2–18; R.W. Scheller, 'Rembrandt en de encyclopedische verzameling,' *Oud Holland*, LXXXIV, 1969, pp. 81–145 (excellent study on collecting in Holland in the 17/the century); More general but still useful are the contributions on the subject made by Rudolf Berliner, 'Zur älteren Geschichte der allgemeinen Museumslehre in Deutschland,' *Münchner Jahrbuch der bildenden Kunst*, 1928, pp. 327–50; as well as D. Murray, *Museums*, Glasgow, 1904. For further bibliographical references see also Balsiger and Scheller, *op.cit.*

[42] Von Schlosser, *op.cit.*, pp. 73–76; Elizabeth M. Hajós,' The Concept of an Engravings Collection in the Year 1665 : Quicchelberg, *Inscriptiones Vel Tituli Theatri Amplissimi*,' *Art Bulletin*, XL, 1958, pp. 151–56; Scheller, *op.cit.*, p. 103; Balsiger, *op.cit.*, pp. 59–60; 303–08; Renate von Busch, *Studien zu deutschen Antikensammlungen des 16. Jahrhunderts*, Diss. Tübingen, 1973, p. 114.

more accessible for visitors, who were attracted because of this encyclopaedic character, in order to learn about nature, history, exotic cultures and so forth.

A brief additional remark about the Vendramin collection should be inserted here. As we saw earlier, the brothers Gerard and Jan (?) Reynst acquired most or all of the works of art and objects during the 1630's from Andrea Vendramin, who actually had collected them. This type of collection, combining paintings, sculptures and *naturalia*, was found more frequently in the North, however, in the *Kunst- und Wunderkammern* of Germany and Austria, where works of art were collected alongside with rare and curious objects.[43] In comparison to other contemporary Italian collections, assembled by private citizens rather than royalty, the Vendramin collection could be considered an exception.[44] In Italy, collectors either favored the works of art by themselves, or the *naturalia*, except for Vendramin, who collected more evenly and assembled a large number of paintings, antique sculptures, manuscripts, coins and vases in addition to objects found in nature. The most extensive and richest such collection in the South was brought together in Rome by the German Jesuit Athanasius Kircher, who as we recall was aware of the Reynst collection, for he had used a few of its Egyptian objects in his treatise on Egyptian art (see

---

[43] See Von Schlosser, *loc.cit.*, as well as Scheller, *op.cit.*, p. 81.

[44] For the various Italian collections of the 16th and 17th centuries, see Von Schlosser, *op.cit.*, pp. 104–10; Salerno, *loc.cit.*; Balsiger, *op.cit.*, pp. 36–503.
The only time Michiel mentioned *naturalia* such as petrified objects, fish, shells in addition to paintings and sculptures was during his visit to the house of Odoni in 1532 (Theodor Frimmel, *Der Anonimo Morelliano, Quellenschriften für Kunstgeschichte und Kunsttechnik*, N.F. 1, 1896, p. 84). One other collector with a similar interest was Bernardino Loredan, whose testament of 1608 lists 'antiques, bronze medals, fossils and pictures' (Logan, *op.cit.*, p. 314). Generally, the collections consisted primarily of Venetian sixteenth-century paintings, at times with an additional work by Raffael, Guido Reni or a Netherlandish master. Often, they also included some pieces of antique sculpture and coins, medals and gems. One such collection was formed by Pietro Bembo in Padua, who owned several busts of Roman Emperors; others who owned a fair number of antique sculptures in addition to paintings were Zuan Ram and Antonio Foscarini in Venice (see Frimmel, *op.cit.*, pp. 92–93 and 104–07).

p. 59). The catalogue of this *museo Kircheriano* in Rome was published in Amsterdam in 1678.[45]

The Vendramin collection never is discussed in the literature on the subject, despite the fact that its content could be reconstructed rather well thanks to the seventeen manuscript catalogues.[46] It certainly deserves to be singled out among early Italian collections because it deviates from the ones assembled by other private collectors of the time like Ceruti and Calceolari (catalogue Verona, 1622),[47] and later by Moscardo (catalogue Padua, 1656),[48] Aldrovandi and Cospi (catalogue Bologna, 1677)[49] or Settala (catalogue Tortona, 1662–77)[50] which were primarily interested in *naturalia* or *artificia rariora*. Among these, only the *museo Cuspiano* and the *museo Moscardo* did include some paintings and antique sculptures, but in addition to the *naturalia* rather than vice versa.

Collectors often were well educated people, so-called gentlemen-*virtuosi*, who bought the works of art out of a genuine interest in the object, in order to learn about it and to further their knowlege.[51] This, however, is hardly true for Gerard and Jan Reynst. As successful, practical merchants they may have considered collecting primarily as a means to attain a higher social respectability, a status symbol which would reflect upon their capabilities. Through this collection they certainly hoped to obtain honor and in time probably even financial gains. How important this idea of *eer* and *gewin* appears to have been in seventeenth-century life in Holland was intimated by Scheller.[52] With this purchase of the Vendramin collection, Gerard Reynst came into the possession of a collection which would have taken him years, if not a lifetime to bring together. Thanks to this collection he obtained the image of a collector, before he really started collecting seriously himself.

There undoubtedly were more such collections in existence in Holland in the seventeenth century, although probably containing

---

[45] Von Schlosser, *op.cit.*, p. 104 and fig. 88 for illustration of title page. Balsiger, *op.cit.*, 93–98 ; 287–90 ; 442–43.

[46] Neither von Schlosser nor Balsiger mentioned the Vendramin-Reynst collection. This may be due to limiting their respective studies to collections accompanied by printed catalogues. The Vendramin catalogues, however, were in many ways more explicit than printed examples, since they were illustrated and they certainly should be taken into consideration.

[47] Von Schlosser, *op.cit.*, p. 106 and fig. 89 for illustration of title-page. Balsiger, *op.cit.*, pp. 121–22 ; 126–34.

[48] Von Schlosser, *op.cit.*, pp. 108–09. Balsiger, *op.cit.*, pp. 331–42.

[49] Von Schlosser, *op.cit.*, p. 108 and

a smaller number of pictures, but no thorough investigation of them has been undertaken. As Scheller pointed out, the publications of old inventories by Bredius were limited primarily to paintings.[53] One other, similar collection has been reconstructed recently, the one assembled by Rembrandt. Rembrandt, according to Scheller, also hoped to obtain through it a higher social status as a patrician and a gentleman-*virtuoso*.[54]

Over the years, the manifold aspects of the Reynst collection receded in favor of the collection of paintings and antique sculptures. The brothers Gerard and Jan Reynst seemed much more interested in collecting works of art as such rather than combining it with *naturalia*. This is also reflected in the very one-sided growth of their collection since they added almost exclusively paintings and sculptures. In these very paintings and sculptures, however, the Reynst collection has survived until today, primarily thanks to the two volumes of engravings after a choice selection among them, the CAELATURAE and the ICONES.

---

fig.90 for illustration of title-page. Balsiger, *op.cit.*, pp.62–64 ; 303–09 ; 149.

[50] Von Schlosser, *op.cit.*, p. 109 and fig. 91 for illustration of title-page. Balsiger, *op.cit.*, pp. 445–52.

[51] Walter E. Houghton Jr., 'The English virtuoso in the Seventeenth Century,' *Journal of the History of Ideas*, 3, 1942, pp. 51–73 and 190–219. Scheller, *op.cit.*, pp. 128–32.

[52] *Ibidem*, pp. 134–37.

[53] Scheller, *op.cit.*, p. 116.

[54] *Ibidem*, pp. 137–41.

IV. CONCLUSION

In concluding, a few words should be said about the significance of the collection that Gerard Reynst had assembled with the assistance of his brother Jan. One striking difference in comparison with other seventeenth-century Dutch collections becomes quite evident: its emphasis on Italian paintings. This aspect must have been well known and the paintings apparently were highly thought of. This can be clearly seen from the reaction of the delegates in the Amsterdam city council. Upon learning that Charles II preferred Italian paintings to contemporary Dutch pictures, they thought right away of the Reynst collection and deemed it worthy to be presented to a king as an official gift on their behalf.

The total number of objects included in this 'cabinet tant renommé de Reinst' was rather astonishing. Besides some 200 primarily Italian paintings (if we assume that the 150 pictures from the Vendramin collection were purchased too), there were some 300 sculptures, 10 tomb monuments, 5 votive reliefs, 9 cinerary urns, a number of votive offerings found in some of the tomb monuments mentioned above, about 60 objects pertaining to religious offerings, some 250 sacrificial vases and flasks, coins dating from 1290 until 1626, a number of Egyptian works of art, some 200 different stones and innumerable specimen of sea life, petrified objects etc., exotic curiosities, books, manuscripts, and finally, miscellaneous items that did not fit any of the above mentioned categories.

Although no systematic investigation into 'cabinets' in Amsterdam of the 17th century has been undertaken, the Reynst collection certainly must have been one of the largest.[1] It was surpassed only with respect to the paintings, since the collection assembled by the Count and Countess of Arundel, housed in Amersfoort, Alkmaar and Amsterdam from 1643 until 1654, was more extensive and of higher quality.[2] Its emphasis on Italian paintings, however, was quite unique. Lugt knew only of three additional collections in Amsterdam between the late 1630's and 1650 that included such

---

[1] That there was an interest in collecting curiosities can also be seen from De Renialme's inventory of 1657, summarily indicated by Bredius, *Künstler-Inventare*, I, p. 239: *Es folgt eine grosse Sammlung von Raritäten, Sachen von Achat, Palmenholz, Kleinodien usw.* (Followed by a large number of curiosities, things of agate, palm wood, precious stones etc.).

[2] The most comprehensive study on the Arundel collection is found in M.F.S. Hervey, *op.cit.* (with earlier bibliography). See also Denys E. Haynes, 'The Arundel Marbles,' *Archaeology*, XXI, 1968, pp. 85-91 and 206-11. For a comparison with contemporary Dutch collections see Lugt, *op.cit.*, pp. 120-22. Based on the inventory, drawn up in 1654 after the death of the Countess of Arundel, the collection

pictures, the ones formed by Van Uffelen, Lopez and Krezter.³ One other name should be added, Reinier van der Wolff⁴ in Rotterdam (died 1679), who owned a number of fine Italian paintings, among them no less than eight by Titian, four by Dosso and two attributed to Giorgione. (Van der Wolff's name came up earlier, when he was asked to appraise the painting bought from De Graeff, included in the 'Dutch Gift' of 1660). Besides paintings, Van der Wolff also collected antique sculpture. Pieces from his collection were later bought by Papenbroek, De Witt and Uilenbroek, who also showed much interest in the antique sculptures formerly with Reynst. Van der Wolff's collection was sold on April 8, 1677 in Rotterdam, after the death of his wife, Maria Dircxdr Pesser. (The sum received totaled 47,000 guilders; one of the Titians alone fetched 9000 guilders).⁵ Dutch collectors, however, seemed to favor their own artists.⁶ This is also true of Rembrandt as a collector, despite his well-known interest in Italian art. His inventory of 1656 listed about fifty paintings by Dutch and Flemish artists, in addition to some seventy-five paintings and sketches by or retouched by him, but only eight by Italian artists: two each by Raphael and Palma Vecchio, one each by Giorgione, Bassano and Lelio Orsi, and two copies after Annibale Carracci.⁷ Based on the many inventories which Bredius published, hardly any seventeenth-century collection in Holland included a sizeable number of Italian paintings. Even the inventory of the art dealer De Renialme, drawn up in 1640 in Amsterdam showed none, while the one drawn up in 1657, only listed ten paintings by Italian artists such as Titian, Tintoretto, Palma Vecchio, Bassano and others, in contrast to over 230 paintings by Dutch and Flemish artists with a few by German and French painters.⁸ The Italian pictures possibly were too expensive for the average collector, since they invariably commanded higher prices than paintings by contemporary Dutch or Flemish artists. One other example may suffice to stress this lack of Italian pictures on the Dutch art market. Among the 1354 items registerd in the sale of paintings belonging to

included 32 paintings by Titian, 14 by Veronese, 12 by Giorgione, Raphael and Tintoretto, and over 40 by Holbein. See F.H.C. Weijtens, *De Arundel-Collectie*, Rijksarchief, Utrecht, 1971 and S.A.C. Dudok van Heel, 'De Kunstverzamelingen Van Lennep met de Arundel-Tekeningen,' *Jaarboek...Amstelodamum*, 67, 1975, pp. 137-48.

³ The material gathered in his article referred to earlier, 'Italiaansche kunstwerken in nederlandsche verzamelingen van vroeger tijden' (1936), has remained the basic source on this subject.

⁴ J.H. Jongkees, 'De verzameling oudheden van Reinier van der Wolff,' *Mededelingen van het Nederlands historisch Instituut te Rome*, XXXI, 1961, pp. 125-26; Van Regteren Altena and Van Thiel, *op.cit.*, pp. 34-35.

members of the St. Luke guild in The Hague that started in April, 1647, just one copy after Michelangelo and an 'Italian piece' could be found.[9] This apparent scarcity would support the assumption that the Reynst brothers purchased the paintings for their collection abroad, preferably in Venice.[10]

In this context, one should briefly recall Rembrandt's statement that he had no time for a study trip to Italy and that besides, he could see the finest Italian paintings right in Holland.[11] This often quoted comment must have been made sometime around 1629, because Constantijn Huygens incorporated it in his autobiographical sketch, started in the spring of 1629.[12] Rembrandt's reason for seeing no necessity to travel to Italy is all the more astonishing if one realizes that it was said about ten years before the justly famous paintings in the Van Uffelen[13] and Lopez[14] collections were in Amsterdam, and some years before many of the Italian paintings in the Reynst collection could be seen in the house on the Keizersgracht. Rembrandt's answer, therefore, may be due partly to youthful exaggeration, but it may also indicate a growing presence of Italian paintings in the Netherlands and reflects an increase in their appreciation.

One of the earliest paintings by Rembrandt that shows a direct borrowing from an Italian picture is his *Descent from the Cross* in Munich, finished in 1633 (Br. 550; Gerson, p. 462). The group of

---

[5] Gerard Hoet, *Catalogus of naamlijst van schilderijen met derzelver prijzen*, II, 's Gravenhage, 1752, pp. 340 ff. and Jongkees, *op.cit.*, p. 125; also mentioned in Dudok van Heel, *op.cit.*, p. 142, note 1 and p. 154, no. 5.

[6] See also Willem Martin, 'Ueber den Geschmack des holländischen Publikums im XVII. Jahrhundert mit Bezug auf die damalige Malerei,' *Monatshefte für Kunstwissenschaft*, I, 1908, pp. 727–53.

[7] Clark, *op.cit.*, pp. 193–211; R.H. Fuchs, *Rembrandt in Amsterdam*, Greenwich, Conn., 1969, pp. 76–80; Scheller, *op.cit.*, p. 87.

[8] Bredius, *op.cit.*, pp. 228–39.

[9] Idem, *Künstler–Inventare*, II, The Hague, 1916, pp. 457–520, especially p. 478, no. 31 and p. 505, no. 36.

[10] Venice seems to have been a favorite place for art purchases. We may recall that Daniel Nijs resided there while negotiating the sale of the Mantua collection (see p. 74 above). Another such transaction was published by L.E. Waterhouse (*loc.cit.*), involving the collection of Bartolommeo della Nave, bought *en bloc* by Basil, Viscount Feilding while ambassador in Venice from 1634–39 on behalf of the Marquess of Hamilton.

[11] J.A.Worp, 'Constantyn Huygens over de schilders van zijn tijd,' *Oud Holland*, IX, 1891, pp. 130–31; A.H. Kan, *De Jeugd van Constantijn Huygens door hemzelf beschreven*, Rotterdam, 1946, pp. 82–83; Seymour Slive, *Rembrandt and His Critics, 1630–1730*, The Hague, 1953, pp. 16–18.

[12] The manuscript is preserved in the Royal Library in The Hague, and was written between May 11, 1629 and April, 1631; cf. Slive, *op.cit.*, p. 9, note 3.

[13] See Lugt, *op.cit.*, pp. 113–14.

[14] *Ibidem*, pp. 114–15. E. Maurice Bloch, 'Rembrandt and the Lopez Collection, *Gazette des Beaux-Arts*, 6th series, XXIX, 1946, pp. 175–86.

the fainting Virgin Mary supported by two assisting women was taken over from Bassano's *Entombment*, painted in 1574 for Sta. Maria in Vanzo, Padua, known in a number of different versions.[15] One such version was in Amsterdam certainly by the middle of the century, when it was engraved by Jeremias Falck while in the Reynst collection and included in the CAELATURAE (cat. no. 5; Pl. P 5). Whether it already was in Reynst's possession by 1631/32, where Rembrandt could have seen it shortly after his arrival in Amsterdam, has to remain open. The possibility does exist, for Jan Reynst is recorded in Venice as early as 1625 and may have started collecting for his brother soon after, as was mentioned earlier (see p. 29). The painting, therefore, could possibly establish the first link between Rembrandt and the Reynst collection. A number of additional points of common interest tend to further corroborate the supposition that Rembrandt indeed was familiar with the 'cabinet ...de Reinst'.

At least three paintings by Rembrandt have been associated with pictures that formerly belonged to Reynst, partly unjustly so.[16] One connection, however, first suggested by Clark[17] and supported by Gerson,[18] seems plausible and should be discussed briefly. The portrait of a *Young Venetian Lady*, similar in attitude to Palma Vecchio's painting in Berlin,[19] reproduced in the Vendramin catalogue DE PICTURIS, fol. 51, may have influenced the

---

[15] Established first by Wolfgang Stechow, 'Rembrandts Darstellungen der Kreuzabnahme,' *Jahrbuch der preuszischen Kunstsammlungen*, 50, 1929, p. 222, fig. 3. See also J.L.A.A. Rijckevorsel, *Rembrandt en de Traditie*, Rotterdam, 1932, p. 153 and Slive, *op.cit.*, p. 25, note 2.

[16] Lugt (*op.cit.*, p. 117) thought that Rembrandt must have known Lotto's *Portrait of Andrea Odoni*, now at Hampton Court, but gave no example of a specific influence upon the artist (see cat. no. 16). Fischel (review of Borenius, p. 29) felt that Rembrandt's painting of *Alexander the Great*, now in the Gulbenkian Foundation in Lisbon (Br. 479; Gerson, p. 388) was inspired by Bordone's painting of a *Male Figure in Armour*, formerly in the Vendramin collection, where it was illustrated in the catalogue DE PICTURIS, fol. 59 (reproduced in Borenius, *op.cit.*, pl. 47 and text p. 34). Later, the painting presumably was purchased for the Reynst collection, and is now lost. This supposition was renewed by Clark (*op.cit.*, p. 140, fig. 131), but it has not been followed by others. One further, even less likely assumption by Fischel (*loc.cit.*, under pl.63; reproduced in Borenius, *op.cit.*, pl. 63 and text p. 38) that a painting of a *Praying St. John*, illustrated in DE PICTURIS on fol. 75, may have been known to Rembrandt because of his somewhat related figure of *St. James*, painted in 1661, presently on loan to the Metropolitan Museum of Art, New York (Br. 617; Gerson, p. 521) has not been accepted in the Rembrandt literature either, and correctly so.

[17] *Op.cit.*, pp. 130. 132 and figs. 123–24.

[18] Horst Gerson, *Rembrandt Paintings*, Amsterdam, 1968, p. 501, no. 339; repeated in A. Bredius, *Rembrandt, The Complete Edition of the Paintings*, revised by H. Gerson, London, 1969, p. 557, no. 116.

[19] Inv. no. 197 A; see *Staatliche Museen, Preussischer Kulturbesitz, The Picture Gallery*, Berlin, 1968, p. 79. Reproduced in Borenius, *op.cit.*, fig. 9; for the illustration in DE PICTURIS see *ibidem*, pl. 39 and text pp. 32–33.

pose of Rembrandt's *Hendrickje Stoffels at an Open Door*, painted ca. 1658/59, in Berlin (Br. 116; Gerson, p. 104). Neither Clark nor Gerson pointed out, however, that the painting most likely was in the Reynst collection in Amsterdam, where Rembrandt could have studied it. This would be another argument in favor of Rembrandt having visited the house on the Keizersgracht.[20] Further facts that lend support to this assertion are Rembrandt's own collection which would certainly make him aware of other such 'cabinets'; secondly Rembrandt's professed interest in Italian, especially Venetian art which might have attracted him to a collection devoted almost exclusively to Venetian paintings; thirdly, the many antique sculptures could have aroused his curiosity since he himself owned a number of busts of Roman emperors and Greek philosophers. Rembrandt may even have been familiar with the often mentioned 'sepultura di Aristotile' and may have used the so-called portrait of Aristotle as a point of reference for his own painting of *Aristotle with the Bust of Homer*, dated 1653, in New York (Br. 478; Gerson, p. 386).[21]

Much of the attempted reconstruction of the former collection of Gerard Reynst had to be based on suppositions and contemporary descriptions. One fact, however, has become clear: it was the largest and richest collection assembled by a private citizen in Amsterdam during the 17th century, where it could be viewed for over thirty years. This 'cabinet tant renommé', formed by the two merchants Gerard and Jan Reynst, grew into one of the attractions of the city of Amsterdam and must have been a source of inspiration and enjoyment for many.

---

[20] The only member of the Reynst family mentioned in Hofstede de Groot's *Urkunden* (*op.cit.*, p. 226), was Jacobus Reynst, a second cousin of Gerard and Jan. On March 3, 1660, he received a complaint from the widow of the 'Keyserskroon' in Amsterdam that Rembrandt had not paid the costs incurred during the auction of the latter's possessions in 1656. Reynst, as commissioner of the Chamber of Insolvency (*desolate boedelkamer*), had to see to it that she was reimbursed.

[21] See J.A. Emmens, *Rembrandt en de regels van de Kunst* (*Utrechtse kunsthistorische studien* x), Utrecht, 1968, especially pp. 169–76; and Julius S. Held, *Rembrandt's Aristotle and Other Rembrandt Studies*, Princeton, N.J., 1969, pp. 3–43.

CATALOGUE OF PAINTINGS AND SCULPTURES IN THE COLLECTION OF GERARD AND JAN REYNST

ABBREVIATIONS USED IN THE CATALOGUE

Baker 1929
C.H.Collins Baker, *Catalogue of the Pictures at Hampton Court*, Glasgow, 1929.

Bathoe
W. Bathoe, *Catalogue of the Collection of Pictures,...Belonging to King James the Second...*, London, 1758.

Block
J.C. Block, *Jeremias Falck*, Danzig, 1890.

Brants
J.P.J. Brants, *Description of the Ancient Sculpture...the Museum of Archaeology of Leiden*, The Hague, 1927.

Brunsting
H. Brunsting, 'Twee gouden eeuwen,' *Archeologie en Historie*, Bussum, 1973, pp. 179–90.

Campori
G. Campori, *Raccolta di Cataloghi ed Inventari inediti*, Modena, 1870.

Clarac
F. de Clarac, *Musée de Sculpture...*III, Paris, 1850.

Hagen
A. Hagen, 'Ueber den Kupferstecher Jeremias Falck', *Kunstblatt*, no. 16, 30 March 1848, pp. 61–64.

Hecquet
R. Hecquet, *Catalogue des estampes gravées d'après Rubens...et... de Visscher*, Paris, 1751.

Heinecken
K.H. von Heinecken, *Idée générale d'une collection complette d'estampes...*, Leipzig/Vienna, 1771.

Hoet
G. Hoet, *Catalogus of naamlyst...*; 's-Gravenhage, 1752.

Hollstein
F.W.H. Hollstein, *Dutch and Flemish Etchings, Engravings and Woodcuts, ca. 1450–1700*, Amsterdam, 1949ff.

Jacobs 1925
Emil Jacobs, 'Das Museo Vendramin und die Sammlung Reynst,' *Repertorium für Kunstwissenschaft*, XLVI, 1925, pp. 15–38.

Janssen
L.F.J. Janssen, *De griekische, romeinsche en etrurische Monumenten van het museum van oudheden te Leyden*, Leiden (1848).

Knorr 1663
Fuchs, 'Aus dem 'Itinerarium' des Christian Knorr von Rosenroth,' *Jaarboek van het Genootschap Amstelodamum*, XIV, 1916, pp. 239–45.

Law 1881, 1898
Ernest Law, *A Historical Catalogue of the Pictures in the Royal Collection at Hampton Court*, London, 1898.

Leplat
B. Leplat, *Suite de divers marbres modernes,...*, (no location), 1733.

Lugt 1936
Frits Lugt, 'Italiaansche Kunstwerken in Nederlandsche verzamelingen van vroeger tijden,' *Oud Holland*, LIII, 1936, pp. 97–135.

Mahon 1949, 1950
Denis Mahon, 'Notes on the 'Dutch Gift' to Charles II,' *Burlington Magazine*, XCI, 1949, pp. 303–05; 349–50; XCII, 1950, pp. 12–18.

Martinioni
Francesco Sansovino, *Venetia città nobilissima...* (with additions by Giustiniano Martinioni), Venice, 1663.

Oudendorp
Franciscus Oudendorp, *Brevis veterum monumentorum ab amplissimo viro Gerard Papenbroekio Academiae Lugduno-Batavae legatorum descriptio*, Lugduno-Batavae, 1746.

Ridolfi
Carlo Ridolfi, *Le Maraviglie dell'Arte*, Venice, 1648, ed. D. von Hadeln, Berlin 1914 and 1924.

Reinach
Salomon Reinach, *Répertoire de la statuaire grecque et romaine*, 6 vols., Paris 1897–1930.

Savini-Branca 1965
Simona Savini-Branca, *Il Collezionismo veneziano nel '600* (*Università di Padova, Pubblicazioni della Facoltà di lettere e filosofia*, XLI), Padua, 1965.

Stimmel
J. G. Stimmel, *Catalogue raisonné du cabinet d'estampes de feu Monsieur (Gottfried) Winckler*, V, Leipzig, 1810, pp. 309–17.

Timmers
J.J.M. Timmers, *Gérard Lairesse*, Amsterdam, 1942.

Uffenbach
Zacharias Conrad von Uffenbach, *Merkwürdige Reisen durch Niedersachsen, Holland und Engelland*, III, Ulm, 1754.

Wussin
Johann Wussin, *Cornel Visscher*, Leipzig, 1865.

Wurzbach
A. von Wurzbach, *Niederländisches Künstler-Lexikon*, 3 vols., Vienna/Leipzig, 1906/11 (reprint 1963).

For abbreviated exhibition titles see under *Exhibitions*.

The engravings of the CAELATURAE and the ICONES are listed first, since they preserve – at least visually – many more works of art formerly in the collection of Gerard Reynst than are extant today.

ROYAL INVENTORIES*

Inventory of Charles II:
(MS. in Surveyor's Office, St. James's Palace, London).
An Inventory of all his Maties Pictures in White-hall ; An Inventory of all his Maties Pictures in Hampton-Court, 2 parts, probably drawn up ca. 1666–67.

Inventory of James II:
(MS. kept in the British Museum, Harl. MS 1890).
Inventory of James II' pictures in 1688 at Whitehall, Windsor, Hampton Court, and in the custody of The Queen Dowager, Catherine of Braganza, at Somerset House:
Inventory of His Majesty's Goods 1688. (Vertue's copy of this MS, in the British Museum, MS 15752, served as basis for edition printed for W. Bathoe, 1758).

Inventory of Queen Anne:
MS. in Surveyor's Office, St. James's Palace, London).
Inventory of Queen Anne's pictures at Kensington, Hampton Court, Windsor, St. James's and Somerset House:

A List of Her Majesties Pictures in Kensington Hampton Court and Windsor Castle
(drawn up by or for Peter Walton, Surveyor and Keeper of the pictures, probably between 1705 and 1710. Additional notes were added, probably ca. 1710–12 by Thomas Coke, Vice-Chamberlain of the Household).

(The above information is based on Mahon, *op.cit.*, 1949, pp. 349–50 and Levey, *op.cit.*, pp. 39–40).

---

* Listed chronologically

ITALIAN PAINTINGS IN THE COLLECTION OF GERARD REYNST, PARTLY FORMERLY IN THE COLLECTION OF JAN REYNST*

FEDERICO BAROCCI
(ca. 1535–1612)

1. *Woman with a Dog*

Painting: No longer traceable in Royal collections.

Provenance: Gerard Reynst (?); Charles II (Inventory, p. 2, no. 14: *Barrotse. An Italian woman laying one hand upon a Dogg and holding a handchercheife in the other. Dutch p^rsent. 3.8 × 3.1*).

Bibliography: Mahon 1950, p. 16, no. 18 (presumably Federico Barocci, no longer identifiable).

JACOPO BASSANO
(1510–1592)

2. *Abraham Leaving Haran*

Etching and Engraving: Cornelis Visscher (Hecquet, no. 2, with reference to Reynst collection; as *God Pormises the Land of Canaan to Abraham's Descendants*; Stimmel, no. 32; Wussin, no. 98, with reference to Reynst collection; idem, p. 273, no. 31 and p. 277, no. 11; Wurzbach, no. 98). 304 × 374 mm. (12 × 14¾ in.)
CAELATURAE 34.

Painting: Lost.

Provenance: Gerard Reynst.

Bibliography: Law 1881, p. 44 and 1898, pp. 54–55 (wrongly identifies Bassano's *Jacob's Journey* at Hampton Court, Inv. no. 142, with this lost painting); Jacobs 1925, p. 31 (confuses this lost painting with Hampton Court, Inv. no. 142); Baker 1929, pp. 4–5, no. 142 (repeats Law's incorrect findings); W. Arslan, *I Bassano*, Bologna, 1931, p. 160 (ca. 1570–80; ill. of engraving, fig. 32) and p. 190 (takes over Law's incorrect statement that the *Departure of Abraham* [*Jacob's Journey*] at Hampton Court was part of the Reynst collection and the 'Dutch Gift' and adds furthermore that it came from the Gabriele Vendramin collection); Mahon 1950, p. 18 (establishes correctly that this lost Bassano painting of *Abraham Leaving Haran* is not identical with *Jacob's Journey* and that the latter therefore was not part of the Reynst collection).

Plate: P 2.

This and the following painting, both lost, probably were considered pendants. Stimmel and Wussin believed that they represented scenes from the life of Abraham. However, it is more likely that the painting reproduced in the following engraving depicts an *Annunci-*

P 2

*Listed alphabetically according to artists

ation to the Shepherds and thus may have formed the corresponding typological New Testament scene to the Old Testament representation of *Abraham Leaving Haran*.

3. *The Annunciation to the Shepherds*

Etching and Engraving: Cornelis Visscher (Hecquet, no. 1, with reference to Reynst collection; as *The Angel Ordering Abraham to Leave his Land*; Stimmel, no. 31; Wussin, no. 99, with reference to Reynst collection; *idem*, p. 273, no. 32 and p. 277, IX, no. 12; Wurzbach, no. 99).
291 × 366 mm. (11½ × 14⅜ in.)
CAELATURAE 33.

Painting: Lost.

Provenance: Gerard Reynst.

Bibliography: W. Arslan, *I Bassano*, Bologna, 1931, p. 162, fig. 33 (as *Annunciation to the Shepherds*, ca. 1570/80); Mahon 1950, p. 18, note 10.

Plate: P 3.

4. *Christ Carrying the Cross*

Engraving: Jeremias Falck (Stimmel, no. 15; Hagen, no. 79, with reference to Reynst collection, after Schiavone; Wussin, p. 273, no. 15 and p. 275, IV, no. 2, as after Schiavone, probably identical with Stimmel, no. 15, and thus probably after Jacopo Bassano; Block, no. 18, after Schiavone, and no. 19, after Veronese; both as after painting in Reynst collection, but neither description corresponds exactly to engraving in CAELATURAE; Wurzbach, no. 2, with reference to Reynst collection; Hollstein, VI, p. 210, nos. 18–19).
355 × 287 mm. (14 × 11 5/16 in.)
CAELATURAE 12.

Copper plate in 1679 with Clement de Jonghe.

Painting: Earl of Bradford, Weston Park, Staffordshire. Oil on canvas, 146 × 132 cm.

Provenance: Gerard Reynst; Charles II (Inventory, p. 10, no. 161: *Bastano. Our Saviour on his knees bearing his Cross & tc. Tenn figures more. Dutch p^rsent*); James II (Catalogue, ms., f. 70 v., no. 46 of paintings in the custody of the Dowager Queen Catherine of Braganza [died 1705]; Bathoe, no. 729: *By Bassano. Christ carrying his Crosse with Eleaven figures in itt one of the Dutch Presents*); not listed in Queen Anne inventory; Viscount Torrington (probably between 1721–27, as deduced from annotation by Coke in the CAELATURAE in the Victoria and Albert Museum); Earl of Bradford.

Exhibition: London 1960, no. 64 (ill.).

Bibliography: Jacobs 1925, p. 30 (reference to catalogue of James II.); George Vertue, 'Inventory of Charles I collection 1687–8,' *The Walpole Society*, XXIV, 1935–36, p. 93 (as Bassano, '... one of the Dutch presents...', with incorrect statement that Reynst bought these

P 3

112 ITALIAN PAINTINGS

paintings at the sale of the collection of king Charles I [see Mahon 1950, p. 303, note 10]); Lugt 1936, p. 117, note 40 (by Bassano); Mahon 1950, no. 15 (establishes for first time provenance for this painting and furnishes date on copy in York museum); P. Zampetti, *Jacopo Bassano* (exh. cat.), Venice (Palazzo Ducale), 1957, under no. 18; E. Arslan, *I Bassano*, Milan (1960), pp. 66–67, 179, fig. 56 (ca. 1540), and p. 386 (copy in York, with bibliography); B. Haak, 'De Kruisdraging door Jacopo Bassano…,' *Bulletin van het Rijksmuseum*, (Amsterdam), VIII, 1960, pp. 63–64; M. Muraro, 'Bassano's 'Way to Calvary',' *Burlington Magazine*, CII, 1960, pp. 53–54, fig. 1 (retraces publications of the painting and the copy); R. Pallucchini, 'La pittura veneta alla mostra 'Italian Art and Britain': Appunti e proposte,' *Eberhard Hanfstaengl zum 75. Geburtstag*, Munich, 1961, p. 75 (not later than 1545); E.K. Waterhouse, 'A note on British collecting of Italian Pictures in the later seventeenth century,' *Burlington Magazine*, CII, 1960, p. 57; K. Garlick, 'Old Masters and Sporting Englishmen: Pictures at Weston Park…,' *Country Life*, 137, May, 1965, p. 1211 (ill.); A. Ballarin, 'Jacopo Bassano e lo studio di Raffaello e dei Salviati,' *Arte Veneta*, XXI, 1967, pp. 77, 94, 98, fig. 112 (ca. 1545); S. J. Freedberg, *Painting in Italy 1500 to 1600*, Hardmondsworth, 1971, p. 372, fig. 241 (1546–47).

Plates: P 4, P 4a.

Bassano's painting in the collection of the Earl of Bradford was correctly identified by Mahon as the picture formerly with Reynst. This is supported further in a comparison of Falck's engraving with the two paintings. (The version in the museum at York, canvas, 147,3 × 137,1 cm., presented by F.D. Lycett Green, Esq., through the National Art Collections Fund, 1955, is a copy after the Bradford painting. See *Catalogue of Paintings*, I, *Foreign Schools 1350–1800*, City of York Art Gallery, York, 1961, p. 5, no. 773 with reference that A. Noack suggested Dirck Barendsz).

Despite the fact that Falck took some liberties when reproducing the painting (Veronica's left hand, for example, or the clothing of the warrior hitting Christ differ from the painting), there are specific instances, where the print definitely follows the Bradford painting and not the one in York, especially in the rendering of the sky and of the women's headgear. Judging from Falck's engraving the painting apparently was cut at the top and at the bottom.

5. *The Entombment of Christ*

Engraving: Jeremias Falck (Stimmel, nos. 16, 39; without name of painter or engraver; added to CAELATURE by Winckler; no. 39 is same print, after letters, as de-

P 4

P 4a

scribed earlier under no. 16; Wussin, p. 279, no. 39 and p. 276, IV, no. 18; Block, no. 24, with reference to Reynst collection; after Caravaggio; Hollstein, VI, p. 210, no. 24).
368×281 mm. ($14\frac{1}{2} \times 11\frac{1}{16}$ in.)
CAELATURAE 14.

Painting: Lost.

Provenance: Gerard Reynst.

Bibliography: Lugt 1936, p. 117, note 40 (known as Tintoretto, but undoubtedly by Jacopo Bassano).

Plate: P 5.

Stimmel listed Falck's engraving twice, once under no. 16 and a second time under no. 39 (after letters). The print is included often in the CAELATURAE and the painting, now lost, most likely was in the collection of Reynst. The composition is based on Bassano's *Entombment* of 1574 in Sta. Maria in Vanzo which is known in a number of versions (modelletto in Vienna, Kunsthistorisches Museum, Inv.no. 5680; preliminary drawing at Christ Church, Oxford, Inv.no. 1334; copy in Munich, Nr. 2786, 1147). As was pointed out earlier Rembrandt used the group of the fainting Virgin Mary and the assisting women in his *Descent from the Cross* of 1633. He, therefore, possibly knew this painting formerly with Reynst and adapted the group for his own composition (see pp. 104–05). Visscher's etching after Tintoretto's *Entombment*, on the other hand, with which Falck's engraving has been confused (see Lugt, *loc. cit.*), is less common and has been found only in the CAELATURAE in Cincinnati and Paris (see cat. no. 31; Pl. P 31).

6. *The Virgin and Child and the Infant St. John*

Engraving and Etching: Theodor Matham (Stimmel, no. 11; Wussin, p. 272, no. 11 and p. 277, VIII, no. 3; Wurzbach, no. 2, with reference to Reynst collection; Hollstein, XI, p. 252, no. 5).
389×283 mm. ($15\frac{5}{16} \times 11\frac{1}{8}$ in.)
CAELATURAE 11.

Painting: Lost.

Provenance: Gerard Reynst.

Bibliography: W. Arslan, 'Un nuovo dipinto di Jacopo Bassano,' *Bolletino d'Arte*, VIII, 1928–29, p. 413, fig. 4 (closest known version is painting formerly in Castle Howard, ca. 1568–72); idem, *op.cit.* 1931, p. 116, fig. 28; Lugt 1936, p. 117, note 40; E. Arslan, *I Bassano*, Milan (1960), p. 109.

Plate: P 6.

P 5

BONIFAZIO DE' PITATI,
CALLED VERONESE
(1487–1553)

7. *The Virgin and Child with Tobias and Saints*

Engraving: Jeremias Falck (Stimmel, no. 20, as after Palma Vecchio, later in English royal collection; Hagen, no. 92, with reference to Reynst collection; Wussin, p. 273, no. 20 and pp. 275–76, no. 15; Block, no. 10, with reference to Reynst collection and incorrect statement that it passed into the collection of Charles I; Wurzbach, no. 15, with reference to Reynst collection; Hollstein, VI, p. 209, no. 10). 273 × 405 mm. (10¾ × 16 in.). CAELATURAE 22.

Painting: Hampton Court Palace (No. 146; Inv.no. 140). Oil on canvas, 162,5 × 249 cm.

Provenance: Gerard Reynst; Charles II (Inventory, p. 10, no. 158: *Old Palma. The coming of the Sheapheards to our Savior.* 5.2 × 7.5); James II (Catalogue, ms., Whitehall, no. 158, as by Palma Vecchio; Bathoe, no. 157); Queen Anne (Inventory, Kensington, no. 56, as by Palma Vecchio).

Bibliography: J. A. Crowe and G. B. Cavalcaselle, *A History of Painting in North Italy*, II, London, 1871, p. 486 (not by Palma); Law 1898, p. 56, no. 146 (ill.; as Bonifazio); Baker 1929, p. 11, no. 146 (School of Bonifazio di Pitati, with reference to Reynst collection); Lugt 1936, p. 101, fig. 27 (as Bonifazio di Pitati); Mahon 1950, p. 15, no. 14 (identified as part of *Dutch Present* in the inventory of Charles II); Bernard Berenson, *Italian Pictures of the Renaissance, Venetian School*, I, London, 1957, p. 42 (as Bonifazio de' Pitati).

Plates: P 7, P 7a.

Although there is no reference to the 'Dutch Gift' in the inventory of Charles II, the painting did belong to the Reynst collection since it was engraved for the CAELATURAE. It must have passed into the Royal collection along with the other Italian paintings from Reynst in 1660.

P 6

116 ITALIAN PAINTINGS

P 7

P 7a

PARIS BORDONE
(1500–1571)

8. *Portrait of a Man Holding a Document*

Painting: Hampton Court Palace (No. 182; Inv.no. 52)
Oil on canvas, 89 × 82,5 cm.

Provenance: Gerard Reynst (?); Charles II (Inventory, p. 11, no. 167: *Paris Burdon. A man in a black cap and black habit reading a greate writing, which he holds in boath his hands. To the Wast. Dutch p^rsent. 3.0 × 2.9*); James II (Catalogue, no. 293, as by Bordone; Bathoe, no. 292).

Bibliography: Law 1898, p. 67, no. 182 (Italian Lawyer? Bordone?); A. Venturi, *Storia dell' Arte Italiana*, IX, 3, Milan, 1928, p. 1032; Baker 1929, p. 13, no. 182 (less likely suggestion that perhaps sold by Commonwealth to De Critz, 1651); Mahon 1950, p. 13, no. 5 (difficult to trace in later royal inventories); Bernard Berenson, *Italian Pictures of the Renaissance, Venetian School*, 1, London, 1957, p. 46 (Bordone; Italian Lawyer?).

Plate: P 8.

Included on the basis of the inventory of Charles II, where the painting is specified as part of the 'Dutch Gift'. It was not engraved for the CAELATURAE.

GIOVANNI BUSI, CALLED CARIANI
(ca. 1480–1548)

9. *Reclining Venus*

Painting: Hampton Court Palace (No. 1103; Inv.no. 1103)
Oil on canvas, 78,5 × 137 cm.

Provenance: Andrea Vendramin, Venice; Jan Reynst; Gerard Reynst; Charles II (Inventory, p. 46, no. 544: *Palma. A naked Venus Lying at Length. Dutch p^rsent. 2.8 × 4.7*); James II (Catalogue, ms., Whitehall, no. 279, as by Titian; Bathoe, no. 278); Queen Anne (Inventory, Kensington, no. 104, as by Paris Bordone).

Exhibitions: London 1946–47, no. 195 (with incorrect provenance from Charles I collection); Amsterdam 1965, no. 2 (ill., as Cariani).

Bibliography: Law 1898, p. x; Detlev von Hadeln, in *Thieme-Becker*, V, Leipzig, 1911, p. 595 (ca. 1520); J. A. Crowe and G. B. Cavalcaselle, *A History of*

P 8

Painting in North Italy, III, New York, 1912, p. 459, note 3; T. Borenius, *The Picture Gallery of Andrea Vendramin*, London, 1923, pp. 5, 31, fig. 8; O. Fischel [review of Borenius], in *Zeitschrift für bildende Kunst*, LVIII, 1924, *Monatsrundschau*, pp. 28–29; Jacobs 1925, pp. 32–33 (identifies painting with a drawing after it in catalogue of Vendramin collection; rejects provenance from Charles I); A. Venturi, *Storia dell'Arte Italiana*, IX, part 3, Milan, 1928, p. 465; Baker 1929, p. 21, no. 1103, pl. V (as Cariani); Ludwig Baldass, 'Ein unbekanntes Hauptwerk des Cariani,' *Jahrbuch der kunsthistorischen Sammlungen in Wien*, N.F. III, 1929, pp. 101 and 109 (ca. 1530–35); T. Borenius, in *Burlington Magazine*, LX, 1932, p. 140; E. G. Troche, 'Giovanni Cariani,' *Jahrbuch der preuszischen Kunstsammlungen*, LV, 1934, p. 113 (ca. 1524–26); Lugt 1936, p. 101, fig. 24; Mahon 1950, p. 14, no. 9 (corrects wrong identification with a painting from the collection of Charles II); L. Gallina, *Giovanni Cariani*, Bergamo, 1954, pp. 124–25, pl. LXVIII; Bernard Berenson, *Italian Pictures of the Renaissance, Venetian School*, I, London, 1957, p. 54 (as late Cariani).

Plate: P 9.

The provenance from the collection of Andrea Vendramin in Venice was established first by Jacobs, who identified the painting through a drawing in the Vendramin catalogue DE PICTURES, fol 43 (see p. 172). Jacobs' supposition that the painting formerly was in the Reynst collection is fully justified since other objects from the Vendramin collection could be traced to the house of Gerard Reynst in Amsterdam (see pp. 55, 57, 59–75).

DOMENICO FETTI
(1589–1623)

10. *The Vision of St. Peter*

Engraving: Jeremias Falck (Stimmel, no. 17, without name of painter or engraver; Robert Naumann, *Archiv für die zeichnenden Künste*, XI, 1856, p. 184, Appendix, no. 1, as Jacob Matham; Wussin, p. 273, no. 17 and p. 274, I, no. 1, without name of painter or engraver; Block, no. 29, by Falck after Liss; Wurzbach, Theodor Matham, no. 7, after Liss; Hollstein, VI, Jeremias Falck, p. 211, no. 29 and XI, after Liss, p. 148, no. 5; *idem*, XI, Theodor Matham, p. 252, no. 10 and XI, after Liss, p. 148, no. 5).
385 × 285 mm. (15⅛ × 11¼ in.).
CAELATURAE 15.

Painting: Probably lost.

Provenance: Gerard Reynst.

Bibliography: Eduard R.V. Engerath, *Kunsthistorische Samm-*

P 9

ITALIAN PAINTINGS

lungen des aller höchsten Kaiserhauses, I, Vienna, 1884, p. 142, no. 198 (formerly in collection of the Duke of Buckingham); J. G. van Gelder, 'Domenico Fettis Vision des Hl. Petrus,' *Die graphischen Künste*, II, 1937, pp. 91–95 (first to relate Falck's engraving to painting in Vienna, but leaves open, whether the latter was in Reynst collection; painting formerly with Reynst either earlier version by Fetti of Vienna painting, or perhaps copy by Liss after Fetti's painting, in 1627–28 with Daniel Nijs in Venice; not a pendant to Liss' *Ecstasy of St. Paul*); Kurt Steinbart, *Johann Liss*, Berlin, 1940, p. 172 (among wrongly attributed paintings; by Fetti, version formerly with Reynst precedes painting in Vienna; engraved by Falck ca. 1655–61).

Plate: P 10.

Fetti's *Vision of St. Peter* in Vienna comes closest in composition to the painting formerly with Reynst, known today only through Falck's engraving. J.G. van Gelder was the first to associate the two. The Vienna version may be identical with the painting listed in the inventory drawn up in 1627 by Duke Vincenzo of Mantua: *Tre quadri di mano del Fetti sopra l'asse, in uno dipinta la visione di S. Pietro…* (A.Luzio, *La Galleria dei Gonzaga…*, Milan, 1913, p. 132, no. 634), sold with the Mantua collection in 1628 to Charles I. In 1634, the picture was given in exchange for a *Leda* by Veronese to the Duchess of Buckingham. This painting was sold later at the Buckingham auction in Antwerp in 1648 (II, no. 8: *The vision of St. Peter, wherein all kinds of animals are seen in a sheet. length 3 f. 0, breadth 2 f. 3*).

There is no record that Gerard Reynst bought Fetti's painting at the Buckingham sale. Since the painting in Vienna, furthermore, was acquired in 1649 (see catalogue 1973, p. 65, pl. 47), we have to assume that Reynst owned a different version of Fetti's *Vision of St. Peter* which might be identical with the one in the collection of Cardinal Alessandro d'Este in Rome (1624) that is lost. This latter opinion is due to Pamela Askew, who believes that Fetti's original composition dates from ca. 1619–22 (Letter from W. Stechow to E. Reynst, September, 1970).

P 10

GIORGIONE (ATTRIBUTED TO)
(ca. 1477–1510)

11. *The Concert*

Engraving: Jeremias Falck
(Stimmel, no. 25, after Giorgione;
Hagen, no. 91, with reference to
Reynst collection; Wussin, p. 273,
no. 25 and p. 275, IV, no. 14;
Block, no. 158, with reference to
Reynst collection; Wurzbach,
no. 14 with reference to Reynst
collection; Hollstein, VI, p. 214,
no. 158).
273 × 386 mm. (10¾ × 15 3⁄16 in.).
CAELATURAE 27.

Painting: Hampton Court Palace
(No. 144; Inv.no. 554).
Oil on canvas, 75 × 97,5 cm.

Provenance: Gerard Reynst;
Charles II (Inventory, p. 46,
no. 534: *A consort of Singing Dutch
p^rsent*); James II (Catalogue, ms.,
Windsor, no. 121, as by Giorgione;
Bathoe, no. 859: *Giorgione, A piece
with four figures to the waste, singing*);
Queen Anne (Inventory, Windsor,
no. 173, as by Giorgione).

Exhibitions: London 1946–47,
no. 202 (as attributed to Giovanni
Bellini; incorrect provenance);
Giorgione e i Giorgioneschi, Venice
(Palazzo Ducale), 1955, no. 42 (ill.);
London 1960, no. 9 (as ascribed to
Giorgione; corrected provenance);
Amsterdam 1965, no. 1 (ill.; as
Giovanni Bellini).

Bibliography: William Hazlitt,
Criticisms on Art, London, 1843,
p. XXIV, no. 551 (*A Concert* by
Giovanni Bellini); G.F. Waagen,
Treasures of Art in Great Britain, II,
London, 1854, p. 368, no. 551
(Prince of Wales's Presence-
Chamber: Giovanni Bellini, *A
Concert*); J. A. Crowe and
G.B. Cavalcaselle, *A History of
Painting in North Italy*, II, London,
1871, p. 532 (possibly a late Lotto);
Law 1898, p. 55, no. 144 (ill.; as
Morto da Feltre? incorrect identi-
fication of the painting with one
listed in the Commonwealth
inventory); Bernard Berenson,
Lorenzo Lotto, London, 1905,
p. VIII (as by Morto da Feltre,
already suggested by Mary Logan,
*Guide to the Italian Pictures at
Hampton Court*, London, 1894);
Lionello Venturi, *Giorgione e il
Giorgionismo*, Milan, 1913, pp. 260–
62, 390 (by an unidentified artist);
Jacobs 1925, p. 31; Roberto Longhi,
Vita Artistica, July 1927, p. 134 (as
Giovanni Bellini, ca. 1515);
A. Venturi, *Storia dell'Arte Italiana*,
IX, part 3, Milan, 1928, p. 560, n. 1
(school of Bellini); Baker 1929,
p. 69, no. 144, pl. XI (school of
Giorgione; repeats incorrect
provenance from Commonwealth
sale); W. Suida, 'Giorgione,
nouvelles attributions,' *Gazette des
Beaux-Arts*, série 6, XIV, 1935, p. 86
(by the young Giorgione, ca.
1500); Lugt 1936, p. 99, fig. 6
(Giorgione ?); Giuseppe Fiocco,
Giorgione, Bergamo, 1941, p. 30
(as Torbido); Antonio Morassi,
Giorgione, Milan, 1942, p. 105
(imitator of Giorgione); Rodolfo
Pallucchini, *La Pittura Veneziana del
Cinquecento*, I, Novara, 1944, p. XVI
(as Torbido); Giuseppe Fiocco,
Giorgione, Bergamo, 2nd ed.,
1948, p. 33, pl. 117 b (Torbido);
Mahon 1950, p. 14, no. 10 (corrects
Law's erroneous provenance);
Carlo Gamba, 'Il mio Giorgione,'
Arte Veneta, VIII, 1954, p. 174 (as
Mancini); Luigi Coletti, *Tutta la
Pittura di Giorgione*, Milan, 1955,
p. 65, pl. 108 (Morto da Feltre);
Giles Robertson, 'The Giorgione
Exhibition in Venice,' *Burlington
Magazine*, XCVII, 1955, p. 277
(appears to be good old copy of lost
original by same hand as Pitti *Three
Ages*); Pietro Zampetti, 'Postille
alla Mostra di Giorgione,' *Arte
Veneta*, IX, 1955, p. 67 (perhaps by
the late Giovanni Bellini); Bernard
Berenson, *Italian Pictures of the
Renaissance, Venetian School*, I,
London, 1957, p. 84, pl. 632
(Master of the three Ages [early
Giorgione?]); Fritz Heinemann,
Giovanni Bellini e i Belliniani (*Saggi
e studi di storia dell'arte*, 6), Venice
(1959), p. 202, no. S. 818 a (as
Torbido; with incorrect provenance); Pietro Zampetti, *L'Opera
completa di Giorgione*, Milan, 1968,
p. 95, no. 37, fig. 37 (perhaps by the
old Giovanni Bellini); Terisio
Pignatti, *Giorgione* (*Profili e saggi di
arte Veneta*, 8), Venice (1969),
p. 121, no. A 18, pl. 135 (as school
of Bellini, close to Giorgione).

Plates: P 11, P 11a.

Law's erroneous suggestion that
the painting might be identical with
A Picture of Musick by Georgion,
mentioned in the Commonwealth
sale, was corrected by Mahon. He
pointed out that this latter painting
was identical with the *concert of five
halflengths* by Titian, listed in the
catalogue of Charles I, now in the
National Gallery in London (see
Bathoe, p. 99, no. 1; reproduced in
the *National Gallery Illustrations,
Italian Schools*, London, 1937,
p. 357, no. 3, as school of Titian;
described by Cecil Gould, *The
Sixteenth-century Venetian School*,
London, 1959, p. 130, no. 3, as by
an imitator of Titian).

GIULIO ROMANO
(ca. 1499–1546)

12. *Portrait of Isabella d'Este*

Engraving: Pieter Holsteyn II (Stimmel, no. 9, by Cornelis Holsteyn after either Correggio or Giulio Romano; Wussin, p. 272, no. 9 and p. 276, v, no. 1; Wurzbach, no. 9, with reference to Reynst collection, after Correggio; Hollstein, IX, p. 77, no. 9, after Parmigianino).
374 × 302 mm. ($14\frac{3}{4} \times 11\frac{7}{8}$ in.).
CAELATURAE 9.

Painting: Hampton Court Palace (No. 306; Inv.no. 76).
Oil on panel, 113 × 89 cm.

Provenance: Gerard Reynst; Charles II (Inventory, p. 1, no. 4: *said to be Raphaels. A Venetian woman in antick habit sitting in a chayre*); James II (Catalogue, ms., Windsor, no. 95; Bathoe, no. 833: *Raphael, An Italian dutchess, at half length*); Queen Anne (Inventory, Windsor, no. 43 in store, as by Raphael); Windsor Castle (Inventory of 1776: *A German Lady by Raphael*); Kensington Palace (Inventory of 1818, no. 597, as by Sebastiano del Piombo).

Exhibitions: London 1946–47, no. 231 (ill.; with incorrect provenance; reference to another version, before 1910 in George Salting collection); London 1960, no. 19.

Bibliography: P. J. Mariette, *Abecedario (Archives de l'Art Français*, VIII), Paris, 1857–58, p. 167 (as Giulio Romano, with reference to drawing for background figure in Crozat collection); Law 1898, p. 112, no. 306 (ill., as Parmigianino ?; engraved in 1653 by Visscher; suggests that it might be identical with Titian's portrait of *Isabella d'Este*, listed in Charles I's inventory); Alessandro Luzio, *La Galleria dei Gonzaga venduta all'Inghilterra nel 1627–28*, Milan, 1913, p. 226 (ill., as *Isabella d'Este* by Parmigianino; corrects Law's wrong identification); Lionel Cust, 'Notes on Pictures in the Royal Collections – XXIX,' *Burlington Magazine*, XXV, 1914, p. 290, fig. A (attributed to Parmigianino; as probably part of the Mantua collection sold to Charles I and bought at the sale of his pictures by Reynst; engraved in 1653 by Visscher, while in Reynst's collection); Jacobs 1925, p. 31; L. Fröhlich-Bume, 'Some unpublished portraits by Parmigianino,' *Burlington Magazine*, XLVI, 1925, pp. 87–88, pls. IA, II, D (as one of the finest portraits by Parmigianino; ca. 1520–30); Lionello Venturi, 'Un ritratto d'Isabella d'Este dipinto da Giulio Romano,' *L'Arte*, XXIX, 1926, pp. 244–45 (by Giulio Romano, ca. 1533); Baker 1929, p. 114, no. 306, pl. XXIV (Parmigianino); Giovanni Copertini, *Il Parmigianino*, Parma, 1932, I, p. 223, fig. 32 (questions attribution to Parmigianino); Frederick Hartt, 'Raphael and Giulio Romano,' *Art Bulletin*, XXVI, 1944, pp. 92–93 (as Giulio Romano);

P 12

P 12a

Jacob Hess, 'Raphael and Giulio Romano,' *Gazette des Beaux-Arts*, série VI, XXXII, 1947, p. 99; A.O. Quintavalle, *Il Parmigianino*, Milan, 1948, p. 101 (as Peruzzi); Sydney J. Freedberg, *Il Parmigianino*, Cambridge, Mass., 1950, pp. 227–28, fig. 157 (attributed to Giulio; incorrect provenance); Mahon 1950, pp. 14–15, no. 12 (rejects Mantua provenance); Frederick Hartt, *Giulio Romano*, New Haven, 1958, I, pp. 82–83, 289, figs. 124, 126–27 (Giulio Romano, ca. 1524–25, perhaps the 'ritratto della Marchesa Isabella con cornice di note'in Mantua collection, mentioned by Daniel Nijs; cf. Luzio, *op.cit.*, p. 151); Bernard Berenson, *Italian Pictures of the Renaissance, Central Italian and North Italian Schools*, London, 1967, I, p. 196 (as Giulio Romano, *Lady Receiving Visitors?*); John Pope-Hennessy, *The Portrait in the Renaissance* (Bollingen Series, XXXV, 12), New York, 1966, p. 165, fig. 182.

Plates: P 12, P 12a.

In his *Abecedario* Mariette already refers to this painting as being in the Reynst collection. He is the first to identify it as a *Portrait of Isabella d'Este* painted by Giulio Romano, an opinion later accepted by Freedberg and Hartt.

Law's suggestion that it might be identical with Titian's portrait of Isabella, listed among the paintings bought by Charles I (Luzio, *op.cit.*, p. 175, no. 92: *Marchioness of Mantua in an old fashioned red velvet apparel...*), initiated the incorrect provenance from the collection of the Duke of Mantua, still mentioned by Freedberg, although Luzio had rejected this identification as early as 1913 (*op. cit.*, p. 226). Mahon came to the same conclusion in 1950.

There is no evidence either that the painting was engraved in 1653 by Cornelis Visscher 'while in Van Reynst's possession', as Law and Cust stated respectively.

GUERCINO (GIOVANNI FRANCESCO BARBIERI) (1591–1666)

13. *Queen Semiramis Receiving News of the Revolt of Babylon*

Engraving: Jeremias Falck (Stimmel, no. 26; Hagen, no. 81, with reference to Reynst collection; Wussin, p. 273, no. 26 and p. 275, IV, no. 4; Block, no. 155, with reference to Reynst collection; incorrectly identified with version in Dresden; Wurzbach, no. 4, with reference to Reynst collection; Hollstein, no. 155). 271 × 382 mm. (10¾ × 15 in.).
CAELATURAE 29.

Painting: Museum of Fine Arts, Boston (Inv. no. 48.1028). Oil on canvas, 112 × 154 cm.

Provenance:
Daniele Ricci, Bologna, 1624; Gerard Reynst; Charles II (?; not recorded in royal inventories); Barbara Villiers, Countess of Castlemaine (Duchess of Cleve-

P 13

land) or Henry Fitzroy, first Duke of Grafton (?); Sir Thomas Hanmer; Duke of Grafton, Euston Hall.

Bibliography: C. C. Malvasia, *Felsina pittrice*, II, Bologna [1678], ed. 1841, (reprint 1967), p. 261 (*Fece al Sig. Daniele Ricci una Semiramide, che fu esposta in Bologna a maraviglia dell' arte ; e questo quadro andò in Inghilterra a quel Re...*); Lugt 1936, p. 117, note 40, fig. 35 (undoubtedly by Guercino; painting of same subject in Dresden perhaps identical with painting for Cardinal Cornaro of 1645); D. Mahon, 'Guercino's Paintings of Semiramis,' *Art Bulletin*, XXXI, 1949, pp. 217–223 (identifies painting with the one engraved by Falck and formerly in Reynst collection; traces picture to Daniele Ricci for whom it was painted in 1624; perhaps part of 'Dutch Gift', with suggestions for subsequent collectors; see under provenance); *idem*, 1950 p. 16 no. 1 (probably part of 'Dutch Gift', but presented by Charles II to Barbara Villiers or to their son, the first Duke of Grafton; not recorded in royal inventories; reference to note by Mariette on Falck's engraving in the Albertina that Guercino's painting was part of the 'Dutch Gift'); N. Brabanti Grimaldi, *Il Guercino*, Bologna [1968] p. 92.

Plates: P 13, P 13a

Mahon was the first to identify this painting with the one formerly in the Reynst collection which was known only through Falck's engraving. (Neither the painting in Dresden, destroyed during World War II and thought by Block to be the prototype for Falck's print, nor the one formerly in the Northbrook collection, identified by Mahon, *loc.cit.* 1949, with the picture painted in 1645 for Cardinal Cornaro, correspond to Falck's engraving). Since Guercino's *Semiramis* was not listed in the royal inventories yet seems to have been in the collection of the Dukes of Grafton, who are direct descendants of Charles II, the painting probably was part of the 'Dutch Gift' (as implied by Mariette) but left the royal collection before ca. 1666–67, when the inventory of Charles II was drawn up.

JAN LISS
(ca. 1597–1629/30)

14. *St. Paul in Ecstasy*

Engraving: Jeremias Falck (Stimmel, no. 18, without name of engraver; Robert Naumann, *Archiv für die zeichnenden Künste*, XI, 1856, p. 184, Appendix, no. 2, as Jacob Matham; Wussin, p. 273, no. 18 and p. 274, I, no. 2; Block, no. 26; Wurzbach, Theodor Matham, no. 8; Hollstein, VI, Jeremias Falck, p. 211, no. 26 and XI, p. 148, no. 4; *idem*, XI, Theodor Matham, p. 252, no. 9, and XI, p. 148, no. 11).
377 × 284 mm. ($14\frac{7}{8} \times 11\frac{3}{16}$ in.).
CAELATURAE 16.

Painting: Staatliche Museen Preussischer Kulturbesitz, Gemäldegalerie, Berlin. (Inv.no. 1858). Oil on canvas, 80 × 58,5 cm.

Provenance: Gerard Reynst; Van de Amory, Amsterdam (sale Amsterdam, 23 June 1722, no. 82: *Paulus Verrukking, door Jan Lis*); A.v. Frey, Berlin (bought ca. 1900 on the Florentine Art Market); bought by the Kaiser-Friedrich-Museum, Berlin in 1919.

Exhibitions: *Werke alter Kunst*, Berlin (Akademie), 1914; *Aufgang der Neuzeit*, Nuremberg, 1952, p. 126, no. N8; *Deutsche Maler und Zeichner des 17. Jahrhunderts*, Berlin (Orangerie des Schlosses Charlottenburg), 1966, no. 49, fig. 49; *Johann Liss*, Augsburg (Rathaus)/Cleveland (Museum of Art), 1975/76, no. A38, fig. 38 and color plate VIII (ca. 1627–29; reference to Reynst collection; extensive bibliography).

Bibliography: Rudolf Oldenbourg, 'Jan Lys,' *Jahrbuch der königlich preuszischen Kunstsammlungen*, XXXV, 1914, p. 162, fig. 15; Eduard Plietzsch, 'Die Ausstellung von Werken alter Kunst...,' *Zeitschrift für bildende Kunst*, N.F. XXV, 1914, p. 232; Wilhelm von Bode, 'Eine neue Erwerbung der Gemäldegalerie: Vision eines Heiligen von Jan Lys,' *Berliner Museen*, XLI, 1919, columns 1–6; *idem*, 'Johan Lys,' *Velhagen und Klasings Monatshefte*, XXXV, 1920, pp. 176–77; Rudolf Oldenbourg, *Giovanni Lys*, Rome, 1921, p. 12; J. Zarnowski, 'Zwei unbekannte Werke Jan Lys' in Russland,' *Belvedere*, VIII, 1925, pp. 93, 95–96; Giuseppe Fiocco, *Die venezianische Malerei des 17. und 18. Jahrhunderts*, Florence and Munich, 1929, p. 22; R. A. Peltzer, 'Johann Liss,' *Thieme-Becker*, XXIII, Leipzig, 1929, p. 286; Kurt Steinbart, *Johann Liss*, Berlin, 1940, pp. 126–29 (ca. 1628–29) 161, 166, 168, pls. 50–51; fig. 56; *idem*, *Johann Liss*, Vienna, 1946, p. 41; Vitale Bloch, 'Lissiana,' *Oud Holland*, LXI, 1946, p. 129; *idem*, 'Liss and his 'Fall of Phaeton',' *Burlington Magazine*, XCII, 1950, p. 282; Kurt Steinbart, 'Das Werk des Johann Liss in alter und neuer Sicht,' *Saggi e memorie di Storia dell' Arte*, II, 1959, p. 186; *Staatliche Museen Berlin, Verzeichnis der ausgestellten Gemälde...*, Berlin, 1962,

P 14

ITALIAN PAINTINGS

P 14a

128　ITALIAN PAINTINGS

p. 48; Rudolf Wittkower, *Art and Architecture in Italy, 1600 to 1750*, Harmondsworth, 1965, p. 67; C. Donzelli and G.M. Pilo. *I Pittori del Seicento Veneto*, Florence [1967], p. 241.

Plates: P 14, P 14a.

Steinbart's supposition that Jan Reynst probably bought the painting after Liss' death in Venice in 1629/30 from the latter's estate may be correct (*op.cit.*, p. 126).

15. *The Brothel*

Engraving: Jeremias Falck (Stimmel, no. 27, without name of painter or engraver; sometimes called the *Brothel* by Rubens; Hagen, no. 84, with reference to Reynst collection; after Liss; Wussin, p. 273, no. 27, after Rubens, without name of engraver and p. 275, IV, no. 7, by Falck after Liss; Block, no. 160, with reference to Reynst collection, after Liss' painting in Cassel; Max Rooses, *L'Oeuvre de P.P.Rubens*, IV, Antwerp, 1890, p. 80: engraved for '*cabinet Rynst*', sometimes attributed to Rubens, but probably after Liss; Wurzbach, no. 7, with reference to Reynst collection; Hollstein, VI, p. 214, no. 160, ill. and XI, p. 148, no. 8).
248 × 375 mm. (9¾ × 14¾ in.).
CAELATURAE 30.

Painting: Staatliche Kunstsammlungen, Cassel (Inv.no. 187).
Oil on canvas, 164,2 × 242 cm.

Provenance: Gerard Reynst; Jan Six, Amsterdam (sale Amsterdam, 6 April 1702, no. 95: *Een Venetiaans Lusthuis, zeer sierlyk en ongemeen curieus*; sold for 49 guilders to Jan Six, Jr.); Pieter Six, Amsterdam (sale Amsterdam, 2 September 1704, no. 16: *Een Bordeel van Jan Lis, heerlyk geschildert*; sold for 465 guilders [Hoet, I, p. 72, no. 16]); first mentioned in the main inventory in Cassel in 1749, no. 521.

Exhibition: *Johann Liss*, Augsburg (Rathaus)/Cleveland (Museum of Art), 1975/76, no. A15, fig. 15 (replica of earlier version in Nuremberg; reference to Reynst collection; list of other versions and extensive bibliography).

Bibliography: Knorr 1663 (Fuchs, pp. 239–40, mentions a *Taberna* which might be this painting); Rudolf Oldenbourg, 'Jan Lys,' *Jahrbuch der königlich preuszischen Kunstsammlungen*, XXXV, 1914, pp. 144–47, fig. 4 (probably identical with painting formerly with Reynst; lists copy by E. van Heemskerk, Leningrad, and copies in private collection near Florence and in Bologna, Pinacoteca, Inv.no. 708); J. Zarnowski, 'Zwei unbekannte Werke von Jan Lys' in Russland,' *Belvedere*, VIII, 1925, pp. 92–93; R. A. Peltzer, 'Johann Liss,' *Thieme-Becker*, XXIII, Leipzig, 1929, p. 286; Kurt Steinbart, *Johann Liss*, Berlin, 1940, pp. 14, 65–82, pls. 27–29; fig. 25 (ca. 1625;

P 15

painting was in Amsterdam before 1646), 152, 158, 162; Vitale Bloch, 'Lissiana,' *Oud Holland*, LXI, 1946, p. 128 (painted in Rome); *Katalog der Gemäldegalerie des Hessischen Landesmuseums Kassel*, Cassel, 1958, pp. 87–88, no. 187 (probably identical with painting formerly with Reynst); Kurt Steinbart, 'Das Werk des Johann Liss in alter und neuer Sicht,' *Saggi e memorie di Storia dell' Arte*, II, 1959, pp. 166–68, fig. p. 167 (ca. 1624, during first Venetian phase); Lisa Oehler, 'Unbekannte Vorzeichnungen zu einigen Gemälden der Kasseler Galerie,' *Kunst in Hessen und am Mittelrhein*, 1/2, 1962, pp. 104–07, fig. 6; Arthur von Schneider, *Caravaggio und die Niederländer*, 2nd edition, Amsterdam, 1967, pp. 49–50, pl. 26 b (painted in Rome or immediately after return to Venice); C. Donzelli and G.M. Pilo, *I Pittori del Seicento Veneto*, Florence (1967), p. 242.

Plates: P 15, P 15a.

Liss' painting is known in two almost identical versions, one in the Gemäldegalerie, Cassel, the other one in the Germanisches National-museum, Nuremberg (oil on canvas, 161 × 259,5 cm; Steinbart, *op.cit.*, pp. 65, 162, pl. 26). In a few details, Falck's engraving corresponds more closely with the painting in Cassel, especially in the rendering of the beam of light and of the dress of the courtesan seated in front of the table. We may, therefore, assume that this was the version formerly with Reynst, a supposition that is strengthened by the recent restoration of the Nuremberg painting which revealed a red plumed beret placed on the bench in front of the table instead of the helmet, found in the Cassel version and in Falck's print. Steinbart suggested that Liss' *Brothel* had been in the Reynst collection in Amsterdam prior to 1646, since parts of it were adapted by Simon de Vos in his own composition of a *Musical Company*, dated 1646, in Vienna (*op.cit.*, p. 79, fig. 31). This supposition is probably correct, since the painting most likely was bought by Jan in Venice, and Jan had died in the summer of 1646. Whether De Vos based his own composition specifically on the painting formerly with Reynst, however has to remain open. Judging from Sandrart Liss apparently painted a number of such scenes (*Teutsche Academie*, ed. Peltzer, pp. 187–88: *mehr hat er schöne Conversationen geharnischter Soldaten mit Venetianischen Courtisanen, da unter lieblichen Seiten- und Kartenspiel bey einem ergötzlichen Trunk jeder nach seinem Gefallen conversirt und im Luder lebt, worinnen die Vielfältigkeit der Affecten, Gebärden und Begierden eines jeden so vernünftig ausgebildet sind, dass diese Werke nicht allein hoch gepriesen, sondern auch von den Kunstliebenden um grossen Wehrt erkauffet worden*).

P 15a

LORENZO LOTTO
(1489–1556)

16. *Portrait of Andrea Odoni*

Etching and Engraving: Cornelis Visscher (Hecquet, no. 30, with reference to Reynst collection; Stimmel, no. 19, without name of painter; Wussin, no. 36, with reference to Reynst collection, by Lotto; *idem*, p. 273, no. 19 and p. 277, IX, no. 4; Wurzbach, no. 36, after Correggio).
308 × 381 mm. (12⅛ × 15 in.).
CAELATURAE 26.
Copper plate in 1679 with Clement de Jonghe.

Painting: Hampton Court Palace (No. 148; Inv. no. 72).
Oil on canvas, 102 × 115,5 cm.
Signed and dated: *Lavrentivs lotvs 1527*.

Provenance: Andrea Odoni, Venice (seen in 1532 by Marcanton Michiel: *El retratto de esso M. Andrea a oglio, meza figura, che contempla li fragmenti marmorei antichi fu de man de Lorenzo Lotto*; see Frimmel, below); Alvise Odoni, Venice; Gerard Reynst; Charles II (Inventory, p. 16, no. 264: *Lorenzo Lottie. A man sitting and holding a small head in one hand with severall statues by him. He being an Antiquery. 3.4 × 3.9*); James II (Catalogue, ms., Whitehall, no. 163, as by Giorgione; Bathoe, no. 162); Queen Anne (Inventory, Windsor, no. 138, as *Portrait of Bandinelli by Coreggio*).

Exhibitions: London 1946–47, no. 217 (ill.; with fictitious date 1653 for engraving; reference to Reynst collection and 'Dutch Gift'); *Mostra di Lorenzo Lotto*, Venice (Ducal Palace), 1953, no. 67 (ill.); London 1960, no. 8 (ill.; with reference to Reynst collection and Dutch Gift); *Konstens Venedig*, Stockholm (Nationalmuseum), 1962–63, no. 81 (as *Portrait of the Sculptor Andrea Odoni* with reference to Reynst collection and fictitious date 1653).

Bibliography: Th. Frimmel, ed., *Der Anonimo Morelliano (Marcanton Michiel's 'Notizia d'opere del Disegno'* [1532]), Quellenschriften für Kunstgeschichte und Kunsttechnik, N.F. 1, Vienna, 1896, p. 84; Giorgio Vasari, *Le Vite*, ed. Novara, 1967, v, p. 43 (*In casa d'Andrea Odoni è il suo ritratto di mano di Lorenzo, che è molto bello* [1541]); Joachim von Sandrart, *Teutsche Academie*, Nuremberg, 1675, ed. R. A. Peltzer, Munich, 1925, p. 417 (*das Contrafät eines Kunst-Liebhabers in seinem studio, von Antonio da Corregio verfärtigt*, listed among the expensive paintings of the time, with reference to Reynst collection); William Hazlitt, *Criticisms on Art*, London, 1843, p. XIV, no. 67 (*The Sculptor Bandinelli* by Correggio); G. F. Waagen, *Treasures of Art in Great Britain*, II, London, 1854, p. 356, no. 67 (by Lotto, erroneously called Bandinelli and attributed to Correggio); J.A. Crowe and G.B. Cavalcaselle, *A History of Painting in North Italy*,

P 16

II, London, 1871, p. 158 and pp. 519–20; Law 1898, p. 57, no. 148 (ill.; with unfounded statement that the painting was in 1653 with Reynst, when engraved by Visscher); Bernard Berenson, *Lorenzo Lotto*, London, 1905, pp. 174–75 (ill.; reference to Reynst collection); Jacobs 1925, p. 31; G.J. Hoogewerff and J.Q. van Regteren Altena, *Arnoldus Buchelius 'Res Pictoriae'* (Quellenstudien zur Holländischen Kunstgeschichte, xv), 's-Gravenhage, 1928, p. 99 (4 September 1639: *pictura quam nuper in auctione Uffeliana ipse emerat 900 flor. quae in catalogo ita exprimitur: Het conterfeitsel van een man op syn antycqs. Halve figuer, seer raer, van Titiaan* ; identified by the authors with the so-called *Ariosto* by Titian in the National Gallery, London, and – less likely – with Lotto's *Andrea Odoni*); Baker 1929, p. 97, no. 148, pl. XV (Lotto, with fictitious date 1653 and reference to Reynst collection); A. Venturi, *Storia dell'Arte Italiana*, IX, part 4, Milan, 1929, pp. 65–66, fig. 58; Lugt 1936, p. 117, fig. 13 (feels Rembrandt must have known the painting); G.M. Richter, *Giorgio da Castelfranco*, Chicago, 1937, p. 225 (Sandrart copied Titian's so-called *Ariosto*, now London, when in Reynst collection, a suggestion rejected by Gould, see below); Mahon 1950, p. 15, no. 13 (corrects Law's erroneous date 1653 for Reynst collection and engraving); A. Banti and A. Boschetto, *Lorenzo Lotto*, Florence [1953], pp. 39 and 81, no. 77, fig. 150 (with incorrect statement that painting passed into collection of Charles II in 1633); Bernard Berenson, *Lorenzo Lotto*, New York, 1956, p. 98, pl. 219 (repeats Law's fictitious date 1653 for Visscher's engraving; reference to Reynst collection and 'Dutch Gift'); idem, *Italian Pictures of the Renaissance, Venetian School*, I, London, 1957, p. 102 (as Lotto); C. Gould, *National Gallery Catalogues, The Sixteenth-Century Venetian School*, London, 1959, p. 112, under no. 635, p. 116, under no. 1944 (doubts Hoogewerff's and Van Regteren Altena's identification of this painting with the so-called *Ariosto* in London; see above; Richter's statement that Sandrart copied the *Ariosto* while in Reynst collection unfounded); P. Bianconi, *All the Paintings of Lorenzo Lotto*, New York, 1963, pp. 19, 79, pl. 124; Margot Seidenberg, *Die Bildnisse des Lorenzo Lotto*, Basel, 1964, pp. 12–13, 58–64, 87–88, 91–92, fig. 3 (with reference to Reynst collection and possible influence on some of Rembrandt's portraits); John Pope-Hennessy, *The Portrait in the Renaissance*, (Bollingen Series, XXXV, 12), New York, 1966, pp. 228, 231, fig. 255; L.O. Larsson, 'Lorenzo Lottos Bildnis des Andrea Odoni in Hampton Court,' *Konsthistorisk Tidskrift*, XXXVII, 1968, pp. 21–33, fig. 1 (establishes that Odoni is primarily represented as a collector; most antique sculptures cannot be identified in the list of his collection; repeats Law's incorrect date 1653); E. Panofsky, *Problems in Titian*, New York, 1969, p. 81, fig. 94; S.J. Freedberg, *Painting in Italy, 1500 to 1600*, Harmondsworth, 1971, p. 207.

Plates: P 16, P 16a.

Lotto's portrait of the Venetian merchant and collector Andrea Odoni figured prominently among the notable sights in Venice and was mentioned by Marcanton Michiel in 1532, and by Vasari in 1541. Larsson (*loc. cit.*) pointed out that Odoni is not represented inmidst prized objects from his collection of antique sculptures but rather portrayed as a collector *per se*. Thus it seems only natural that the painting should have passed on to Jan Reynst, another merchant with a similar interest.

17. *Portrait of a Gentleman*

Engraving: Cornelis van Dalen II (Stimmel, no. 3, as *Portrait of Giorgione*, after Sebastiano del Piombo; Wussin, p. 272, no. 3 and p. 275, III, no. 2; Wurzbach, no. 2, with reference to Reynst collection, after Titian; Hollstein, V, p. 107, no. 109, with reference to Reynst collection, after Lotto).
412 × 295 mm. (16¼ × 11⅝ in.).
CAELATURAE 3.

Painting: Hampton Court Palace (No. 114; Inv.no. 486).
Oil on canvas, 51 × 37,5 cm.

Provenance: Gerard Reynst; Charles II (Inventory, p. 83, no. 116: *A man in a blacke gowne with his hayre behind his eares. Dutch p^{r}sent. 1.10 × 1.6*); James II (Catalogue, ms., no. 87: *Giorgione, A man's head in a black habit, with his hair behind his ears* ; Bathoe, no. 952); possibly Queen Anne (Inventory, Kensington, no. 31, as by Giorgione).

Exhibitions: London 1946–47, no. 228 (as Lotto; with reference to Reynst collection and Dutch present; engraving incorrectly identified as by Vorsterman after Titian); Amsterdam 1965, no. 4 (ill.; as Lotto).

Bibliography: J.A. Crowe and G.B. Cavalcaselle, *A History of Painting in North Italy*, II, London, 1871, p. 158 (Giorgione, in James II collection), and p. 159 (Follower of

ITALIAN PAINTINGS

P 17

P 17a

Giorgione, Bergamasque, related to Lotto); Law 1898, p. 44–45, no. 144 (ill.; as Lotto, with reference to Reynst collection and 'Dutch Gift'); B. Berenson, *Lorenzo Lotto*, London, 1905, pp. 15–16, 257 (1508–09; reference to Reynst collection); Jacobs 1925, p. 31; G.J.Hoogewerff and J.Q. van Regteren Altena, *Arnoldus Buchelius 'Res Pictoriae'* (Quellenschriften zur Holländischen Kunstgeschichte, xv), 's-Gravenhage, 1928, p. 100 (4 September 1639: *Eminebant inter omnes duae tabulae satis magnae, quas nullo quantumvis magno pretio carere vellet ;... ; posterior, manu Corregii in qua vir admodum vivide expressus, quam tabulam vel quovis pretio ambivit Britanniae rex, sed frustra, cum vel multa millia offerrentur*; tentatively identified by the authors with this painting by Lotto); Baker 1929, pp. 96–97, no. 114 (Lotto; repeats Law that there is an engraving of same subject by Vorsterman as after Titian); Lugt 1936, p. 100, fig. 12 (as Lotto); Mahon 1950, p. 14, no. 11 (corrects Law and Baker: not engraved by Vorsterman); A. Banti and A. Boschetto, *Lorenzo Lotto*, Florence 1953, p. 107 (Oliverio?); B. Berenson, *Lorenzo Lotto*, New York, 1956, p. 16, pl. 49 (1508; last portrait of Treviso period; part of Reynst collection, engraved by Vorsterman); B. Berenson, *Italian Pictures of the Renaissance, Venetian School*, I, London, 1957, p. 102 (early Lotto); P. Bianconi, *All the Paintings of Lorenzo Lotto*, II, New York, 1963, p. 101 (attributed to Lotto); John Pope-Hennessy, *The Portrait in the Renaissance* (Bollingen Series, xxxv, 12), New York, 1966, p. 129, fig. 141.

Plates: P 17, P 17a.

LORENZO LOTTO
(ATTRIBUTED TO)

18. *The Adoration of the Shepherds*

Engraving: Jeremias Falck (Stimmel, no. 12, according to annotation by Winckler after L. Cotto? (probably should read Lotto); Wussin, p. 272, no. 12 and p. 276, IV, no. 16, as doubted by Hagen; Block, no. 6, after Veronese; Hollstein, VI, p. 209, no. 6).
306 × 371 mm. ($12\frac{1}{16} \times 14\frac{3}{4}$ in.).
CAELATURAE 20.
Copper plate in 1679 with Clement de Jonghe.

Painting: Lost.

Provenance: Jan Reynst; Gerard Reynst.

Bibliography: Ridolfi, I, p. 145 (*Trattenendosi Lorenzo [Lotto] in Venetia, lauorò parimente opere diverse, e tra quelle la nascita di Christo, fingendo l'attione di notte tempo, con Angeli intorno al Presepe,*

P 18

in compagnia della Vergine Madre illuminati da splendori, che escono dal Bambino, rarissima Pittura, hor è nelle case del Signor Giouanni Reinst Gentil'huomo Olandese in Amsterdamo); Lugt 1936, p. 117, note 40 (undoubtedly by Lotto); Mahon 1949, p. 303, note 9; Savini-Branca 1965, pp. 70, 269 (as Lotto).

Plate: P 18.

According to Ridolfi, the painting was by 1646 at the latest in one of Jan Reynst's houses in Amsterdam.

MARCO D'OGGIONO (ATTRIBUTED TO)
(ca. 1475–ca. 1530)

19. *The Infant Christ and St. John, Embracing*

Painting: Hampton Court Palace (No. 64; Inv.no. 391).
Oil on panel, 62 × 46,5 cm.

Provenance: Gerard Reynst (?); Charles II (Inventory, Whitehall, p. 20, no. 335: *Leonard De Vince. Two Boyes naked. A Landskip. Dutch Present. 2.1 × 1.6*); not securely identifiable in James II collection (Bathoe, nos. 386 and 561, lists two paintings of this subject by Parmigianino); Queen Anne (Inventory, Hampton Court, no. 38, as Leonardo).
Bibliography: Law 1898, pp. 21–22, no. 64 (ill.; as Marco d'Oggionno; identifies painting with no. 26 in Charles I's catalogue); Baker 1929, p. 111, no. 64 (attributed to Marco d'Oggiono, as part of 'Dutch Gift'); Lugt 1936, p. 99, fig. 3 (as Marco d'Oggiono); Mahon 1950, pp. 17–18, among paintings whose provenance from 'Dutch Gift' is questionable (identifies two paintings of this subject in royal collection, one from Charles I's collection [Bathoe, no. 26], the other one bought in 1662 from William Frizell [list, no. 49]); Bernard Berenson, *Italian Pictures of the Renaissance, Central Italian & North Italian Schools*, I, London, 1968, p. 242 (as Marco d'Oggiono, variant of painting in Mond collection, London).

Plate: P 19.

The fact that two painting of this subject are listed in the royal inventories, one given to Leonardo, the other to Parmigianino, both with rather impeccable provenances other than from the Reynst collection, makes one doubt the reference to the Dutch Gift in Charles II's inventory, as Mahon rightly observed (*loc.cit.*). A fair number of copies of this Leo-

nardesque composition does exist, however, which might still leave the possibility open that there may indeed have been a third version in the royal collections which no longer is traceable today. (The composition is also known from a print by J. Chr. Le Blon; Hollstein, II, p. 102, no. 8, as after Correggio).

PARMIGIANINO
(FRANCESCO MAZZOLA)
(1503–1540)

20. *Athena*

Engraving: Cornelis Visscher (Hecquet, no. 28, with reference to Reynst collection; after La Bosse; Stimmel, no. 1; Wussin, no. 136, with reference to Reynst collection; as *Artemisia*; idem, p. 272, no. 1 and p. 277, IX, no. 1; Wurzbach, no. 136, with reference to Reynst collection; *Artemisia*, after Guercino).
364 × 257 mm. ($14\frac{5}{16} \times 10\frac{1}{8}$ in.).
CAELATURAE 1.
Copper plate in 1679 with Clement de Jonghe.

Painting: Windsor Castle (Inv.no. 1138).
Oil on canvas, 63,5 × 46 cm.

Provenance: Gerard Reynst; Charles II (Inventory, p. 19, no. 315: *Parmizano. Minerva's head wth one hand on her breast. Dutch p^{r}sent*); James II (Catalogue, ms., f. 66 v., no. 98 of the second numbered group of the ms., probably placed in the King's Closet at Whitehall; Bathoe, no. 632); Queen Anne (Inventory, Windsor, no. 103: *Parmeggiano*).

Exhibition: London 1946–47, no. 215 (with reference to preliminary drawings for gold plaque).

Bibliography: William Hazlitt, *Criticisms on Art*, London, 1843, p. XXXVI, no. 102 (as *Minerva*, by Parmigianino); G. Copertini, *Il Parmigianino*, Parma, 1932, p. 118, pl. CLVIII, a (by imitator of Parmigianino); Lili Fröhlich-Bume, 'Unpublished Drawings by Parmigianino,' *Old Master Drawings*, IX, 1934–35, pp. 55–56 (publishes two double-sided preliminary drawings for the medallion on Athena's chest without connecting them to the painting); C. H. Collins Baker,

P 20

P 20a

Catalogue of the Principal Pictures in the Royal Collection at Windsor Castle, London (1937), p. 250, pl. LXXXII (as Parmigianino); S.J.Freedberg, *Parmigianino*, Cambridge, Mass., 1950, p. 226 (as *Minerva*?, attributed to Parmigianino, probably identifiable with painting of a *Pallade armata* attributed to Parmigianino in the 1662 inventory of the studio Muselli in Verona; no reference to Reynst collection; suggests that the painting may have been derived from the Parmigianino drawing in the Pierpont Morgan Library, New York [see Popham, below]); Mahon 1950, p. 13, no. 7 (lists references in royal inventories); B. Berenson, *Italian Pictures of the Renaissance, Central Italian and North Italian Schools*, I, London, 1968, p. 319 (as *Bust of* [?] *Minerva* by Parmigianino); A.E. Popham, *Catalogue of the Drawings of Parmigianino*, New Haven, 1971, p. 26 (as by Parmigianino, painted in the second Parmese period, based on preliminary drawing at the Pierpont Morgan Library, New York, of ca. 1530–31 [cat. no. 316, pl. 364]; lists four, possibly five preparatory drawings for Minerva's medallion, present whereabouts unknown [cat. no. 799]).

Plates: P 20, P 20a.

Freedberg's suggestion (*loc.cit.*) that this painting might be identical with *una Pallade Armata*, a painting attributed to Parmigianino in the 1662 inventory of the studio Muselli in Verona, cannot be sustained since it was part of the 'Dutch Gift' of 1660 and was specifically listed as such in the inventory of Charles II.
The figure should be identified as *Athena* rather than *Artemisia* since one of the preliminary drawings for the medallion bears the inscription ATENE also taken over in the painting. This supposition is further strengthened by Popham's suggestion (*loc.cit.*) that a third drawing with a *Seated Figure of Athena* (on the same mount as the other two preparatory drawings for the medallion) may have been an alternative study for this same plaque.

SCHOOL OF RAPHAEL

21A. *The Virgin and Child and St. Anne*

Engraving: Jeremias Falck (Heinecken, p. 83; Stimmel, nos. 10 and 34 as Jacob Matham after Raphael, done twice; painting part of 'Dutch Gift'; Hagen, no. 78, with reference to Reynst collection, as after Andrea del Sarto; Wussin, p. 272, no. 10 and no. 34 and p. 276, VII, nos. 1 and 3, with reference to 'Dutch Gift'; Block, no. 9, with reference to Reynst collection, after Del Sarto; Wurzbach, no. 1, with reference to Reynst collection; Hollstein, VI, p. 209, no. 9). 363×288 mm. $(14\frac{5}{16} \times 11\frac{3}{8}$ in.$)$.
CAELATURAE 10.
Copper plate in 1679 with Clement de Jonghe?

Painting: Lost.

Provenance: Gerard Reynst.

Bibliography: Lugt 1936, p. 117, note 40 (Perino del Vaga? same composition in Galleria Borghese, Rome).

Plate: P21A.

21B. *The Virgin and Child and St. Anne*

Engraving: Unidentified engraver; see above.
CAELATURAE 18.

Plate: P21B.

Two engravings, by different artists, after the same painting.

SCHOOL OF RAPHAEL (?)

22. *Christ on a Lamb, the Virgin and St. Joseph*

Painting: Not securely identifiable.

Provenance: Gerard Reynst; Charles II (Inventory, p. 23, no. 390: *Said to be Raphael. Our Savio^r on a Lambe ye B Virgin & S^t. Joseph. Dutch p^{r}sent.* 1.0×0.9); James II (Catalogue, ms., f. 70 r, no. 33 of paintings in the custody of the Dowager Queen Catherine of Braganza: *By Raphell. Our Lady, and Christ with a Lamb and Joseph one of the Dutch Presents*; Bathoe, no. 716); not in Queen Anne inventory or subsequent royal inventories.

Bibliography: Joachim von Sandrart, *Teutsche Academie*, Nuremberg, 1675, ed. R.A. Peltzer, Munich, 1925, p. 417 (*ein Marien-Bild in der Grösse nur eines Papir-Bogens, von Raphael gemahlt*, with reference to Reynst collection);

P21A

J. D. Passavant, *Raphael d'Urbin*, II, Paris, 1840, pp. 55–56; G. F. Waagen, *Treasures of Art in Great Britain*, II, London, 1854, p. 476, no. 12 (as Raphael); Jacobs 1925, p. 30; George Vertue, 'Inventory of Charles I collection 1687–8,' *The Walpole Society*, XXIV, 1935–36, p. 93 (with reference to 'Dutch Gift', but incorrect provenance from collection of Charles I); Mahon 1950, p. 16, no. 17 (leaves possible identification with Raphael's painting in the Prado, once in the custody of Catherine of Braganza, open).

The painting probably relates to Raphael's *Holy Family with the Lamb* of 1507, in the Prado, Madrid (O. Fischel, *Raphael*, Berlin, 1962, p. 37, fig. 55), as Passavant already suggested. Since Raphael's composition is known in a number of versions and since the one formerly with Reynst was not engraved for the CAELATURAE, a secure identification remains elusive.

GUIDO RENI
(1575–1642)

23. *Susannah and the Elders*

Engraving and Etching: Cornelis Visscher (Hecquet, no. 3, with reference to Reynst; after painting by Reni in the collection of the Duke of Orleans; Stimmel, no. 24; Wussin, no. 100, with reference to Reynst collection; *idem*, p. 273, no. 24 and p. 277, IX, no. 7; Wurzbach, no. 100).
290 × 373 mm. ($11\frac{7}{16} \times 14\frac{11}{16}$ in.).
CAELATURAE 28.
Copper plate in 1697 with Clement de Jonghe.

Painting: Not securely identifiable among the known versions (Florence, Uffizi; see below; Hannover, Niedersächsische Landesgalerie, Inv.no. VAM 930; London, National Gallery, Inv. no. 196; Nantes, Musée des Beaux-Arts, Inv.no. 85; engraving by P. Beljambe for the *Galerie du Duc d'Orléans*, Paris, 1786, no. VIIE; possibly after Uffizi version).

Provenance: Gerard Reynst.

Bibliography: Knorr 1663 (Fuchs, p. 239, mentions a *Historia Susannae*); Lugt 1936, p. 117, note 40 (as Guido Reni; lists paintings in Vienna, Florence, Hannover, and in the Galerie d'Orléans, which all are slightly different from engraving in the CAELATURAE); Otto Kurz, 'Guido Reni,' *Jahrbuch der kunsthistorischen Sammlungen Wien*, N.F., XI, 1937, p. 217 (with reference to 17th century versions after the lost original of ca. 1621, among them the Reynst painting; Mahon 1949, p. 304, note 18 (reference to Knorr's *Historia Susannae*); Count Antoine Seilern, *Flemish Paintings & Drawings at 56 Princes Gate London SW 7, Addenda*, London, 1969, p. 33, under no. 311 (traces Uffizi version, ex Orléans, to the collection of Queen Christina of Sweden); Cesare Garboli and Edi Baccheschi, *Guido Reni*, Milan,

P 21B

140 ITALIAN PAINTINGS

P 23

P 23a

1971, no. 101 (for various versions of the painting); Michael Levey, *National Gallery Catalogues, The Seventeenth and Eighteenth Century Italian Schools*, London, 1971, p. 192, no. 196 (for various versions).

Plates: P 23, P 23a.

Several versions of Reni's *Susannah and the Elders* are still extant today, although the original itself apparently is lost (perhaps the painting once in the Imperial collection in Vienna, known from the engraving in the *Theatrum Pictorium* of 1660, Plate 38). Both Hecquet and Stimmel, stated that Visscher's engraving was made after the painting formerly owned by the Duke of Orléans (*Galerie du Palais Royal*, I, Paris, 1786, Guido Reni, VII; engraved by P. Beljambe). The Orléans painting has been identified with the version in the Uffizi which shows marked differences, however, with Visscher's engraving, especially in the rendering of the fountain and in the headgear of Susannah. Closer to Visscher's print are the version in London and in particular the one in Nantes (Pl. P 23a). Too many alterations between Visscher's print in the CAELATURAE and these various paintings remain, however, to allow for a secure identification.

GUIDO RENI (ATTRIBUTED TO)

24. *St. Bartholomew*

Engraving: Theodor Matham (Stimmel, no. 6, as *St. Peter*, after Reni; Wussin, p. 272, no. 6 and p. 277, VIII, no. 1; Hollstein, XI, p. 252, no. 11, as *St. Paul*). 376×282 mm. ($14\frac{13}{16} \times 11\frac{1}{8}$ in.). CAELATURAE 6. Copper plate in 1679 with Clement de Jonghe.

Painting: Lost.

Provenance: Gerard Reynst.

Bibliography: Knorr 1663 (Fuchs, p. 240: mentions an *effigies S. Bartholomaei* which might refer to this painting).

Plate: P 24.

The saint probably should be identified as St. Bartholomew since he is represented with a flaying knife. A *Head of St. Bartholomew* was seen by Knorr von Rosenroth in 1663, during his visit to the

P 24

P 25

P 25a

Reynst collection in Amsterdam. Annotations in pencil on Matham's print attribute the lost painting to Guido Reni, in a few instances to Jacopo Palma. No painting by either of these two artists was found that corresponded to Matham's engraving.

25. *St. John the Evangelist*

Stipple engraving: Jan Lutma II (Stimmel, no. 33, without name of painter; Wussin, p. 273, no. 33 and p. 276, VI, no. 1; Dimitri Rovinski, *L'Oeuvre gravé des élèves de Rembrandt et des maîtres qui ont gravé dans son goût*, St. Petersburg, 1894, column 74, no. 55, as after Rembrandt (?); Hollstein, XI, p. 111, no. 2, ill.).
348 × 289 mm. (15 1/8 × 11 3/8 in.).
CAELATURAE 17.

Painting: Museum der bildenden Künste, Leipzig (Inv. no. 193). Oil on canvas, 67 × 57 cm.

Provenance: Gerard Reynst.

Plates: P 25, P 25a.

Reni's painting was identified by Dr. Susanne Heiland, who kindly brought it to my attention.

26. *Allegory of Painting*

Engraving: Theodor Matham (Stimmel, no. 7, after Reni; Wussin, p. 272, no. 7 and p. 277, VIII, no. 2; Hollstein, XI, p. 253, no. 43).
332 × 280 mm. (13 1/8 × 11 in.).
CAELATURAE 7.
Copper plate in 1679 with Clement de Jonghe.

Painting: Presumably lost.

Provenance: Gerard Reynst; perhaps James II (Catalogue, ms., f. 50v., Whitehall, no. 168: *By Gwedoe. A Paintresse with a Cupid putting a Laurell upon her head*; Bathoe, no. 167); perhaps Queen Anne (Kensington, no. 106 or Somerset House, no. 10).

Bibliography: Jacobs 1925, p. 31 (refers to Reynst collection and James II catalogue, but the painting is no longer traceable); Mahon 1950, p. 17, no. ii (gives full history of two paintings of this subject and composition in the royal collections, both catalogued as 'after Reni'; leaves identification with painting ex Reynst open); Michael Levey, *The Later Italian Pictures in the Collection of Her Majesty the Queen*, London (1964), p. 93, cat. nos. 582–83 (unclear whether either of the two copies in royal collection can be identified with painting formerly with Reynst).

Plate: P 26.

Acording to Mahon and Levey (*loc.cit.*) there were two painted versions of this subject in the royal collections by the early 18th century (listed in the Queen Anne Inventory, ca. 1706–10). An annotation in pencil by Coke

P 26

(*Guido/att Kensington*) on the engraving in the CAELATURAE in the Victoria and Albert Museum identifies the version at Kensington with the picture formerly in the Reynst collection. Two paintings of this subject are still in the royal collections, both poor copies. Since it is unlikely that either of them passed as an original in the 17th century, one should not exclude the possibility that Guido's original version once was in the collection but left it at a later date, while two copies after it remained.

ANDREA SCHIAVONE
(1522–1563)

27. *Presentation in the Temple*

Engraving: Jeremias Falck (Stimmel, no. 13; Wussin, p. 272, no. 13 as *Sacrifice in the Tempel* and p. 275, no. 18; Block, no. 8, with reference to Reynst collection; Hollstein, VI, p. 209, no. 8). 273×372 mm. ($10\frac{3}{4} \times 14\frac{11}{16}$ in.). CAELATURAE 21.
Copper plate in 1679 with Clement de Jonghe?

Painting: Lost.

Provenance: Gerard Reynst.

Plate: P 27.

28. *Christ Before Pilate*

Painting: Hampton Court Palace (No. 289; Inv. no. 522). Oil on canvas, $96,5 \times 153,5$ cm.

Provenance: Jan Reynst; Gerard Reynst; Charles II (Inventory, p. 4, no. 54: *Andreas Shevone. Our Saviour before Pilat to the Waste. Dutch p^{r}sent. 3.4 × 5.2*); James II (Catalogue, ms., Windsor, no. 6; Bathoe, no. 743); Queen Anne (Inventory, Windsor, no. 7).

Bibliography: Ridolfi, I, pp. 250–51 (*Pilato, che per discolparsi del Sangue di Christo si laua le mani, e da vn seruo gli viena gettata l'acqua con vase d'oro, e gli stà innanzi il Saluatore in atto humilissimo, annodato da dure ritorte, tenuto da Soldati… Mà i due primi quadri della Vergine, e del Pilato, furono de migliori, & più celebri che si facesse, ritenendo certo che di gratia, e di finimento, non sempre accostumato dall' Autore: …nella Galeria del Signor Gio. Reinst…* [under Schiavone]); Law 1898, p. 107, no. 289; L. Fröhlich-Bum, 'Andrea Meldolla, genannt Schiavone,' *Jahrbuch der kunsthistorischen Sammlungen des allerhöchsten Kaiserhauses*, XXXI, 1913, p. 201 (reference to Ridolfi and paintings in the Accademia, Venice and at

P 27

29. The Judgment of Midas

Painting: Hampton Court Palace (No. 175; Inv.no. 470). Oil on canvas, 165 × 195 cm.

Provenance: Jan Reynst; Gerard Reynst; Charles II (Inventory, p. 11, no. 169: *Andrea Shevone. The Judgment of Minos, Dutch present. 4.11 × 6.3*); James II (Catalogue, ms., Windsor, no. 27: *Minos*; Bathoe, no. 764); Queen Anne (Inventory, Kensington, no. 66).

Exhibitions: London 1946–47, no. 208 (fictitious date 1653 for presence in Reynst collection); Amsterdam 1965, no. 6 (ill.).

Bibliography: Ridolfi, I, p. 251 (*Tre fauole d'Ouidio con figure poco men del viuo; ... ; il Giuditio di Mida con Apollo, che con gentil maniera suona la lira, e Pane il flauto* [under Schiavone]... *nella Galeria del Signor Gio. Reinst...*); Law 1898, p. 66, no. 175 (no reference to Charles II inventory); L. Fröhlich-Bum, 'Andrea Meldolla, genannt Schiavone,' *Jahrbuch der kunsthistorischen Sammlungen des allerhöchsten Kaiserhauses*, XXXI, 1913, pp. 185–87, 216, fig. 47 (early; possibly identical with painting mentioned by Ridolfi, *op. cit.*, p. 257, as in collection Walter van der Voort, Venice); Baker 1929, p. 136, no. 175, pl. XXVII (reference to Reynst collection and Dutch Present); A. Venturi, *Storia dell' Arte Italiana*, IX, part 4, Milan, 1929, pp. 709–10, fig. 505; Lugt 1936, p. 101, fig. 25; R. Bishop, Hampton Court), p. 216; Baker 1929, pp. 136–37, no. 289; Lugt 1936, p. 101, fig. 26; Mahon 1950, p. 13, no. 2 (corrects Baker's unlikely reference to Commonwealth sale; Bernard Berenson, *Italian Pictures of the Renaissance, Venetian School*, I, London, 1957, p. 160, no. 289 (as Schiavone); Savini-Branca 1965, p. 269.

Plate: P 28.

The painting is specifically listed as *Dutch present* in the inventory of Charles II and thus may be identified with the picture listed by Ridolfi as in the *Galeria* of Jan Reynst, although he leaves open whether it was still in Venice or already in Amsterdam. It was not engraved for the CAELATURAE.

P 29

Paintings of the Royal Collections, London/Bombay/Sidney, 1937, p. 163 (as school of Schiavone); Mahon 1950, p. 13, no. 6 (suggests that Schiavone's painting was already in Amsterdam at time of Ridolfi's description, since there are discrepancies with painting at Hampton Court); Bernard Berenson, *Italian Pictures of the Renaissance, Venetian School*, I, London, 1957, p. 160, no. 175, pl. 1170 (Schiavone); C. Gould, *National Gallery Catalogues, The Sixteenth Century Venetian School*, London, 1959, under no. 1884, p. 75, note 1 (possibly from Reynst collection); Savini-Branca 1965, p. 269.

Plate: P 29.

The identification of this painting with the one described by Ridolfi as in Jan Reynst's collection is not absolutely certain (Apoll is not playing a lyre, Pan is not playing a flute, and Minerva is not mentioned specifically). Charles II's inventory, on the other hand, specifies *Dutch present*. Mahon (loc.cit.) suggested that Ridolfi's inaccuracies were caused by the fact that the painting already was in one of Jan's houses in Amsterdam. It was not engraved for the CAELATURAE. Unless further evidence is found to the contrary, the Hampton Court *Judgment of Midas* should be left among the group of paintings which formerly belonged to the Reynst collection.

BERNARDO STROZZI
(1581–1644)

30. *The Old Courtesan*

Engraving: Jeremias Falck (Stimmel, no. 8, by Cornelis Visscher after Liss; Hagen, no. 90, with reference to Reynst collection, after Liss; Wussin, p. 272, no. 8, and p. 275, IV, no. 13, by Falck; Block, no. 156, with reference to Reynst collection, after Liss; Wurzbach, no. 13, with reference to Reynst collection; Hollstein, VI, p. 214, no. 156 and XI, p. 148, no. 6; *idem*, XI, Theodor Matham, p. 253, no. 42).
370 × 304 mm. ($14\frac{9}{16} \times 12$ in.).
CAELATURAE 8.
Copper plate in 1679 with Clement de Jonghe.

Painting: Pushkin Museum, Moscow.
Oil on canvas, 135 × 109 cm.

Provenance: Gerard Reynst; collection Giraud (confiscated during revolution); Museum Rumianzov, Moscow (1924).

Bibliography: Rudolf Oldenbourg, 'Jan Lys,' *Jahrbuch der königlich preuszischen Kunstsammlungen*, XXXV, 1914, pp. 162, 166, fig. 17 (as Liss); R. Longhi, review of Oldenbourg, *L'Arte*, XX, 1917, p. 303 (as Strozzi); J. Zarnowski, 'Zwei unbekannte Werke Jan Lys' in Russland,' *Belvedere*, VIII, 1925, pp. 93–94, figs. 2, 3 (formerly with Reynst; by Liss, from Giraud collection; reference to B. Wipper, 'Ueber genuesische Bilder in Moskau,' *Wissenschaftliche Nachrichten*, 1922, no. 2 [in Russian] where the painting is attributed to Strozzi); Victor Lasareff, 'Beiträge zu Bernardo Strozzi,' *Münchner Jahrbuch der bildenden Kunst*, N.F., VI, 1929, pp. 19–21, fig. 3 (formerly with Reynst; Strozzi; 1620's); R.A. Peltzer, 'Johann Liss,' *Thieme-Becker*, XXIII, Leipzig, 1929, p. 286 (Liss?); J. G. van Gelder, 'Domenico Fettis Vision des Hl. Petrus,' *Die graphischen Künste*, II, 1937, p. 93 (Strozzi); Kurt Steinbart, *Johann Liss*, Berlin, 1940, pp. 178–79 (in ca. 1650 with Gerard Reynst, probably bought by Jan in Venice; Strozzi, Venetian period); Anna Maria Matteucci, 'L'Attività Veneziana di Bernardo Strozzi,' *Arte Veneta*, IX, 1955, p. 140 (Genoese period; publishes version in collection Modiano, Bologna, fig. 156): Luisa Mortari, 'Su Bernardo Strozzi,' *Bolletino d'Arte*, XL, 1955, pp. 321, 327, 331 (formerly with Reynst); *Catalogue Moscow, Museum Pushkin*, Moscow, 1961, p. 176, no. 221; Luisa Mortari, *Bernardo Strozzi*, Rome, 1966, pp. 59–60, fig. 285 (another version of Moscow painting in Modiano collection, Bologna which is better); C. Donzelli and G.M. Pilo, *I Pittori del Seicento Veneto*, Florence (1967), pp. 384, 387 (Strozzi).

Plates: P 30, P 30a.

Zarnowski was the first to associate Strozzi's painting in Moskow with Falck's engraving in the CAELATURAE. On the basis of the close correspondence between the two he concluded that this was the painting formerly with Reynst. This provenance of the Moscow painting has been accepted ever since. In the meantime another, apparently better version has been located in a private collection in Bologna (Matteucci, *loc. cit.*). Both painted versions correspond so closely to Falck's engraving, however that it is impossible to identify the prototype. (The painting formerly with Reynst cannot be identical with the one mentioned as being in the house of Vincenzo Imperiale: *1 Vecchia che si specchia del Prete palmi 5 altezza palmi 7 larghezza* [letter of October 22, 1661, to Charles II; A. Luzio, *La Galleria dei Gonzaga...*, Milan, 1913, p. 307], because the measurements do not correspond and the painting, furthermore, probably was in Amsterdam by 1646).

TINTORETTO (JACOPO ROBUSTI) (1518–1594)

31. *Pietà*

Etching: Cornelis Visscher (Hecquet, no. 4, after Tintoretto, with reference to Reynst collection; Stimmel, no. 16, without name of painter and etcher; Wussin, no. 105, with reference to Reynst collection, and p. 277, no. 3; Wurzbach, no. 105, with reference to Reynst collection).
388 × 280 mm. ($15\frac{5}{16}$ × 11 in.).
CAELATURAE 14 (alternate)

Painting: Lost.

Provenance: Possibly Gerard Reynst.

Bibliography: Lugt 1936, p. 117, under note 40 (does not know this etching, believes it to be same as engraving after Bassano's painting, see cat. no. 5).

Plate: P 31.

Hecquet, probably based on the CAELATURAE in Paris, gives the earliest reference to the Reynst collection. Visscher's etching, however, is only rarely included in the CAELATURAE; in its place one finds more often Falck's engraving of an *Entombment* after a lost painting by Bassano (see cat. no. 5). Both prints represent the same subject and therefore are not easily distinguishable.
While it seems likely that a version of Bassano's *Entombment* was in the Reynst collection, we can be less certain about Tintoretto's *Pietà*. For the time being, however, it is retained in the group of paintings formerly with Reynst.

148 ITALIAN PAINTINGS

P 30

P 31

ITALIAN PAINTINGS

P 30a

32. Portrait of a Dominican Friar

Engraving: Cornelis van Dalen II (Stimmel, no. 4, as *Portrait of Sebastiano del Piombo, Dominican Friar*, after Titian or Tintoretto; Stimmel's confusing title is taken over in all future references; Wussin, p. 272, no. 4, as *Portrait of Sebastiano del Piombo*, after Titian, and p. 275, III, no. 3; Wurzbach, no. 4, with reference to Reynst collection; Hollstein, v, p. 107, no. III, as *Sebastiano del Piombo* after Tintoretto or Titian).
398×290 mm. ($15\frac{11}{16} \times 11\frac{7}{16}$ in.)
CAELATURAE 4.

Painting: Hampton Court Palace (No. 78; Inv.no. 772).
Oil on canvas, 75×58 cm.

Provenance: Gerard Reynst; Charles II (Inventory, p. 8, no. 103: *Tintoretto. A Church-man without hands. A Dutch Present. 2.5 × 2.0*); not securely identifiable in catalogue of James' II collection; Queen Anne (Inventory, Kensington, no. 153: *A Dominican ffrier in his habit* [by] *Tintoret*).

Exhibitions: London 1946–47, no. 203 (as Tintoretto, with incorrect date 1653 for engraving); Amsterdam 1965, no. 7 (ill.; as Tintoretto).

Bibliography: Law 1898, p. 28, no. 78 (characteristic example of Tintoretto); F.P.B. Osmaston, *The Art and Genius of Tintoret*, II, London, 1915, p. 180; Jacobs 1925, p. 31; Baker 1929, p. 154, no. 78 (as Venetian school); Lugt 1936, p. 101, fig. 23 (as Venetian school); E. von der Bercken, *Die Gemälde des Jacopo Tintoretto*, Munich, 1942, pp. 52 and 111, no. 138 (1540's); Mahon 1950, p. 13, no. 3; Bernard Berenson, *Italian Pictures of the Renaissance, Venetian School*, I, London, 1957, p. 173 (as Tintoretto); C. Bernari and P. de Vecchi, *L'Opera completa del Tintoretto*, Milan, 1970, p. 134, no. F 18, (ill.; attributed to Tintoretto).

Plates: P 32, P 32a.

TITIAN (TIZIANO VECELLIO)
(ca. 1480–1576)

33. Portrait of a Man
(so-called *Pietro Aretino*)

Engraving: Cornelis van Dalen II (Stimmel, no. 5, supposedly *Aretino*; Wussin, p. 272, no. 5 as *Aretino*, after Titian, but more likely after Tintoretto, and p. 275, III, no. 4; Wurzbach, no. 1, with reference to Reynst collection; Hollstein, v, p. 107, no. 108, with reference to Reynst collection).
377×278 mm. ($14\frac{7}{8} \times 10\frac{15}{16}$ in.).
CAELATURAE 5.

Painting: Lost.

Provenance: Gerard Reynst; William III (probably sale Amsterdam, Het Loo, 26 July 1713, no. 20; bought for 160 guilders by Mourik); Nicolaas Antoni Flinck (sale Rotterdam, 4 November 1754, no. 2:
Het Pourtrait van Pietro Aretino, door Titiaan, waar in men met een opslag

P 32

ITALIAN PAINTINGS

P 32a

P 33

van een OOg beschouwen kan de groote smaak van dien Meester: hebbende dit eertyds berust in 't Kabinet van de Heer Reynst, en naderhand in 't Kabinet van Koning Willem de derde, zynde hoog 2 voet 10 duim, breet 2 voet 2 duim. 71–0 [Hoet, III, p. 101]).

Bibliography: Lugt 1936, p. 117, note 40 (undoubtedly by Titian), and p. 132, note 71 (first reference to Flinck sale and collection of William III); Mahon 1950, p. 17 no. III (Quotes Coke's annotation in CAELATURAE in the Victoria and Albert Museum; leaves possibility open that painting never left Holland).

Plate: P 33.

According to a handwritten note by Thomas Coke, Vice-Chamberlain of H.M. Household from 1706–1727, under Van Dalen's engraving in the copy of the CAELATURAE in the Victoria and Albert Museum (Plate 11), the painting was then at Kensington and attributed to Paris Bordone. Since the picture cannot be identified securely in the royal inventories, it seems doubtful that it was part of the 'Dutch Gift'. Its appearance in the Flinck sale with the reference that it belonged to the collection of William III leaves the possibility open that it never left Holland.

34. *Portrait of a Man* (so-called *Sannazaro*)

Engraving: Cornelis van Dalen II (Stimmel, no. 2, *Boccaccio*; engraving sometimes given to Vorsterman; Wussin, p. 272, no. 2 and p. 274, III, no. 1; Wurzbach, no. 3, with reference to Reynst collection; Hollstein, V, p. 107, no. 110, *Boccaccio*).
378 × 280 mm. ($14\frac{15}{16} \times 11$ in.).
CAELATURAE 2.
Copper plate in 1679 with Clement de Jonghe.

Painting: Hampton Court Palace (No. 149; Inv. no. 68).
Oil on canvas, 84 × 71 cm.

Provenance: Gerard Reynst; Charles II (Inventory, p. 2, no. 21: *Titian. A man in black without a band having a booke in his hand with one finger in it. Dutch p^rsent. 2.8 × 2.4.*); James II (Catalogue, ms., Whitehall, no. 441, as Giorgione; Bathoe, no. 440); probably Queen Anne (Inventory, Kensington, no. 61: *A mans head in black with a Book in his right Hand*, as by Giorgione; see Mahon, *loc.cit.*).

Exhibitions: London 1946–47, no. 199 (ill.; with wrong provenance, incorrect reference to engravings and fictitious date); London 1960, no. 20.

Bibliography: G.F. Waagen, *Treasures of Art in Great Britain*, II, London, 1854, p. 357, no. 79 (refers to incorrect identification as *Alexander de' Medici*); J.A. Crowe and G.B. Cavalcaselle, *Life and Times of Titian*, II, London, 1877, p. 465; Law 1898, pp. 57–58, no. 149 (ill.; ca. 1513–15; with incorrect statements that the painting was

P 34

engraved in 1653 by C. Visscher when in Reynst collection); G. Gronau, *Titian*, London, 1904, pp. 43, 279 (ill.; ca. 1511; with incorrect provenance); O. Fischel, *Tizian*, Stuttgart/Leipzig, 1924, p. 23 (ill.; ca. 1511); Jacobs 1925, p. 31; L. Cust, in *Apollo*, VII, 1928, p. 49; Baker 1929, pp. 143–44, no. 149, pl. XXX (as *Jacopo Sannazaro*); W. Suida, *Le Titien*, Paris, 1935, pp. 33, 157, pl. XXIX, b; Lugt 1936, p. 100, fig. 18; R. Bishop, *Paintings of the Royal Collections*, London/Bombay/Sidney, 1937, p. 167 (ill.); Mahon 1950, pp. 12–13, no. 1 (corrects Law's and Jacobs' wrong identification of this painting with the one in James' II collection, Bathoe, no. 134, which, according to Mahon, is by the school of Tintoretto, at Windsor; also corrects Law's fictitious date of 1653 for the engraving; shows that Crowe and Cavalcaselle wrongly stated that the painting was engraved by Peter de Jode, and published by Bonenfant as *Giovanni Boccaccio*); J. Lauts, in *Venezia e Europa, Atti del XVIII congresso internazionale di Storia dell'arte*, Venice, 1956, p. 79 (as having been known to Rembrandt in Amsterdam about 1640); Bernard Berenson, *Italian Pictures of the Renaissance, Venetian School*, I, London, 1957, p. 188 (early Titian, Sannazaro?); C. Cagli and F. Valcanover, *L'Opera completa di Tiziano*, Milan, 1969, no. 94, fig. 94 (1518–20); R. Pallucchini, *Tiziano*, I, Florence, 1969, p. 254, figs. 134–35 (ca. 1518–20); H.E. Wethey, *The Paintings of Titian*, II, *The Portraits*, London, 1971, no. 93, pl. 19 (as so-called *Sannazaro*; ca. 1512).

Plates: P 34, P 34a.

The painting was not mentioned by Ridolfi when he described the Titians in the Reynst collection which might indicate that by then it already was in Amsterdam. Lauts' reference that Rembrandt saw the painting about 1640 in Amsterdam is unfounded. Mahon (*loc.cit.*) correctly points out that Law's statement that the engraving was made in 1653, when the painting was in the Reynst collection, is highly conjectural. (Baker changed it to the even more unlikely year 1633).

TITIAN (ATTRIBUTED TO)

35. *Holy Family with St. John and St. Elizabeth*

Engraving and Etching: Cornelis Visscher (Hecquet, no. 8, with reference to Reynst collection; Stimmel, no. 22, painting attributed to Palma; Wussin, no. 103, with reference to Reynst collection, supposedly after Palma; engraving sometimes attributed to Theodor Matham, but certainly by Visscher; *idem*, p. 273, no. 22 and p. 277, IX, no. 6; Wurzbach, no. 103, with reference to Reynst collection, after Palma).
388 × 388 mm. (14½ × 15 1/16 in.).
CAELATURAE 25.

Painting: Lost.

Provenance: Perhaps Andrea Odoni, Venice (*El quadro della nostra donna nel paese, cun el Christo fanziuollo et S. Giovan fanziuollo, et S[anta]... fu de mano de Titiano*); Jan Reynst; Gerard Reynst; Charles II (Inventory, p. 11, no. 166: *Titiano. Our Saio^r with his feete on a Cusheon the B: Virgin St John and St Elizabeth. Dutch p^{r}sent.* 4.0 × 5.3); James II (Catalogue, ms., f. 70v, no. 48 of paintings in the custody of the Dowager Queen Catherine of Braganza: *Titian. Our Lady, and Christ St John, and Elizabeth one of the Dutch Presents*; Bathoe, no. 731).

Bibliography: Theodor Frimmel, *Der Anonimo Morelliano (Marcanton Michiel's 'Notizia d'opere del Disegno'* [1532]), *Quellenschriften für Kunstgeschichte und Kunsttechnik*, N.F. I, Vienna, 1896, p. 84 (as seen in 1532 in house of Odoni; Gould [see below] suggests identification of this painting with the one later with Reynst); G.F. Waagen, *Treasures of Art in Great Britain*, II, London, 1854, p. 479 (with incorrect provenance from Charles I's collection); Jacobs 1925, p. 31; George Vertue, 'Inventory of Charles I collection 1687–88,' *Walpole Society*, XXIV, 1935–36, p. 93 (as Titian, one of Dutch presents, incorrectly referred to as bought at sale of Charles I's collection); Mahon 1950, p. 15, no. 16 (since no longer traceable in subsequent royal inventories, the painting may have been disposed of after Queen Catherine's death or removed by her to Portugal); C. Gould, *National Gallery Catalogues, The Sixteenth-Century Venetian School*, London, 1959, pp. 111–12, under no. 635 (with reference to Odoni and Reynst collections).

Plate: P 35.

ITALIAN PAINTINGS

TITIAN (SCHOOL OF)

36. *The Virgin and Child with Tobias and the Angel*

Engraving and Etching: Cornelis Visscher (Hecquet, no. 6, with reference to Reynst collection; Stimmel, no. 21; Wussin, no. 101, with reference to Reynst collection; *idem*, p. 273, no. 21 and p. 277, IX, no. 5; Wurzbach, no. 101, with reference to Reynst collection. 280 × 383 mm. (11 × 15$\frac{1}{8}$ in.).
CAELATURAE 23.

Painting: Hampton Court Palace (No. 112; Inv.no. 465).
Oil on panel, 81,2 × 143 cm.

Provenance: Jan Reynst; Gerard Reynst; Charles II (Inventory, p. 46, no. 532: *Titian. The best Madonna wth a Tobias in it. Dutch p^{r}sent. 2.9 × 3.9*); James II (Catalogue, ms., Whitehall, no. 432; Bathoe, no. 431); Queen Anne (Inventory, Kensington, no. 29).

Exhibition: Amsterdam 1965, no. 5 (ill.; as Scarsellino).

Bibliography: Ridolfi, I, p. 201 (*Il Signor Giouanni Reinst Gentilhuomo Olandese, hà medesimamente di questa eccellente mano* [Titian]...; & *altra tela con Maria Vergine vicina ad vn fiorito rosaio, che contempla il figliuolino posto tra le molli herbette, e lungi è l'Angelo Raffaello in camino, vna delle singolari fatiche di Titiano*); P.J. Mariette, in *Abecedario*, ed. *Archives de l'Art français*, X, Paris, 1858–59, p. 307 (*une des plus belles pièces qui ayent été exécutées d'après le Titien*, with reference to Reynst collection, presently at Kensington; with the arms of the La Tour family, painted for a member of the Tassi family); William Hazlitt, *Criticisms on Art*, London, 1843, p. XXII, no. 466 (as Titian); G.F. Waagen, *Treasures of Art in Great Britain*, II, London, 1854, p. 484 (as Titian, incorrectly listed as from Charles I's collection); J.A. Crowe and G.B. Cavalcaselle, *Life and Times of Titian*, II, London, 1877, p. 464 (as Sante Zago); Leupe, in *De Nederlandsche Spectator*, 1878, p. 83 (singled out in letter of 16/26 November 1660, as having pleased Charles II especially); Law 1898, p. 43, no. 112 (ill.; School of Titian; reference to Reynst collection and 'Dutch Gift'); Jacobs 1925, p. 30; L. Cust, in *Apollo*, VII, 1928, p. 50 (ill.; with incorrect provenance from Charles I's collection); Baker 1929, pp. 146–47, no. 112 (School of Titian; coat of arms of the Torriani family); Lugt 1936, p. 111, fig. 40 (school of Titian); Mahon 1950, p. 14, no. 8; C. Gould, *National Gallery Catalogues, The Sixteenth-Century Venetian School*, London, 1959, p. 112, under no. 635 (as other version of painting in the National Gallery); Savini-Branca 1965, p. 270; H.E. Wethey, *The Paintings of Titian*, I, *The Religious Paintings*, London, 1969, p. 105, under no. 59.

Plates: P 36, 36a.

P 35

P 36

P 36a

ITALIAN PAINTINGS

Ridolfi does not specify whether the painting, at the time of his writing, was in Jan Reynst's collection in Venice or in Amsterdam. His inexact description of the background figures as 'the archangel Raphael' led Mahon (*loc.cit.*) to suggest that the painting possibly might have been in Amsterdam. (The painting was also put into print by J. Chr. Le Blon; Hollstein, II, p. 106, no. 15, ill.).

PAOLO VERONESE (PAOLO CALIARI) (1528–1588)

37. *The Resurrection of Christ*

Engraving and Etching: Cornelis Visscher (Hecquet, no. 5, with reference to Reynst collection; Stimmel, no. 14; Wussin, no. 106, reference to Reynst collection; *idem*, p. 273, no. 14 and p. 277, no. 2; Wurzbach, no. 106, with reference to Reynst collection). 405 × 309 mm. (16 × $12\frac{3}{16}$ in.).
CAELATURAE 13.
Copper plate in 1679 with Clement de Jonghe.

Painting: Presumably lost.

Provenance: Jan Reynst; Gerard Reynst; Van de Amory, Amsterdam (sale Amsterdam, 23 June 1722, no. 40: *De Hemelvaert Christi door Paulo Veronese, uit het Cabinet van Rynst, door Cornelis Visser in Prent, h. 1 en een half v., br. 1 v. 2 d. 40.–0* [Hoet, I, p. 262]).

Bibliography: Ridolfi, I, p. 340 (*Ammiransi nella Galeria del Signor Giouanni Reinst in Venetia, di cui altroue habbiamo fauellato...; il Saluatore risorto in picciolo quadro inuolto in vn panno lino, à cui fan corona molti spiritelli volanti, con vn breue sostenuto da alcuni di loro con l'ali iscritto: Ego, & pater vnum sumus* [by Veronese]; Knorr 1663 (Fuchs p. 240, refers to a *Transfiguratio Christi* and a *Salvator Mundi*; either painting might correspond to this picture); Savini-Branca 1965, p. 270.

Plate: P 37.

38. *The Marriage of St. Catherine*

Engraving: Theodor Matham (Stimmel, no. 23, by Jacob Matham; Wussin, p. 273, no. 23 and p. 276, VII, no. 2; Wurzbach, no. 10, with reference to Reynst collection. Hollstein, XI, p. 252, no. 21, with reference to Reynst collection). 292 × 416 mm. ($11\frac{1}{2}$ × $16\frac{7}{16}$ in.).
CAELATURAE 24.

P 37

158 ITALIAN PAINTINGS

P 38

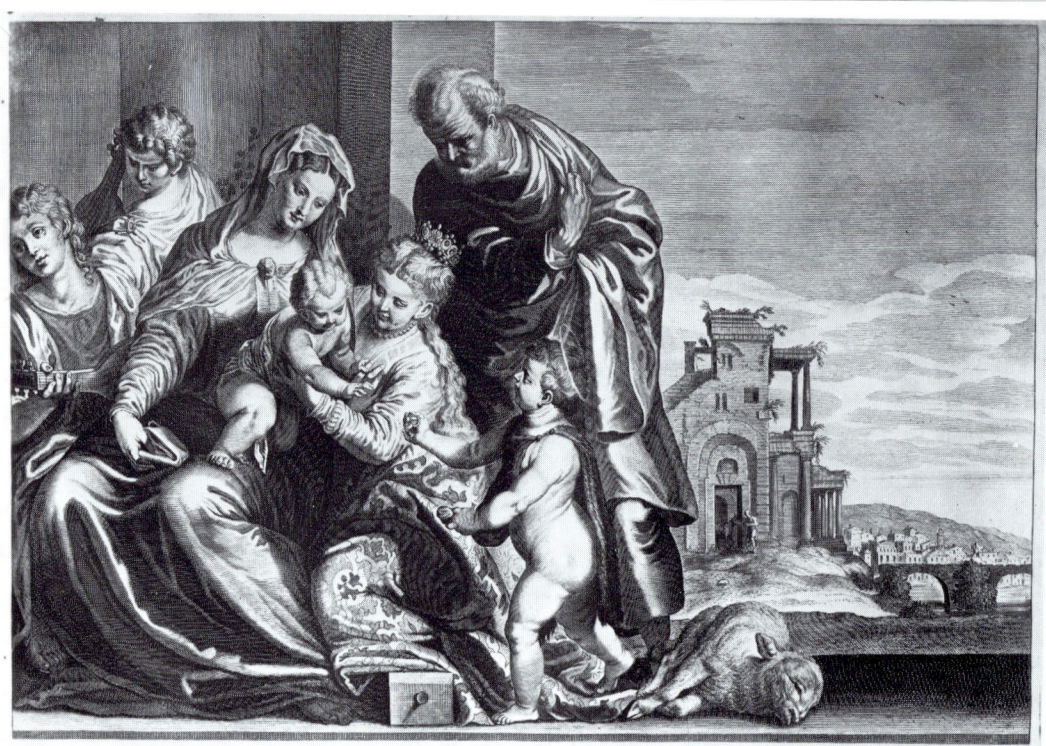

Painting: Hampton Court Palace (No. 178; Inv.no. 96).
Oil on canvas, 147,5 × 198 cm.

Provenance: Jan Reynst; Gerard Reynst; Charles II (Inventory, p. 11, no. 165: *Poleveronese. Our Saviour, the B: Virgin St John St Katherine with a Lambe and other figurs by. Dutch p^{r}sent. 4.9 × 6.5*); not in James II catalogue; Queen Anne (Inventory, Hampton Court, no. 27).

Exhibitions: London 1946–47, no. 194 (ill.; reference to Reynst collection); Amsterdam 1965, no. 8 (ill.).

Bibliography: Ridolfi, 1, p. 340 (*Mandò questi ancora alle sue case in Amsterdamo di tanto Autore [Veronese] vn quadro con Santa Caterina, che si sposa à Christo, con Angeli lietissimi, che festeggiano le reali nozze col suono de' leuti, e vi si mirano casamenti lontani, che fù delle opere più pregiate di Paolo*); Joachim von Sandrart, *Teutsche Academie*, Nuremberg, 1675, ed. R.A. Peltzer, Munich, 1925, p. 417 (*die Vermählung Christi mit S. Catharina von Verone*, as from Reynst collection); Law 1898, p. 66, no. 178 (as Veronese?); Jacobs 1925, p. 31 (identical with Bathoe, no. 1154; see Mahon, below); Baker 1929, pp. 155–56, no. 178, pl. XXXII; Lugt 1936, p. 102, fig. 33; Mahon 1950, p. 13, no. 4 (rightly states that engraving probably was done by Theodor Matham, since Jacob Matham died in 1631; corrects Jacobs' mistaken identification of this painting with Bathoe, no. 1154); L. Vertova, *Veronese*, Milan/Florence, 1953 (ill.; ca. 1560); Bernard Berenson, *Italian Pictures of the Renaissance, Venetian School*, I, London, 1957, p. 132, pl. 1065 (as Veronese); Savini-Branca 1965, p. 270; G. Piovene and R. Marini, *L'Opera completa del Veronese*, Milan, 1968, p. 105, no. 89, fig. 89 (ca. 1563).

Plates: P 38, P 38a.

In this instance Ridolfi specifically states that the painting was in Jan Reynst's Amsterdam collection which indicates that works of art, although apparently bought in Venice, were in part transferred to Holland previous to Jan's death in 1646.

ITALIAN PAINTINGS

UNIDENTIFIED ARTIST

39. *Christ and the Virgin*

Painting: Lost.

Provenance: Gerard Reynst; Charles II (Inventory, p. 46, between nos. 546 and 547: *O^r Savio^r & y^e B Virgin. Dutch p^{r}sent*).

Bibliography: Mahon 1950, p. 16, no. 19 (untraceable in royal collection).

P 38a

PIETER VAN LAER
(ca. 1592–1642)

40. *The Ambush*

Etching and engraving: Cornelis Visscher (Hecquet, no. 18, with reference to Reynst collection; Stimmel, no. 30; Wussin, no. 169, with reference to Reynst collection; *idem*, p. 273, no. 30 and p. 277, IX, no. 10; Wurzbach, no. 169, with reference to Reynst collection; Hollstein, X, p. 10, no. 17).
408 × 492 mm. ($16\frac{1}{16} \times 19\frac{3}{8}$ in.).
CAELATURAE 19.
Copper plate in 1679 with Clement de Jonghe.

Painting: Banca Sannitica, Naples. Oil on canvas, 74 × 98 cm.

Provenance: Gerard Reynst; perhaps with Adriaen Paets (sale Rotterdam, 26 April 1713: *Een zeer capitael en schoon stuk, van Pieter van Laer, alias Bamboots, zynde een Bataille...* [Hoet, I, p. 156, no. 27]); Van Diemen, Amsterdam (until 1 September 1875, when it came into the Mansi collection through marriage); Marchese Mansi, Lucca.

Bibliography: Knorr 1663 (Fuchs, pp. 239–40, mentions an *Iter mercatorum* which might refer to this painting); E.W. Moes, 'Een te weinig opgemerkte bron voor het leven van Pieter van Laer,' *Oud Holland*, XII, 1894, p. 104 (reference to Van Laer's paintings in Reynst collection); *idem*, 'De Hollandsche afkomst der collectie Mansi te Lucca,' *Oud Holland*, XXVI, 1908, p. 67; G. J. Hoogewerff, in *Thieme-Becker*, XXII, Leipzig, 1928, pp. 196–97 (with references to Reynst, Van Diemen and Mansi collections; main work from his later Dutch period); *idem*, 'Pieter van Laer en zijn vrienden,' *Oud Holland*, L, 1933, pp. 116–17 (identifies painting with the *Bataill* in the collection of Adriaen Paets, sold in 1713 to Van Diemen, an opinion only hesitantly accepted by Janeck [see below]); G. Briganti, 'Pieter van Laer e Michelangelo Cerquozzi,' *Proporzioni*, III, 1950, p. 189, fig. 1 (as one of the two securely attributed paintings done in Italy); Axel Janeck, *Untersuchung über den holländischen Maler Pieter van Laer, genannt Bamboccio*, Diss. Würzburg, 1968, pp. 73–74, cat. A 14 (securely attributed, done in Italy, since the second state of Visscher's print is inscribed *Romae*), pp. 232–37 (ca. 1636–37).

Plates: P 40, P 40a.

Kramm (*De levens en werken...*, III, Amsterdam, 1857, p. 927) writes that he remembers having read somewhere, although he is unable to recall the source that Gerard Reynst had travelled to Italy with Pieter van Laer and died there. This certainly is untrue, since Gerard Reynst drowned in Amsterdam, in 1658. Kramm continues that the journey must have taken place before 1639, the year Van Laer returned to Amsterdam.

P 40

DUTCH PAINTINGS

According to him, the three paintings by Van Laer formerly in the possession of Gerard Reynst were painted in Italy and passed on to Reynst through the intermediary of Sandrart, sometime before 1639. These latter suppositions by Kramm have yet to be confirmed, although it apparently is true that the paintings originated in Rome, for all three engravings are inscribed in later states as *P. de Laer Pinxit Romae.*

Kramm's statement would be more appropriate for Jan Reynst, who certainly travelled to Italy and died there. Furthermore, Jan may have known Sandrart from the latter's stay in Venice in 1628–29, where he met Renieri and Liss and also made the acquaintance of Lucas van Uffelen (see *Teutsche Academie*, ed. Peltzer, pp. 25, 232).

41. *The Large Limekiln*

Etching and engraving: Cornelis Visscher (Hecquet, no. 18, with reference to Reynst collection; Stimmel, no. 28; Wussin, no. 171, reference to Reynst collection; *idem*, p. 273, no. 28 and p. 277, IX, no. 8; Wurzbach, no. 171, with reference to Reynst collection; Hollstein, X, p. 10, no. 19). 318 × 367 mm. (12½ × 14 7/16 in.). CAELATURAE 31. Copper plate in 1679 with Clement de Jonghe.

Painting: Formerly Prince of Liechtenstein, Castle Feldsberg, Bohemia (lost since after World War II).

Oil on canvas (?), ca. 95 × 132 cm.

Provenance: Gerard Reynst; perhaps Joachim von Sandrart (mentions as in his collection by *Peter van Laar / ali. Bambots: eine Tafel / darinn etliche Italianische Spitzbuben zu Rom das Spiel 'Alamore' spielen / bey einem Kalchofen / des Autoris beste Arbeit*); Johann van Tongeren (sale The Hague, 24 March 1692: *Een groote Kalckoven van Pieter van Laer, alias Bamboots* [Hoet, I, p. 12, no. 17]); Quirijn van Biesum (sale Rotterdam, 18 October 1719: *Een Kalkoven, waerin A l'amour Speelders, van denzelven* [Pieter van Laer] [Hoet, I, p. 232, no. 110]); Marinus de Jeude (sale The Hague, 18 April 1735: *Een Kapitael Stuk zeer uytmuntend en uytvoerig geschildert, door Pieter* van Laer [alias Bamboots] *verbeeldende een Italiaensche Kalkbranderye, zeer ryk geordonneert* [Hoet, I, p. 432, no. 25]); Richard Pickfatt (sale Rotterdam, 12 April 1736: *Een Kalk-Oven, waar in A l'Amour-Speelders, van denzelven* [Van Laer] [Hoet, I, p. 468, no. 37]); Heer van Zwieten (sale The Hague, 12 April 1741: *De Vermaarde en zeer bekende kalk-oven met veel bywerk, door Bamboots* [Hoet, II, p. 24, no. 168]); Magazijn-catalogus Willem Lormier, The Hague (*Pieter van Laer alias Bamboots. Een Kalk Oven, Koeyen, beelden, en bywerk* [Hoet, II, p. 420]); sale 's Gravenhage, 4 July 1763: *Een Kalk-Oven, en verders veel Figuren en bywerk, door Pieter Baambott, gezegt van Laer* [Terwesten, p. 317, no. 52]; J. Palthe (sale Leiden, 20 March 1770, no. 90: *Een schoon stuk, verbeeldende de Kalk-Ovens, door de Laar, waar van de Print uitgaat*); sale Leiden, 26 August 1788, no. 76: *Een Kalkoven in een Landschap met Koeien, Beelden, en Bywerk*.

Bibliography: Joachim van Sandrart, *Teutsche Academie*, Nuremberg, 1675, ed. R.A. Peltzer, Munich, 1925, p. 330 (reference to painting in his collection which might be identical with the one with Reynst); Arnold Houbraken, *De Groote Schouburgh der Nederlantsche Konstschilders en Schilderessen*, Amsterdam, 1718–20, ed. P.T.A. Swillens, Maastricht, 1943, I, p. 286 (mentions that among Van Laer's important paintings the so-called *Lime-Kiln* was put into print); G.J. Hoogewerff in *Thieme-Becker*, XXII, Leipzig, 1928, p. 197 (late work; pendant to the *Shot with the Pistol*, cf. cat. no. 42); idem, 'Pieter van Laer en zijn vrienden,' *Oud Holland*, XLIX, 1932, p. 9 (reference to Cornelis van den Bergh's portrait of 1765 of Pieter van Laer, copied after Sandrart's illustration for the *Academie*); idem, *Oud Holland*, L, 1933, pp. 115–16 (as Van Laer's most important painting, formerly with Reynst), p. 251 (identifies painting in Sandrart's collection with Van Laer's *Small Lime-Kiln* or the *Alla-Mora-Players*), p. 253 (pendant to the *Ambush*, cat. no. 40, Pl. P 40a); A. Blankert, *Nederlandse 17e eeuwse italianiserende Landschapschilders* (exhibition catalogue), Utrecht (Centraal Museum), 1965, under no. 37; Axel Janeck, *Untersuchung über den holländischen Maler Pieter van Laer*, Diss. Würzburg, 1968, especially pp. 32 and 54, footnote 28 (painting in Sandrart's collection probably is the *Large Limekiln*), pp. 92–95, cat. A IV 5 (extensive references to provenance and bibliography), pp. 230–32 (painted in Rome, ca. 1636–37, under influence of Andries Both).

Plate: P 41.

42. *The Shot with the Pistol*

Etching and Engraving: Cornelis Visscher (Hecquet, no. 18, with

P 41

DUTCH PAINTINGS

P 42

P 42a

reference to Reynst collection; Stimmel, no. 29; Wussin, no. 170, with reference to Reynst collection; *idem*, p. 273, no. 29 and p. 277, IX, no. 9; Wurzbach, no. 170, with reference to Reynst collection; Hollstein, X, p. 10, no. 18).
315 × 387 mm. ($12\frac{7}{16}$ × $15\frac{1}{4}$ in.).
CAELATURAE 32.
Copper plate in 1679 with Clement de Jonghe.

Painting: Hermitage, Leningrad (Inv.no. 6931).
Oil on canvas, 62,5 × 78,5 cm.

Provenance: Gerard Reynst; sale Amsterdam, 10 November 1762, no. 33: *Een capitael stuk, verbeeldend een Spelonk, of Ruine, daarin eenige vegtende Ruiters en weinig verder een koets met Paarden...*; Puschkin, Ekaterinsky Palace (until 1931).

Bibliography: Arnold Houbraken, *De Groote Schouburgh der Nederlantsche Konstschilders en Schilderessen*, Amsterdam, 1718–20, ed. P.T.A. Swillens, Maastricht, 1943, I, p. 286 (mentions among the important works a *stuik – roovery in een roots* which was put into print); G.J. Hoogewerff in *Thieme-Becker*, XXII, Leipzig, 1928, p. 197 (late work; pendant to the *Large Lime-Kiln* [?]); *idem*, 'Pieter van Laer en zijn vrienden,' *Oud Holland*, XLIX, 1932, p. 16 and *Oud Holland*, L, 1933, p. 253 (late work, 1639–42, perhaps pendant to *Bataille*; known only in Visscher's and Stopendaal's prints); *Hermitage, Catalogue of Paintings*, II, Leningrad/Moscow, 1958, p. 210, no. 6931; Axel Janeck, *Untersuchung über den holländischen Maler Pieter van Laer*, Diss. Würzburg, 1968, especially p. 50 (believes that Houbraken knew the Reynst collection), pp. 72–73, cat. A 13 (references to provenance and bibliography), and pp. 232–234 (painted in Rome, ca. 1636–37, under the influence of Andries Both).

Plates: P 42, P 42a.

PAINTINGS IN THE COLLECTION OF JAN REYNST, MENTIONED BY RIDOLFI*

GENTILE BELLINI
(1429–1507)

43. *Madonna and Saints*

Painting: Lost.

Provenance: Jan Reynst.

Bibliography: Ridolfi, I, p. 62 (*Habbiamo veduto ancora di questo Autore* [Gentile Bellini] *vna mezza figura di Maria Vergine con più Santi intorno, conforme la diuotione di coloro per quali la dipinse, dal Sig. Giouanni Reinst gentil'huomo Olandese*); Savini-Branca 1965, p. 269.

BONIFAZIO DE' PITATI, CALLED VERONESE
(1487–1553)

44. *The Adoration of the Magi*

Painting: Lost.

Provenance: Jan Reynst; Nicolò Renieri, Venice.

Bibliography: Ridolfi, I, p. 289 (*Nella Galeria del Signor Giouanni Reinst ammiransi in gran tela, la visita de' Magi, oue l'Autore, oltre l'hauer diuisata l'attione con ogni singolarità, hauui vsato vn esquisito colorito, e formata la Vergine, e que' Regi cosi maestosi, e naturali, che si può quell'opera pareggiare, a qual si sia fatica di celebre mano* [Bonifazio]); G. Martinioni, *Venetia città nobilissima*, Venice, 1663, p. 378 (as in house of Nicolò Renieri: *Di Bonifacio hà l'Adoratione de Magi con quantità di figure*); G. Campori, *Raccolta di cataloghi ed inventari inediti*, Modena, 1870, p. 445; Savini-Branca 1965, pp. 55, note 33, 265, 269.

HANS ROTTENHAMMER
(1564–1625)

45. *Madonna in Adoration of the Child and Angels*

Painting: Lost.

Provenance: Jan Reynst.

Bibliography: Ridolfi, II, p. 85 (*Tra le singolari Pitture del Signor Giouanni Reinst trouasi vna picciola Madonna, che adora il bambino con molti Angeli assistenti mirabile cosa dell'Autore* [Giovanni Rothamer]); R.A. Peltzer, 'Hans Rottenhammer,' *Jahrbuch der kunsthistorischen Sammlungen des allerhöchsten Kaiserhauses*, XXXIII, 1916, p. 350 (lost; painted in Italy, 1589–1606); Savini-Branca 1965, p. 269.

ANDREA SCHIAVONE
(1522–1563)

46. *Christ Presented to the People*

Painting: Lost.

Provenance: Jan Reynst.

Bibliography: Ridolfi, I, p. 251 (*Galeria del Signor Gio. Reinst, Ha in oltre di questa mano vn quadro di braccia quattro, oue nella sommità d'vna scala Pilato mostra il Saluatore al popolo : & à piè vi sono degli Hebrei, che gridano il Crucifige*); Savini-Branca 1965, p. 269.

47. *The Madonna and Child with St. John and St. Joseph*

Painting: Lost.

Provenance: Jan Reynst; Nicolò Renieri, Venice (sale Venice, 4 December 1666, no. G. 26: *Vn Quadro di mano d'Andrea Schiauon, con il ritratto d'vna Madonna posta à sedere, con il Figlio in braccio, con appresso S. Gioseppe, & vn S. Gio : Battista in ginocchione auanti, figure intiere, con vn bellissimo Paese con dentro vna figurina, bellissimo pezzo, le figure sono quasi al naturale, largo quarte tredici, alto 10. in circa in vn Cornisone tutto intagliato, e tutto indorato*).

Bibliography: Ridolfi, I, pp. 250–51 (*La Vergine tra le solitudini col'-fanciullino, S. Giouan Battista, e S. Giuseppe* [under Schiavone]; *Mà i due primi quadri della Vergine, e del Pilato, furono de' migliori, & più celebri che si facesse, ritenendo certo che di gratia, e di finimento, non sempre accostumato dall'Autore : parte de'*

---

*Listed alphabetically according to artists

quali si conseruano nella Galeria del Signor Gio. Reinst, gentilhuomo Olandese in Venetia, & altri nelle sue case in Amsterdamo); M. Boschini, *La Carta del Navegar Pitoresco*, Venice, 1660, p. 313 (*Ma stupenda Pitura per la mente / Me vien adesso desto Autor famoso* [Schiavone], / *Un quadro venerando, e maestoso. / Che certo el stima tal ogni intendente. / Questo è Giesù, che in brazzo de Maria / Discore co'l Parente Zambatista. / Ghè Sant' Isepo, venerando in vista. / In tun Paese, ch'è tuto alegria. / … E possessor de questo xè'l Renieri*); G. Martinioni, *Venetia città nobilissima*, Venice, 1663, p. 378 (as in house of Nicolò Renieri: *Di Andrea Schiauone vna Madonna*); G. Campori, *Raccolta di cataloghi ed inventari inediti*, Modena, 1870, p. 447; Savini-Branca 1965, pp. 55, note 33, 267, 269.

48. *The Battle Between Lapiths and Centaurs*

Painting: Lost.

Provenance: Jan Reynst; Nicolò Renieri, Venice (sale Venice, 4 December 1666, no. G. 27: *Vn Quadro compagno del sudetto di mano d'Andrea Schiauon, oue è dipinto il ratto delle donne di Lapiti fatto dalli Centauri, oue si vede vn combattimento furioso, & vn miscuglio bizzaro, con parte di quelle donne che scappano dalla tauola, & altre rapite sopra le Groppe delli Centauri, tutte figure intiere con Cornice compagna al sopradetto*).

Bibliography: Ridolfi, I, p. 251 (*Tre fauole d'Ovidio con figure poco men del viuo ; il ratto de' Centauri delle Donne de' Lapiti, e vi apparise vn rouinoso miscuglio di donne gridanti conculcate da que Semicauali* [under Schiavone]); G. Martinioni, *Venetia città nobilissima*, Venice, 1663, p. 378 (as in house of Nicolò Renieri: *Di Andrea Schiauone…*; *Il Ratto delle donne de Lapiti fatto dalli Centauri, doue si vede un combattimento furioso, & un miscuglio bizzaro*); G. Campori, *Raccolta di cataloghi ed inventari inediti*, Modena, 1870, p. 447; Savini-Branca 1965, pp. 55, note 33, 267, 269.

49. *Perseus Liberating Andromeda*

Painting: Lost.

Provenance: Jan Reynst.

Bibliography: Ridolfi, I, p. 251 (*…nella Galeria del Signor Gio. Reinst… Tre fauole d'Ouidio con figure poco men del viuo;…; Andromeda ignuda, legata al sasso, liberata da Perseo*), Savini-Branca 1965, p. 269.

50. *Several Heads of Philosophers*

Painting: Lost.

Provenance: Jan Reynst.

Bibliography: Ridolfi, I, p. 251 (*…nella Galeri del Signor Gio. Reinst…; & alcune teste di Filosofi raramente colorite*); Savini-Branca 1965, p. 269.

LAMBERT SUSTRIS
(1515/20–ca. 1595)

51. *The Death of the Pharaoh in the Red Sea*

Painting: Lost.

Provenance: Jan Reynst; Nicolò Renieri, Venice (?).

Bibliography: Ridolfi, I, p. 226 (*…[Lamberto], il Signor Giouanni Reinst,…, hà la sommersione di Faraone*); R.A. Peltzer, 'Lambert Sustris von Amsterdam,' *Jahrbuch der kunsthistorischen Sammlungen des allerhöchsten Kaiserhauses*, XXXI, 1913, pp. 224, 245 (reference to Ridolfi and Reynst collection); Savini-Branca 1965, p. 57, note 39 and pp. 104, 106 (points to painting of same subject in the Renieri collection, there listed as by Lucas van Leyden), p. 270.

A picture representing the death of Pharaoh and his army was sold on December 4 1666 as lot G. 55 from among paintings owned by Nicolò Renieri (*Vn Quadro di mano di Luca d'Olanda oue è dipinto Moisè col Popolo Ebreo, che hà passato il Mar rosso, e nel detto Mare si vede sommergere Faraone con tutta l' Armata, con quantità di figure fatto in tela à guazzo, largo braccie 3 alto due in circa con Cornice nera à filo d'oro*). Since other paintings formerly with Jan Reynst appeared in that sale (i.e. cat. nos. 44, 47, 48, 52, 53, 57) it might well be possible that this work is identical with the *Sommersione del Faraone* which Ridolfi mentioned as with Jan Reynst (as already observed by Savini-Branca, *loc.cit.*). Martinioni does not refer to the painting.

JACOPO TINTORETTO
(JACOPO ROBUSTI)
(1518–1594)

52. *The Group Portrait of the Pellegrini Family*

Painting: Presumably lost.

Provenance: Jan Reynst; Nicolò Renieri, Venice (sale Venice, 4 December 1666, no. G. 16: *Vn Quadro di mano del Tintoretto vecchio largo quarte 39. e vn quarto, alto 13. e meza, oue sono dipinti otto ritratti grandi al naturale intieri, cioè vna famiglia nobile, li tre più vecchi, cioè due homini, & vna donna, sono à sedere vicino ad vna Tauola coperta d'vn Tapeto Persiano, & appresso sono due Gentil donne grandi in piedi, le quali stanno à vedere tre Gentil houmini, quali ritornano dalla caccia con cani, e Liureri, portando seco Lepri, & altre prede fatte, con adietro bellissimo paese, con il seruo che và alla caccia in Cornice toccata d'oro*); Lord Barrymore, England.

Bibliography: Ridolfi, II, p. 55 (*Ma è singolarissimo un lungo quadro, che si ammira nella Galeria del Signor Caualier Gio. Reinst descritto, entroui ritratti interi della famiglia Pellegrina, che siedono con alcune Matrone ad una tauola in un giardino alla cui presenza compariscono i loro figli giouinetti, tolti anch'eglino dal naturale, uenuti dalla caccia con cani à mano, e serui che han lepri in collo, che oltre la bellezza, e l'eccellenza del colorito formano vn gentile, e pellegrino componimento, nella quale medesimamente si veggono numerosi quadri con inuentioni, diuerse paesi, & altre cose, oltre le descritte, de' più celebri Pittori dell'età nostra* [under Tintoretto]); G. Martinioni, *Venetia città nobilissima*, Venice, 1663, p. 378 (as in house of Nicolò Renieri: *Vedesi del Tintoretto un gran Quadro di braccia sei in circa, nel quale sono otto ritratti, grandi al naturale intieri, d'una Famiglia Nobile, tre de più vecchi, cioè, due Gentil'huomini, &*

nva Gentildonna stanno à sedere ad vna tauola coperta di Tapetto Persiano, & appresso sono due Gentildonne giouani in piedi, le quali mirano tre gentilhuomini, che ritornano dalla Caccia con Cani, portando in seco Lepri, & altre prede. Questa è pittura singolare); G. Campori, *Raccolta di cathaloghi ed inventari inediti*, Modena, 1870, pp. 444–445; F.P.B. Osmaston, *The Art and Genius of Tintoret*, II, London, 1915, pp. 25, 181–82 (as in Collection Lord Barrymore; no reference to earlier provenances); Jacobs 1925, p. 24, note 2 (not mentioned by Ridolfi in 1642 under *vita di Tintoretto*); Savini-Branca 1965, pp. 55, note 33, 268, 270.

The painting was possibly acquired by Jan between 1642 and 1646, since it does not yet figure in Ridolfi's *Vita di Giacopo Robusti detto il Tintoretto...*, Venice, 1642.

TITIAN (TIZIANO VECELLIO)
(ca. 1480–1576)

53. *St. Francis in Adoration of the Cross*

Painting: Lost.

Provenance: Jan Reynst; Nicolò Renieri, Venice.

Bibliography: Ridolfi, I, p. 201 (*Il Signor Giouanni Reinst Gentil-huomo Olandese, ha medesimamente di questa eccellente mano* [i.e. Titian]...; *picciola figura di San Francesco in vn paese posto vicino d'alcuni sassi, che mira con molto affetto la Croce, che tiene in mano, e stilla dagli occhi amorose lagrime*); G. Martinioni, *Venetia città nobilissima*, Venice, 1663, p. 377 (as in house of Nicolò Renieri); G. Campori, *Raccolta di*

cataloghi ed inventari inediti, Modena, 1870, p. 443; Savini-Branca 1965, pp. 55, note 33, 57, 268, 270.

No *St. Francis* by Titian is listed in the sale of paintings from Renieri's collection on December 4, 1666, and the picture, therefore, presumably was dispersed between 1663 and 1666.

54. *Portrait of a Senator*

Painting: Lost.

Provenance: Jan Reynst.

Bibliography: Ridolfi, I, p. 201 (*Il Signor Giouanni Reinst Gentil-huomo Olandese, hà medesimamente di questa eccellente mano* [i.e. Titian] *l'effigie d'vn vecchio Senatore con veste aurata di somma bellezza:...*); Savini-Branca 1965, p. 270.

PAOLO VERONESE (PAOLO CALIARI)
(1528–1588)

55. *Sacrifice of Abraham*

Painting: Lost.

Provenance: Jan Reynst.

Bibliography: Ridolfi, I, p. 340 (*Ammiransi nella Galeria del Signor Giouanni Reinst in Venetia,* [di cui altroue habbiamo fauellato]...; *e'l sacrificio d'Abraamo* [by Veronese]); Savini-Branca 1965, p. 270.

56. *The Good Samaritan*

Painting: Not securely identifiable

Provenance: Jan Reynst.

Bibliography: Ridolfi, I, p. 340 (*Ammiransi nella Galeria del Signor Giouanni Reinst in Venetia,* [di cui *altroue habbiamo fauellato]... ; la parabola del Samaritano tipo della Christiana pietà, [senza la quale ogn'opera è vana,] che sceso dal poledro nel mezzo d'vna boscaglia gli medica le ferite, infondendogli l'oglio, & il vino, attione molto bene dispiegata per la languidezza espressa nel ferito e per lo affetto del Samaritano)*; Savini-Branca 1965, p. 270.

A painting of this subject attributed to the studio of Veronese in the Gemäldegalerie in Dresden (Inv. no. 230, canvas, 167×253 cm., bought in 1746 from the Archducal Gallery in Modena), has tentatively been identified by Hadeln (*loc.cit.*, note 2) with the picture formerly in the Reynst collection. Since Ridolfi's description is rather sketchy and since the painting was not engraved for the CAELATURAE, a secure identification is impossible unless further evidence is found.

57. *Portrait of a Man and of a Woman from the Soranza Family*

Painting: Lost.

Provenance: Jan Reynst; Nicolò Renieri, Venice.

Bibliography: Ridolfi, I, p. 340 (*Ammiransi nella Galeria del Signor Giouanni Reinst in Venetia, [di cui altroue habbiamo fauellato] due ritratti di sposi di casa Soranza* [by Veronese]); G. Martinioni, *Venetia città nobilissima*, Venice, 1663, p. 378 (as in house of Nicolò Renieri: *Del medesimo Paolo* [Veronese] *tiene ancora due altri Quadri di due ritratti Nobili di Casa Soranza, cioè, di sposo, e sposa ; La sposa sta à sedere con vn Cagnolino nelle mani, figura intiera grande al naturale; il sposo stà anch'egli à sedere in Romana con bellissima attitudine, e dietro à lui, si vede gratioso ordine di Architettura)*; G. Campori, *Raccolta di cataloghi ed inventari inediti*, Modena, 1870, p. 442; Savini-Branca 1965, pp. 55, note 33, 268, 270; G. Piovene and R. Marini, *L'Opera completa del Veronese*, Milan, 1968, Milan, 1968, p. 137 (mentioned in Ridolfi as in Reynst collection).

According to Ridolfi (1646) with Jan Reynst in Venice, by 1663 (Martinioni) in Renieri's collection, but it does not figure in the sale of Renieri's paintings on December 4, 1666, and thus presumably left the latter's collection between 1663 and 1666.

PAINTING IN THE COLLECTION OF GERARD REYNST, APPRAISED ON 12 MAY, 1672

JACOPO PALMA, THE YOUNGER

58. *Dance of Naked Children*

Painting: Jhr. Jan Six van Hillegom, Amsterdam.

Provenance: Gerard Reynst; Gerrit Uylenburgh; Jan Six (sale Amsterdam, 6 April 1702, no. 7: *Tien Danssende Kinderen, van Palma Giovene, leevens groote* ; sold to Jan Six jr. for 110.– fl.).

Bibliography: A. Bredius, 'Italiaansche schilderijen in 1672,' *Oud-Holland*, 4, 1886, p. 279 (*Item. Een dans van naeckte kindertiens, geseyt van* JACOMO PALMA, *heel naeckt en ontbloot van die qualiteyten, die in de voorsz. meester geresideert hebben, soo in couleur, teyckening, actie, ordonnantie, als dagingh, waarom niet waerdigh geacht met de naem van soo braven meester verciert te werden*); idem, Oud Holland, 34, 1916, p. 90; J. Six, 'La famosa Accademia di Eeulenborg,' *Jaarboek der Koninklijke Akademie van Wetenschappen te Amsterdam*, 1925–26, 1926, p. 240, ill.

Plate: P 58.

The only painting formerly in the collection of Gerard Reynst that is still in Holland today. Constantijn Huygens had mentioned in his letter of May 23, 1672, to von Pölnitz that these Italian paintings Uylenburgh tried to sell to the Elector of Brandenburg were part of the 'cabinet' Reynst (see pp. 91–95).

P 58

LODOVICO, AGOSTINO,
AND ANNIBALE CARRACCI

59. A large number of *drawings*, no longer identifiable today.

Provenance: Jan and Gerard Reynst.

Bibliography: C.C. Malvasia, *Felsina Pittrice*, I, Bologna, 1678, p. 467 (*Io mi atterrisco, mi confondo, quando penso solo all' infinità de' loro disegni passati per le mie mani,..., li tanti... de' Signori Reinst,...*); Savini-Branca 1965, p. 269.

PAINTINGS FORMERLY IN THE COLLECTION OF ANDREA VENDRAMIN, VENICE, PROBABLY BOUGHT EN BLOC BY GERARD REYNST (BASED ON BORENIUS)*

1. (Fol. 13):
Giovanni Bellini, *Bust of Christ* (for the type see painting in the Academia de San Fernando, Madrid; Bernard Berenson, *Italian Pictures of the Renaissance, Venetian School*, I, London, 1957, p. 32, fig. 250).

2. (Fol. 14):
Giovanni Bellini, *The Virgin and Child* (closely related to painting in the Fogg Art Museum, Cambridge, Mass.; *Borenius*, fig. 3; school piece).

3. (Fol. 15):
Giorgione, *David with the Head of Goliath* (similar to painting in Prado; mentioned by Ridolfi as in collection of Andrea Vendramin; see p. 68, note 69).

4. (Fol. 15):
Giovanni Bellini, *Portrait of a Lady*.

5. (Fol. 16):
Palma Vecchio, *Portrait of a Man*.

6. (Fol. 16):
Unidentified artist, *Portrait of a Lady* (manner of Palma Vecchio).

7. (Fol. 17):
Lambert Sustris (?), *Tower of Babel*.

8. (Fol. 18):
Rocco Marconi, *Susanna brought before Daniel*.

9. (Fol. 18):
Unidentified artist, *Portrait of a Young Lady* (manner of Bartolomeo Veneto).

10. (Fol 19):
Raphael, (alleged) *Self-Portrait* (unconvincing attribution).

11. (Fol. 19):
Giorgione, *Two Lovers* (variant of composition known in a number of versions; possibly later in Six collection; see p. 71, note 74).

12. (Fol 20):
Unidentified artist, *St. Sebastian*.

13. (Fol. 20):
Giorgione, *A Family of Fauns Preparing for a Sacrifice*.

14. (Fol. 21):
Palma Vecchio, *The Magdalen*.

15. (Fol. 21):
Giorgione, *Nymph and Faun before an Altar* (possibly companion piece to 13).

16. (Fol. 22):
Unidentified artist, *Woman with a Vase*.

17. (Fol. 22):
Giovanni Bellini, *Portrait of a Man*.

18. (Fol. 23):
Giorgione, '*La Diuitia*' (later in collection of Gerrit Uylenburgh and appraised in 1672; see p. 95).

19. (Fol. 23):
Palma Vecchio, *St. John the Evangelist*.

20. (Fol. 24):
Titian, *Ecce Homo*.

21. (Fol. 24):
Unidentified artist, *St. Sebastian*.

22. (Fol. 25):
Titian, *The Virgin and Child with St. Catherine of Alexandria and a Bishop introducing a Donor.* (probably by a follower of Titian).

23. (Fol. 25):
Unidentified artist, *The Virgin and Child*.

24. (Fol. 26):
Giorgione, *The Judgment of Paris* (probably by a follower).

25. (Fol. 26):
Giorgione, *Portrait of a Man*.

26. (Fol. 27):
Giorgione, *Landscape*.

27. (Fol. 27):
Giorgione, *Youth holding a Flute* (variant known in many versions; see *Borenius*, fig. 6).

28. (Fol. 28):
Unidentified artist, *A Concert by Moonlight* (reminiscent of Savoldo).

29. (Fol 28):
Giovanni Bellini, *Portrait of a Lady*.

30. (Fol. 29):
Palma Vecchio, *Group of Three Figures*.

31. (Fol. 29):
Giovanni Bellini, *Portrait of a Lady*.

32. (Fol. 30):
Veronese, *Ecce Homo*.

33. (Fol. 30):
Unidentified artist, *Portrait of a Woman*.

34. (Fol. 31):
Unidentified artist, *Portrait of a Man* (manner of Lotto).

35. (Fol. 31):
Unidentified artist, *Portrait of a Youth* (manner of Bellini).

36. (Fol. 32):
Giorgione, *Bust of Christ* (similar to painting in private collection; see *Borenius*, fig. 6; more likely follower of Bellini).

37. (Fol. 32):
Unidentified artist, *B. Lorenzo Giustiniani*.

38. (Fol. 33):
Unidentified artist, *Portrait of a Man* (manner of Franciabigio).

39. (Fol. 33):
Unidentified artist, *Portrait of a Young Woman*.

40. (Fol. 34):
Bellin Bellini, *The Virgin and Child*.

41. (Fol. 34):
Lucas van Leyden, *Portrait of a Man*.

42. (Fol. 35):
Palma Vecchio, *Portrait of a Lady*.

43. (Fol. 35):
Unidentified artist, *Portrait of a Man* (manner of Bellini).

---

* In order of illustrations in the manuscript of DE PICTURIS

44. (Fol. 36):
Unidentified artist, *A Courtesan* (manner of Bordone).

45. (Fol. 36):
Unidentified artist, *Portrait of a Lady*.

46. (Fol. 37):
Unidentified artist, *Portrait of a Lady* (manner of Palma Vecchio).

47. (Fol. 37):
Unidentified artist, *Portrait of a Lady* (manner of Carpaccio).

48. (Fol. 38):
Sebastiano del Piombo, *Half-length Portrait of a Youth*.

49. (Fol. 38):
Titian, *Self-Portrait* (not convincing).

50. (Fol. 39):
Unidentified artist, *Portrait of a Man* (manner of Bartolomeo Veneto).

51. (Fol. 39):
Unidentified artist, *Adam and Eve*.

52. (Fol. 40):
Pordenone, *Horse Tamer*.

53. (Fol. 41):
Unidentified artist, *Landscape with a Faun Leaning over a Nymph* (manner of Giorgione).

54. (Fol. 41):
Titian, *Portrait of the Sculptor Martino dal Sfriso*.

55. (Fol. 42):
Unidentified artist, *Satyr playing with a Hare*.

56. (Fol. 42):
Unidentified artist, *Portrait of a Lady*.

57. (Fol. 43):
Cariani, *Reclining Venus* (now at Hampton Court, formerly in Reynst collection; see cat. no. 9).

58. (Fol. 43):
Unidentified artist, *Head of a Youth* (follower of Giorgione).

59. (Fol. 44):
Pordenone, *Warrior on a Careering Horse*.

60. (Fol. 44):
Unidentified artist, *Portrait of a Man*.

61. (Fol. 45):
Palma Vecchio, *Christ* (rather Rocco Marconi).

62. (Fol. 45):
Unidentified artist, *Portrait of a Youth* (manner of Bellini).

63. (Fol. 46):
Licinio, *Portrait of a Man*.

64. (Fol. 46):
Unidentified artist, *Bust of a Young Woman*.

65. (Fol. 47):
Titian, *Portrait of a Man*.

66. (Fol. 47):
'Bellini', *Portrait of a Youth* (follower of Giovanni Bellini).

67. (Fol. 47):
Giorgione, *Portrait of a Man* (unconvincing).

68. (Fol. 48):
Unidentified artist, *Portrait of a Man* (manner of Cariani).

69. (Fol. 48):
Gentile Bellini, *Portrait of a Youth*.

70. (Fol. 48):
Unidentified artist, *Portrait of a Young Lady*.

71. (Fol. 49):
Unidentified artist, *Portrait of a Man*.

72. (Fol. 49):
Unidentified artist, *Portrait of a Youth*.

73. (Fol. 49):
Giovanni Bellini, *Portrait of a Boy*.

74. (Fol. 50):
Lucas van Leyden, *Lucretia*.

75. (Fol. 50):
Unidentified artist, *Portrait of a Man*.

76. (Fol. 51):
Unidentified artist, *Portrait of a Young Woman* (similar to Palma Vecchio's *Portrait of a Young Woman* in Berlin, Inv.no. 197 A; see *Borenius*, fig. 9; see also p. 105).

77. (Fol. 51):
Unidentified artist, *Portrait of a Young Woman*.

78. (Fol. 52):
Unidentified artist, *Portrait of a Venetian Dignitary* (Giovanni Bellini or Catena?).

79. (Fol. 52):
Giorgione, (alleged) *Self-Portrait* (unconvincing).

80. (Fol. 53):
Palma Vecchio, *Portrait of a Geometer*.

81. (Fol. 53):
Unidentified artist, *Portrait of a Man*.

82. (Fol. 54):
Vincenzo Catena, *The Virgin and Child with St. Joseph and the Infant St. John* (unconvincing; more likely follower of Titian).

83. (Fol. 54):
Giorgione, *Man in Armour*.

84. (Fol. 55):
Vittore Carpaccio, *The Virgin and Child between St. Anthony the Abbot and a Female Saint*.

85. (Fol. 55):
'Bellini', *St. Francis*.

86. (Fol. 56):
Bonifazio dei Pitati, *Portrait of a Man*.

87. (Fol. 56):
Unidentified artist, *Portrait of a Man*.

88. (Fol. 57):
Unidentified artist, *Portrait of a Youth*.

89. (Fol. 57):
Unidentified artist, *Portrait of a Man*.

90. (Fol. 58):
Paris Bordone, *Caricatured Portrait*.

91. (Fol. 58):
Unidentified artist, *Portrait of a Man*.

92. (Fol. 59):
Paris Bordone, *Youthful Roman Warrior* (see p. 105, note 16).

93. (Fol. 59);
Unidentified artist, *Portrait of a Lady* (companion piece to 95).

94. (Fol. 60):
Paris Bordone, *Half-length Portrait of a Woman* (see p. 71, note 74).

95. (Fol. 60):
Unidentified artist, *Portrait of a Man*.

96. (Fol. 61):
Unidentified artist, *Half-length Portrait of a Young Woman* (manner of Bordone).

97. (Fol. 61):
Unidentified artist, *Fauns and Nymphs Dancing* (manner of Schiavone; companion piece to 99).

98. (Fol. 62):
Licinio, *A Courtesan*.

99. (Fol. 62):
Unidentified artist, *Concert in a Landscape* (manner of Schiavone).

100. (Fol. 63):
Unidentified artist, *Portrait of a Man* (school of Palma Vecchio).

101. (Fol. 63):
Unidentified artist, *Lovers in a Landscape* (same series as 103, 105, 107).

102. (Fol. 64):
Gentile Bellini, *Portrait of a Man*.

103. (Fol. 64):
Unidentified artist, *Nude Woman* (probably *Venus*) *in a Landscape*.

104. (Fol. 65):
'Campagnola', *Portrait of a Man*.

105. (Fol. 65):
Unidentified artist, *Lot and his Daughters*.

106. (Fol. 66):
Unidentified artist, *Portrait of a Young Man* (manner of Bellini).

107. (Fol. 66):
Unidentified artist, *Concert in a Landscape*.

108. (Fol. 67):
Unidentified artist, *Portrait of a Woman* (manner of Carpaccio).

109. (Fol. 67):
Unidentified artist, *Portrait of a Young Woman in Turban* (manner of Bonifazio).

110. (Fol. 68):
Unidentified artist, *Portrait of a Youth* (manner of Licinio).

111. (Fol. 68):
Unidentified artist, *Portrait of a Man*.

112. (Fol. 69):
Unidentified artist, *Portrait of a Man* (manner of Licinio).

113. (Fol. 69):
Andrea Schiavone, *Figures Bathing*.

114. (Fol. 69):
Andrea Schiavone, *The Killing of the Calydonian Boar* (same series as 116, 117, 119, 120, 122, 123).

115. (Fol. 70):
Unidentified artist, *Faun Leaning over a Nymph*.

116. (Fol. 70):
Andrea Schiavone, *Callisto Surrounded by Other Nymphs*.

117. (Fol. 70):
Andrea Schiavone, *Apollo Flaying Marsyas*.

118. (Fol. 71):
Unidentified artist, *Deucalion and Pyrrha* (manner of Andrea Schiavone).

119. (Fol. 71):
Andrea Schiavone, *The Contest of Apollo and Marsyas* (formerly R. Neumann, Berlin; Borenius, BM, 1932, fig. 1C).

120. (Fol. 71):
Andrea Schiavone, *The Contest of Apollo and Pan*.

121. (Fol. 72):
Unidentified artist, *Portrait of a Man*.

122. (Fol. 72):
Andrea Schiavone, *The Birth of Adonis*.

123. (Fol. 72):
Andrea Schiavone, *The Judgment of Paris* (formerly R. Neumann, Berlin; Borenius, BM, 1932, fig. 1A).

124. (Fol. 73);
Unidentified artist, *Nymphs Surprised while Bathing* (manner of Tintoretto).

125. (Fol. 73):
Unidentified artist, *Subject from Classical History*.

126. (Fol. 73):
Unidentified artist, *Subject from Classical History* (companion piece to 123).

127. (Fol. 74):
Unidentified artist, *Portrait of a Woman with Eastern Headdress*.

128. (Fol. 74):
Unidentified artist, *Mythological Subject*.

129. (Fol. 74):
Unidentified artist, *Mythological Subject*.

130. (Fol. 75):
Unidentified artist, *St. John the Baptist*.

131. (Fol. 75):
Unidentified artist, *Ecce Homo*.

132. (Fol. 76):
Lorenzino, *Cupid Asleep*.

133. (Fol. 76):
Unidentified artist, *Mythological Subject*.

134. (Fol 76):
Unidentified artist, *Mythological Subject*.

135. (Fol. 77):
Unidentified artist, *The Virgin and Child with a Young Monastic Saint and a Child*.

136. (Fol. 77):
Unidentified artist, *Ariadne Abandoned by Theseus*.

137. (Fol. 77):
Unidentified artist, *Nymph and Satyr*.

138. (Fol. 78):
Tintoretto, *Portrait of a Venetian Senator* (now Winterthur, Coll. O. Reinhart; formerly probably in Six collection; see *Borenius*, fig. 11; see p. 96, note 37).

139. (Fol. 78):
Unidentified artist, *Cupid and Psyche*.

140. (Fol. 78):
Unidentified artist, *The Story of the Infant Paris?*

141. (Fol. 79):
Unidentified artist, *The Virgin and Child between SS. Catherine and Peter*.

142. (Fol. 79):
Unidentified artist, *The Virgin and Child*.

143. (Fol. 80):
Tintoretto, *Lamentation over the Dead Christ* (related to painting in Brera, Milan [Inv.no. 149]; see *Borenius*, fig. 12).

144. (Fol. 80):
Unidentified artist, *Portrait of a Youth Holding a Flute* (manner of Giorgione).

145. (Fol. 81):
Unidentified artist, *Portrait of a Young Woman* (similar to Titian's *Flora*, Uffizi, Florence; Inv.no.626).

146. (Fol. 81):
Unidentified artist, *A Courtesan*.

147. (Fol. 82):
Unidentified artist, *A Sibyl*.

148. (Fol. 82):
Unidentified artist, *Bacchus as a Child* (manner of Bellini).

149. (Fol. 83):
Unidentified artist, *Allegory of Spring*.

150. (Fol. 83):
Unidentified artist, *Allegory of Summer*.

151. (Fol. 83):
Unidentified artist, *Portrait of a Man*.

152. (Fol. 84):
Unidentified artist, *Allegory of Autumn*.

153. (Fol. 84):
Unidentified artist, *Allegory of Winter*.

154. (Fol. 84):
Unidentified artist, *Profile Portrait of a Youth*.

155. (Fol. 84):
Unidentified artist, *Profile Portrait of a Youth*.

156–202 (Fol. 85):
List of additional 47 paintings of inferior quality and therefore not reproduced and not included in Index.

203–301 (Fol. 86):
List of additional 20 portrait paintings used as decoration around the house.
65 drawings and watercolors of animals.
14 drawn copies after a series of Apostles, Christ and the Virgin Mary by Dürer.

SCULPTURES IN THE COLLECTION OF GERARD REYNST, REPRODUCED IN THE ICONES*

1. *Torso of Male Figure Draped in Himation*

Engraving: Anonymous.
327 × 196 mm.
ICONES 1: *Consul*.

Sculpture:
Rijksmuseum van Oudheden, Leiden (Inv.no. Pb. 91). White marble of fine grain. Original height of entire statue according to inventory of 1818–24: 1,56 m. Torso: 0,75 m. Head and neck: 0,28 m. Lower part of body to knee and left hand, part of right forearm and right shoulder modern. Head of different marble and too small, does not belong to statue; stuccoed parts of beard removed.

Provenance: Andrea Vendramin, Venice (see Oudendorp); Gerard Reynst; Jan Six, Amsterdam (sale Amsterdam, 6 April 1702, no. 35: *Een Consul, leevens groote*); Nicolaas Antoni Flinck, Rotterdam (see below); Gerard van Papenbroek, Amsterdam (*cat*. no. 91).

s 1a

s 1
s 1b

* Listed according to plate numbers in the standard editions of the ICONES

Bibliography: Knorr 1663 (Fuchs, p. 240: *In Conclavi secundo… Consul Romanus togatus*); Uffenbach (1710, p. 316: *Ferner ein Bild in ganzer Statur, aber nur vier Fuss hoch mit einer toga consulari ; sehr schön, der Kopf aber war neu aufgesetzt*); Oudendorp, II, no. 5 (reference to Reynst ICONES and Vendramin collection; lower part of statue of different stone); Janssen, I, no. 80 (head modern); Clarac, III, p. CCLXVI (with Gerard Reynst); Brants, no. 16, pl. IX, 16 (insignificant, probably late work); Timmers, no. 128; J.G. van Gelder, 'Jan de Bisschop's Drawings after Antique Sculpture,' *Studies in Western Art, Acts of the Twentieth International Congress of the History of Art*, III, Princeton, N.J., 1963, p. 53, figs. 9–10 (reference to Reynst collection); Brunsting, p. 189, no. 35.

Plates: s 1, s 1a, b.

2. *Torso of a Roman Military Statue*

Engraving: Anonymous. 326 × 195 mm.
ICONES 2: *T. Hostilius*.

Sculpture: Rijksmuseum van Oudheden, Leiden (Inv.no. Pb. 92). Grayish-white marble of very coarse grain.
Torso: 1,32 m. Head: 0,28 m. Arms and legs missing. Head from a different statue (cavities to attach a metal wreath ?); neck stucco.

Provenance: Andrea Vendramin, Venice (cat. 1,4: *Tullo Hostilio*); Gerard Reynst; Jan Six, Amsterdam (sale Amsterdam, 6 April 1702, no. 36: *T. Hostilius, Antiq. dito*); Gerard van Papenbroek, Amsterdam (cat.no. 92).

Bibliography: Arnoldus Buchelius, 'Res Pictoriae', 4 September 1639 '…het een was maar tot de helfte, habitu armato, was grooter als d'andere, credebatur een Tiberius Imperator…'; see G.J. Hoogewerff and J.Q. van Regteren Altena, in *Quellenstudien zur Holländischen Kunstgeschichte*, XV, 's-Gravenhage, 1928, p. 98; Knorr 1663 (Fuchs, p. 240: *In Conclavi secundo… Tullus Hostilius, pectore tenus*); Oudendorp, II, no. 6 (reference to Reynst ICONES and Vendramin collection; head too small, from different statue); Janssen, I, no. 78; Clarac, III, p. CCLXVI (with Gerard Reynst); Reinach, II, p. 587, no. 2 (ill.; with reference to Reynst ICONES); Jacobs, p. 26 and note 1 (concordance of ICONES with *De Sculpturis*); Brants, no. 25, pl. XIII, 25 (torso belongs to middle of 2nd century AD; head does not belong to statue; 3rd century AD); Timmers, nos. 129 and 151; I. Q. van Regteren Altena and P.J.J. van Thiel, *De portretgalerij van de Universiteit van Amsterdam en haar stichter Gerard van Papenbroeck 1673–1743*, Amsterdam, 1964, p. 35; Brunsting, p. 189, no. 36.

Plates: s 2, s 2a.

S 2
S 2a

3. *Statuette of Venus (Capitoline type)*

Engraving: Anonymous.
324 × 194 mm.
ICONES 3: *Venus*.

Sculpture: Rijksmuseum van Oudheden, Leiden (Inv.no. Pb. 96 [torso]; Pb. 97 [head]). Yellowish-white marble with dark blue veins. Torso: 0,56 m.; head: 0,20 m.; left arm and hand: 0,195 m; right arm and hand holding cloth: 0,29 m.; right leg and inner part of foot: 0,46 m; end part of toe: 0,075 m.; left leg: 0,34 m.; left foot (broken): 0,17 m. Head, arms, legs were removed since they did not fit the torso, but still preserved (only partly antique); head more likely of an Amazone (according to Reuvens modern); nose lost; tree and plinth no longer extant, perhaps originally added in stucco.

Provenance: Andrea Vendramin, Venice (*cat.* I, 44 B); Gerard Reynst; Jan Six, Amsterdam (sale Amsterdam, 6 April 1702, no. 9: *Een kleine Antique Venus*); Gerard van Papenbroek, Amsterdam (*cat.* nos. 96, 97).

Exhibition: *Klassieke kunst uit particulier bezit*, Leiden (Rijksmuseum van Oudheden), 1975, no. 24, figs. 3–5 (reference to Reynst collection).

Bibliography: Knorr 1663 (Fuchs, p. 240: *In Conclavi secundo... Venus pudica nuda, manum inguinibus praetendens*); Oudendorp, II, no. 10 (reference to Reynst ICONES and Vendramin collection); Janssen, I, no. 91 (arms and legs modern); Clarac, III, p. CCLXVI (with Gerard Reynst); J.J. Bernoulli, *Aphrodite*, Leipzig, 1873, p. 235, no. 62; Jacobs, p. 26, note 1 (concordance of ICONES with *De Sculpturis*) and p. 33, note 3 (ex Papenbroek); Brants, no. 9, pl. VI, 9, 9a (small replica of type of Capitoline Venus); Timmers, no. 130; *Rijksmuseum van Oudheden te Leiden, Gids voor de verzameling van griekse en romeinse beeldhouwwerken*, 's-Gravenhage, 1966, p. 17, no. 59 (quite good Roman copy); Brunsting, p. 189, no. 9.

Plates: s 3, s 3a, b.

4. *Statue of Fortune*

Engraving: Anonymous.
326 × 200 mm.
ICONES 4: *Flora*.

Sculpture: Rijksmuseum van Oudheden, Leiden (Inv.no. Pb. 90 [draped figure]; Pb. 90 a [head]; formerly Pb. 151, identified by H. Brunsting; Pb. 90 b [hands]). Grayish-white, fine grain marble. Figure, including cornucopia and tip of drapery: 1,18 m.
Head: 0,23 m; left hand: 0,165 m. Head (Venus?) and left hand belong to different statues. Feet, lower part of drapery and plinth are modern.

Provenance: Andrea Vendramin, Venice (*cat.* I, 8: *Flora*); Gerard

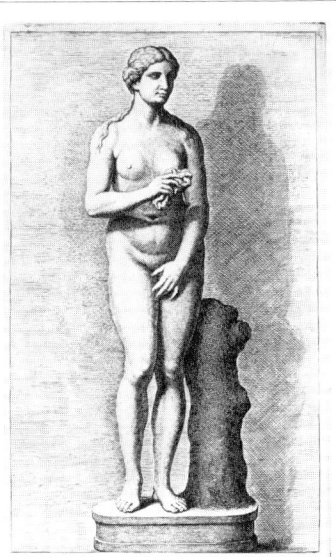

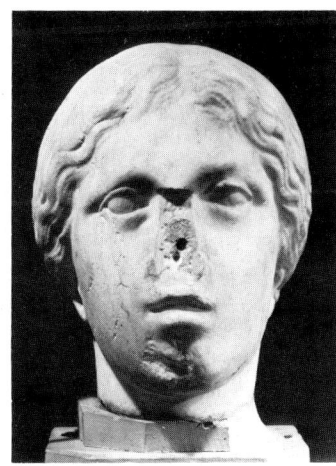

s 3
s 3a, b

S 4a Reynst; Jan Six, Amsterdam (sale Amsterdam, 4 April 1702, no. 19: *Flora, dito heel Beeld Antiq.*); Gerard van Papenbroek, Amsterdam (*cat*. no. 90).

Bibliography: Knorr 1663 (Fuchs, p. 240: *In Conclavi secundo... Flora Nuda*); Oudendorp, II, no. 4 (reference to ICONES and Vendramin collection; not *Flora*; either *Abundance* or *Good Fortune*); Janssen, I, no. 72 (as *Abundantia*, left arm antique, but from different sculpture; feet, part of drapery and plinth modern); Clarac, III, p. CCLXVI (with Gerard Reynst); Reinach, II, p. 251, no. 4 (ill.); Jacobs, p. 26 and note 1 (concordance of ICONES with *De Sculpturis*) and p. 33, note 3 (ex Papenbroek); Brants, no. 33, pl. XVII, 33 (*Standing Fortune*; poor, decorative work); Timmers, no. 131; Brunsting, p. 189, no. 19.

Plates: S 4, S 4a, b.

5. *Statuette of a Venus with a Dolphin*

Engraving: Anonymous.
331 × 202 mm.
ICONES 5.

Sculpture: Rijksmuseum van Oudheden, Leiden (Inv.no. Pb. 111). White, fine grain marble. Torso: 0,59 m. Head: 0,16 m. Legs with dolphin: 0,34 m.

Arms (antique?) do not belong to statue; shell modern; torso and head possibly also modern; head more likely an Apollo.

Provenance: Andrea Vendramin, Venice (*cat*. I, 12); Gerard Reynst; Jan Six, Amsterdam (?; sale Amsterdam, 4 April 1702, no. 20: *Een Venus, kleender dito*); Gerard van Papenbroek, Amsterdam (*cat*. no. 111).

Bibliography: Knorr 1663 (Fuchs, p. 240: *In Conclavi secundo... Venus alia*); Oudendorp, II, no. 23 (reference to Reynst ICONES and Vendramin collection; shell in left hand recently added); Clarac, III, p. CCLXVI (with Gerard Reynst); Timmers, no. 132; Brunsting, p. 189, no. 20.

Plates: S 5, S 5a, b.

Not included in Janssen (*op.cit.*) because Reuvens believed neither the torso nor the head were antique.

S 4
S 4b

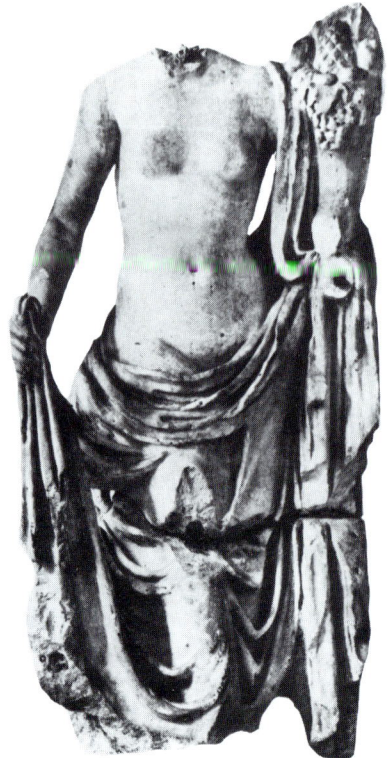

SCULPTURES IN THE COLLECTION OF GERARD REYNST

6. *Statue of a Muse*

Engraving: Anonymous.
325 × 199 mm.
ICONES 6: *Abundantia*.

Sculpture: Rijksmuseum van Oudheden, Leiden (Inv. no. Pb. 93 [torso and fragments]; Pb. 94 [head]). Yellowish-white marble of middle grain. Torso: 1,04 m (with plinth: 1,155 m.).
Head: 0,21 m. Right arm with dish: 0,48 m. left arm: 0,455 m; fragments of neck: 0,09 m; upperbody: 0,23 m. and drapery: 0,18 m; 0,15 m. 0,08 m. Upper part of body, arms and base are modern, still preserved; head (more likely a Venus or Apollo) does not belong to statue and probably is not antique.

Provenance: Andrea Vendramin, Venice (*cat.* 1,5: *Dea Abondanza*); Gerard Reynst; Jan Six, Amsterdam (sale Amsterdam, 6 April 1702, no. 18: *Abundantia Antiq., zynde een Vrouwe Beeld, by na leevens groote, ongemeen fraai*); Nicolaas Antoni Flinck, Rotterdam (?; see below); Gerard van Papenbroek, Amsterdam (*cat.* nos. 93, 94).

Bibliography: Knorr 1663 (Fuchs, p. 240: *In Conclavi secundo... Abundantia cum cornu copiae*); Uffenbach (1710, p. 316: *Hierbey war auch eine tanzende Matrone von*

S 5a
S 6a

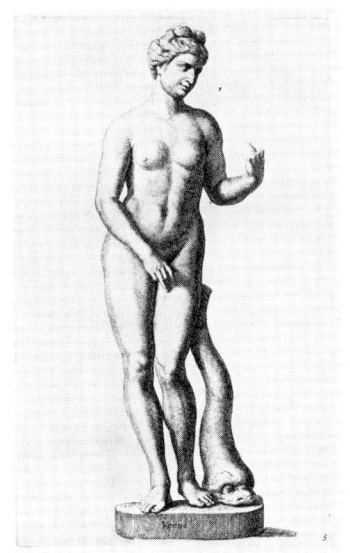  

S 5
S 5b
S 6

eben der Grösse); Oudendorp, II, no. 7 (reference to Reynst ICONES and Vendramin collection; more likely *Piety* or *Religion*) and II, nos. 59, 60; Janssen, I, no. 69 (*Muse*; upper part of body and plinth modern); Clarac, III, p. CCLXVI (with Gerard Reynst); Reinach, II, p. 655, no. 6 (ill.; *Muse* ?); Carl Watzinger, 'Das Relief des Archelaos von Priene,' *BWPr*, XXIII, Berlin, 1903, pp.9–10; Jacobs, p. 26 and note 1 (concordance of ICONES with *De Sculpturis*) and p. 33, note 3 (ex Papenbroek); Brants, no. 18, pl. IX, (*Muse* ?; school of Philiscos of Rhodos, beginning of 3rd century BC); Rudolf Horn, 'Stehende weibliche Gewandstatuen in der hellenistischen Plastik,' *Mitteilungen des deutschen archaeologischen Instituts, Roemische Abteilung*, 2. *Ergänzungsheft*, 1931, p. 85, pl. 31,3; Timmers, no. 133; I.Q. van Regteren Altena and P.J.J. van Thiel, *De portret-galerij van de Universiteit van Amsterdam en haar stichter Gerard van Papenbroeck 1673–1743*, Amsterdam, 1964, p. 35; Brunsting, p. 189, no. 18.

Plates: s 6, s 6a.

7. *Statuette of the Hercules Farnese*

Engraving: Anonymous. 327 × 198 mm.
ICONES 7: *Hercules*.

Sculpture: Rijksmuseum van Oudheden, Leiden (Inv.no. Pb. 87). Yellowish-white marble of middle grain for torso. H. 0,86 m. Head of different marble. Left arm, possibly legs up to knee, feet, club with cloth and plinth modern. Large part of right hand with apples as well as part of left hand missing.

Provenance: Andrea Vendramin, Venice (*cat.* 1, 26); Gerard Reynst; possibly Jan Six, Amsterdam (sale Amsterdam, 6 April 1702, no. 22: *Keizers Zoon als een Hercules, leevens groote, schoon Antiq.*); Gerard van Papenbroek, Amsterdam (*cat.* no. 87).

Bibliography: Knorr 1663 (Fuchs, p. 240: *In Conclavi secundo... Hercules cum exuviis leonis pectore tenus ; artificio insigni elaboratus*); Oudendorp, II, no. 1 (reference to Reynst ICONES and Vendramin collection); Janssen, I, no. 64; Clarac, III, p. CCLXVI (with Gerard Reynst); Reinach, IV, p. 127, no. 4 (ill.); Jacobs, p. 26 and note 1 (concordance of ICONES with *De Sculpturis*) and p. 33, note 3 (ex Papenbroek); Brants, no. 11, pl. VII, 11 (small copy of Hercules Farnese, Naples; late Roman work); Timmers, no. 134; I.Q. van Regteren Altena and P.J.J. van Thiel, *De portret-galerij van de Universiteit van Amsterdam en haar stichter Gerard van Papenbroek 1673–1743*, Amsterdam, 1964, p. 35; *Rijksmuseum van Oudheden te Leiden, Gids voor de verzameling van griekse en romeinse beeldhouwwerken*, 's-Gravenhage, 1966, p. 17, no. 57

s 7
s 7a

(Roman copy); Brunsting, p. 189, no. 22 (possibly identical with statue formerly in Six collection).

Plates: s 7, s 7a.

8. *Partly Draped Female Figure Standing*

Engraving: Anonymous.
327 × 199 mm.
ICONES 8: *Iulia*.

Sculpture: Lost.

Provenance: Andrea Vendramin, Venice (*cat.* I, 33); Gerard Reynst.

Bibliography: Knorr 1663 (Fuchs, p. 240: *In Conclavi secundo... Iulia pectore tenus*); Clarac, III, p. CCLXVI (with Gerard Reynst); Jacobs, p. 26, note 1 (concordance of ICONES with *De Sculpturis*); Timmers, no. 135.

Plate: s 8.

9. *Statue of Adonis*

Engraving: Anonymous.
327 × 200 mm.
ICONES 9: *Adonis*.

Sculpture: Lost.

Provenance: Andrea Vendramin, Venice (*cat.* I, 19); Gerard Reynst.

Bibliography: Knorr 1663 (Fuchs, p. 240: *In Conclavi secundo... Adonis nudus*); Clarac, III, p. CCLXVI (with Gerard Reynst); Jacobs, p. 26, note 1 (concordance of ICONES with *De Sculpturis*) and p. 33, note 3 (erroneously listed as formerly in Uilenbroek collection); Timmers, no. 136.

Plate: s 9.

10. *Statuette of Trajan*

Engraving: Anonymous.
331 × 201 mm.
ICONES 10: *Gladiator*.

Sculpture: Staatliche Museen, East Berlin (Inv.no. Sk. 355).
White marble. H. 0,815 m.
Restored are tip of nose, neck, both arms which originally were lowered, hands holding sword and sheath, penis, whole left leg, right leg from middle of thigh onward. Head, heavily cleaned, does not belong to statue.

Provenance: Andrea Vendramin, Venice (*cat.* I, 17: *Gladiatore*); Gerard Reynst; Great Elector of Brandenburg.

Bibliography: Knorr 1663 (Fuchs, p. 240: *In Conclavi secundo... Gladiator nudus dextra pugionem, sinistra vaginam tenens*); L. Beger, *Thesauri Regii et Electorialis Brandenburgici...*, III, Coloniae Marchicae (1696), pp. 341–42 (ill.; as *Trajan*); E. Gerhard, *Berlin's antike Bildwerke*, I, Berlin, 1836, p. 98, no. 158 (as *Trajan*); Clarac, III, p. CCLXVI (with Gerard Reynst); (Alexander Conze), *Königliche Museen zu Berlin, Beschreibung der antiken Skulpturen...*, Berlin, 1891, p. 144, no. 355 (*Trajan*); Reinach, II, p. 571, no. 9 (ill.; *Trajan?*); Jacobs, p. 26, note 1 (concordance

s 8
s 9

182 SCULPTURES IN THE COLLECTION OF GERARD REYNST

S 10a

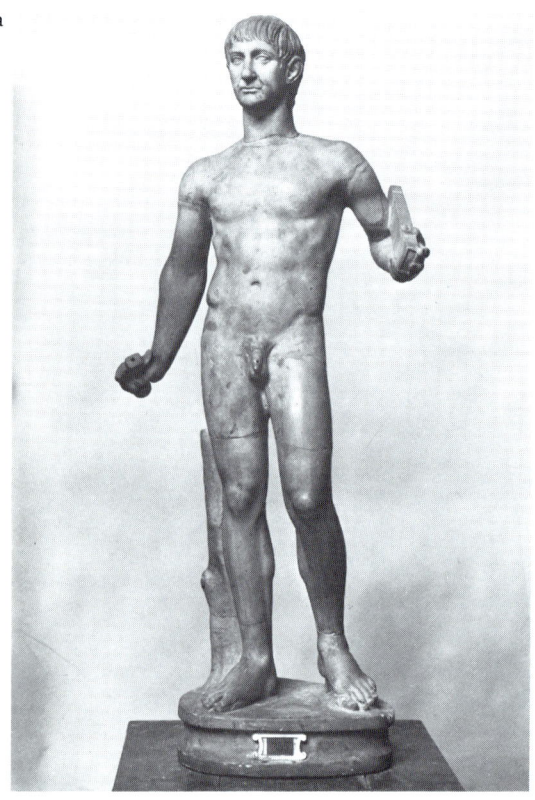

S 10
S 11
S 12

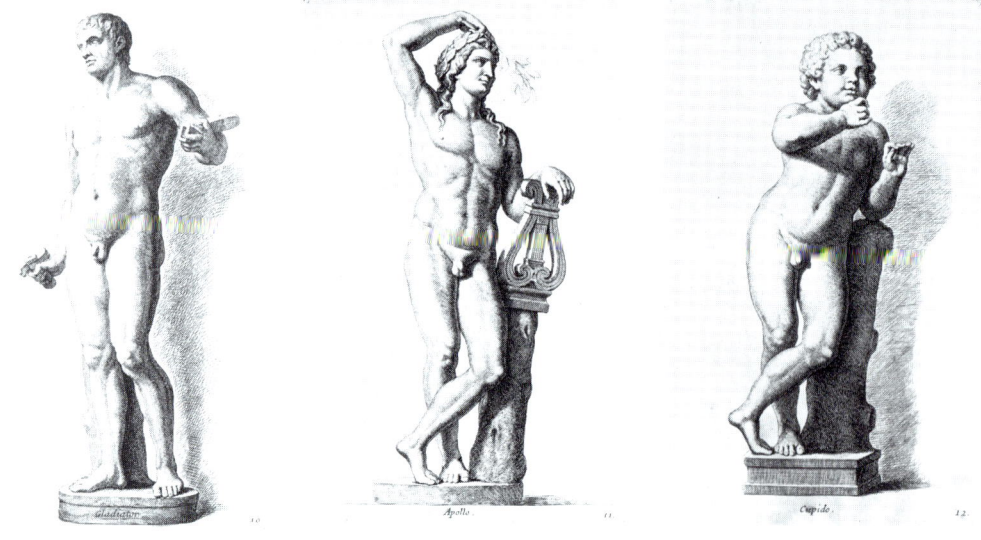

of ICONES with *De Sculpturis*) and p. 35 (to Berlin via Uylenburgh); C. Blümel, *Staatliche Museen zu Berlin, Römische Bildnisse*, Berlin, 1933, p. 15, no. R 34 (ill.; as *Trajan*; references to Vendramin and Reynst collections; in 1671 via Uylenburgh to Berlin); W.H. Gross, *Bildnisse Trajans*, Berlin, 1940, p. 128, no. 32 (*Trajan*); Timmers, no. 137.

Plates: S 10, S 10a.

According to Blümel the head represents a portrait of Trajan, which is mounted on a worthless torso of a male nude; according to Gross, the statuette itself also is antique.

11. *Statue of Apollo with his Lyre, Holding a Branch of Laurel*

Engraving: Anonymous.
328 × 199 mm.
ICONES 11: *Apollo*.

Sculpture: Lost.

Provenance: Andrea Vendramin, Venice (*cat.* I, 13); Gerard Reynst; Jan Six, Amsterdam (sale Amsterdam, 6 April 1702, no. 23: *Apollo met een Harp, heel Beeld, ongemeen fraai*).
Bibliography: Knorr 1663 (Fuchs, p. 240: *In Conclavi secundo... Apollo nudus cum arcu*); Clarac, III, p. CCLXVI (with Gerard Reynst); Jacobs, p. 26, note 1 (concordance of ICONES with *De Sculpturis*); Timmers, no. 138; Brunsting, p. 189, no. 23.

Plate: S 11.

12. *Statue of Cupid*

Engraving: Anonymous.
326 × 193 mm.
ICONES 12: *Cupido*.

Sculpture: Lost.

Provenance: Gerard Reynst.

Bibliography: Knorr 1663 (Fuchs, p. 240: *In Conclavi secundo... Cupido nudus*); Clarac, III, p. CCLXVI (with Gerard Reynst); Timmers, no. 139.

Plate: S 12.

13. *Statues of Dionysus and a Faun*

Engraving: Anonymous.
330 × 201 mm.
ICONES 13: *Hermaphroditus*.

Sculpture: Lost.

Provenance: Gerard Reynst; Jan Six, Amsterdam (sale Amsterdam, 6 April 1702, no. 21: *Bachanten, Twee heele Beelden by een, half leevens groote Antiq.*).

Bibliography: Knorr 1663 (Fuchs, p. 240: *In Conclavi secundo... Hermaphroditus nudus cum Salmace mutata, vestibus virgineis, pudendo masculino*); Clarac, III, p. CCLXVI (with Gerard Reynst); Timmers, no. 140; Brunsting, p. 189, no. 21.

Plate: S 13.

S 13
S 14

14. *Draped Female Figure Standing*

Engraving: Anonymous.
331 × 201 mm.
ICONES 14.

Sculpture: Lost.

Provenance: Gerard Reynst.

Bibliography: Clarac, III, p. CCLXVI (with Gerard Reynst).

Plate: s 14.

15. *Statue of Male Nude with Chlamys*

Engraving: Anonymous.
330 × 201 mm.
ICONES 15.

Sculpture: Lost.

Provenance: Gerard Reynst; Laurens van Campen, Amsterdam (sale Amsterdam, 17 July 1724, no. 19: *Een gladiator van Grieks Marmer, voor dezen gekomen uit het Cabinet van de Heer Reinst*); Gosuin Uilenbroek, Amsterdam.

Bibliography: Sigebert Havercamp, *Museum Uilenbroekianum*, Amsterdam, 1741, p. 289, no. 17 (with reference to Reynst collection; as *gladiator*, Parian marble, 2 feet 2 inches high [ca. 0,66 m.]); Clarac, III, p. CCLXVI (with Gerard Reynst); Timmers, no. 142.

Plate: s 15.

16. *Draped Female Figure Standing, with a Pitcher*

Engraving: Anonymous.
328 × 200 mm.
ICONES 16.

Sculpture: Lost.

Provenance: Andrea Vendramin, Venice (*cat.* I, 23 A); Gerard Reynst; Jan Six, Amsterdam (sale Amsterdam, 6 April 1702, no. 28: *Een staand Vrouwtje, met een Waterkruik, Antiq.*); Gosuin Uilenbroek, Amsterdam.

Bibliography: Sigebert Havercamp, *Museum Uilenbroekianum*, Amsterdam, 1741, p. 288, no. 15 (with reference to Reynst collection; Parian marble, 3 feet high [ca. 0,91 m.]); Clarac, III, p. CCLXVI (with Gerard Reynst); Jacobs, p. 26, note 1 (concordance of ICONES with *De Sculpturis*) and p. 33, note 3 (Uilenbroek collection); Timmers, no. 143; Brunsting, p. 189, no. 28.

Plate: s 16.

17. *Partly Draped Female Figure Standing*

Engraving: Anonymous.
327 × 199 mm.
ICONES 17.

Sculpture: Lost.

Provenance: Andrea Vendramin, Venice (*cat.* I, 31); Gerard Reynst.

Bibliography: Clarac, III, p. CCLXVI (with Gerard Reynst); Reinach, II, p. 406, no. 3 (ill.; reference to Reynst collection); Jacobs, p. 26,

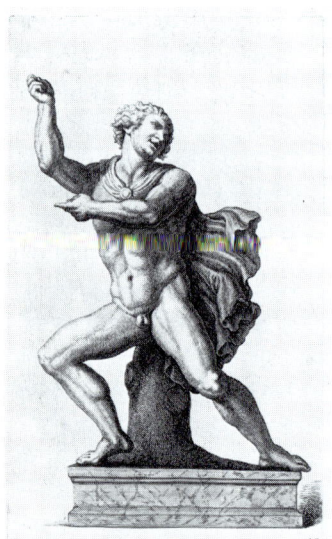
s 15

s 16

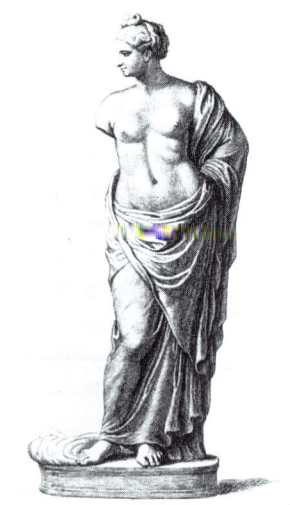
s 17

note 1 (concordance of ICONES with *De Sculpturis*); Timmers, no. 144 (as Aphrodite).

Plate: S 17.

18. *Draped Female Figure Standing*

Engraving: Anonymous.
334 × 202 mm.
ICONES 18.

Sculpture: Lost.

Provenance: Andrea Vendramin, Venice (*cat*. I, 22 B); Gerard Reynst; Jan Six, Amsterdam (sale Amsterdam, 6 April 1702, no. 29: *Een dito vrouwtje zonder handen een weerga*).

Bibliography: Clarac, III, p. CCLXVI (with Gerard Reynst); Jacobs, p. 26, note 1 (concordance of ICONES with *De Sculpturis*); Timmers, no. 145; Brunsting, p. 189, no. 29.

Plate: S 18.

19. *Bust of Hercules*

Engraving: Anonymous.
326 × 195 mm.
ICONES 19: *Hercules*.

Sculpture: Lost.

Provenance: Andrea Vendramin, Venice (*cat*. II, 21); Gerard Reynst.

Bibliography: Knorr 1663 (Fuchs, p. 240: *In Conclavi secundo... Hercules alius pector*); Jacobs, p. 26, note 1 (concordance of ICONES with *De Sculpturis*); Timmers, no. 154.

Plate: S 19.

20. *Male Portrait Bust*.

Engraving: Anonymous.
327 × 195 mm.
ICONES 20: *I. Caesar*.

Sculpture: Lost.

Provenance: Andrea Vendramin, Venice (*cat*. II, 19); Gerard Reynst; possibly Jan Six, Amsterdam (sale Amsterdam, 6 April 1702, no. 2: *Julius Caesar, dito Borststuk, Antiq.*).

Bibliography: Knorr 1663 (Fuchs, p. 240: *In Conclavi secundo... Caesar pectore tenus artificiosissime sculptus, Graecum opus*); Jacobs, p. 26, note 1 (concordance of ICONES with *De Sculpturis*); Timmers, no. 155; Brunsting, p. 189, no. 2 (possibly formerly with Jan Six).

Plate: S 20.

21. *Female Portrait Bust*

Engraving: Anonymous.
324 × 193 mm.
ICONES 21: *Agrippina Major*.

Sculpture: Staatliche Museen, East Berlin (Inv.no. Sk. 447). Gray-white, coarse marble. Head: 0,245 m. Tip of nose restored; surface badly worn. Bust modern.

Provenance: Gerard Reynst; Great Elector of Brandenburg.

Bibliography: Knorr 1663 (Fuchs, p. 240: *In Conclavi secundo... Agrippina major pector, praestantis-*

S 18
S 19
S 20

S 21a

S 21
S 22
S 23

Agrippina Major.
21.

Poppea Sabina.
22.

Augustus.
23.

sima, *vetustate paululum exesa*); E. Gerhard, *Berlin's antike Bildwerke*, I, Berlin, 1836, p. 113, no. 244 (unknown matron); (Alexander Conze), *Königliche Museen zu Berlin, Beschreibung der antiken Skulpturen,...*, Berlin, 1891, p. 175, no. 447 (Roman matron; ca. 200 AD); C. Blümel, *Staatliche Museen zu Berlin, Römische Bildnisse*, Berlin, 1933, p. 46, no. R 111 (ill.; reference to Reynst collection; head of old Roman woman; ca. middle 3rd century AD); Timmers, no. 156.

Plates: S 21, S 21a.

22. *Female Portrait Bust*

Engraving: Anonymous.
326 × 194 mm.
ICONES 22: *Poppea Sabina*.

Sculpture: Lost.

Provenance: Andrea Vendramin, Venice (*cat.* II, 33); Gerard Reynst; Jan Six, Amsterdam (sale Amsterdam, 6 April 1702, no. 8: *Poppeä Sabina, Borstbeeld, dito*).

Bibliography: Knorr 1663 (Fuchs, p. 240: *In Conclavi secundo... Poppaea Sabina*); Jacobs, p. 26, note 1 (concordance of ICONES with *De Sculpturis*); Timmers, no. 157; Brunsting, p. 189, no. 8.

Plate: S 22.

23. *Male Portrait Bust*

Engraving: Anonymous.
324 × 194 mm.
ICONES 23: *Augustus*.

Sculpture: Lost.

Provenance: Andrea Vendramin, Venice (*cat.* II, 5: *Ottaviano*); Gerard Reynst; perhaps Jan Six, Amsterdam (sale Amsterdam, 6 April 1702, no. 14: *Een Jonge Keizer Aug., dito*).

Bibliography: Knorr 1663 (Fuchs, p. 240: *In Conclavi secundo... Augustus pector.*); Jacobs, p. 26 and note 1 (concordance of ICONES with *De Sculpturis*); Timmers, no. 158; Brunsting, p. 189, no. 14.

Plate: S 23.

24. *Portrait Bust of Tiberius*

Engraving: Anonymous.
324 × 191 mm.
ICONES 24: *Tiberius*.

Sculpture: Lost.

Provenance: Gerard Reynst.

Bibliography: Knorr 1663 (Fuchs, p. 240: *In Conclavi secundo... Tiberius pector.*); Timmers, no. 159.

Plate: S 24.

25. *Female Bust*

Engraving: Anonymous.
326 × 196 mm.
ICONES 25: *Calphurnia*.

Sculpture: Staatliche Museen, East Berlin (Inv.no. Sk. 619). White marble. Head: 0,456 m; antique part: 0,315 m. Nose, part of left eyebrow, breast

S 24
S 25
S 25a

restored. Face heavily worked over. Hair abraided at top right.

Provenance: Andrea Vendramin, Venice (*cat.* II, 52); Gerard Reynst; Great Elector of Brandenburg.

Bibliography: Knorr 1663 (Fuchs, p. 240: *In Conclavi secundo... Calphurnia*); Jacobs, p. 26, note 1 (concordance of ICONES with *De Sculpturis*); E. Gerhard, *Berlin's antike Bildwerke*, I, Berlin, 1836, p. 89, no. 138; (Alexander Conze), *Königliche Museen zu Berlin, Beschreibung der antiken Skulpturen, ...*, Berlin, 1891, p. 240, no. 619; C. Blümel, *Staatliche Museen zu Berlin, Römische Kopien griechischer Skulpturen des vierten Jahrhunderts v.Chr.*, Berlin, 1938, p. 15, no. K 218 (ill.; formerly in Vendramin and Reynst collections; probably bought by the Great Elector in Holland; head belonged to replica of so-called *Pothos* statue; dry work); Timmers, no. 160.

Plates: S 25, S 25a.

26. *Female Portrait Bust*

Engraving: Anonymous.
324 × 191 mm.
ICONES 26: *Octavia*.

Sculpture: Lost.

Provenance: Andrea Vendramin, Venice (*cat.* II, 38); Gerard Reynst.

Bibliography: Knorr 1663 (Fuchs, p. 240: *In Conclavi secundo... Octavia*.); Jacobs, p. 26, note 1 (concordance of ICONES with *De Sculpturis*); Timmers, no. 161.

Plate: S 26.

27. *Portrait Bust of so-called Vitellius*

Engraving: Anonymous.
326 × 196 mm.
ICONES 27: *Vitellius*.

Sculpture: Lost.

Provenance: Andrea Vendramin, Venice (*cat.* II, 1). Gerard Reynst.

Bibliography: Knorr 1663 (Fuchs, p. 240: *In Conclavi secundo... Vitellius*.); Jacobs, p. 26, note 1 (concordance of ICONES with *De Sculpturis*); Timmers, no. 162; A.N. Zadoks-Josephus Jitta, 'A Creative Misunderstanding,' *Nederlands Kunsthistorisch Jaarboek*, XXIII, 1972, pp. 7–8 (ill.; surmises that portrait was in Jan Reynst's collection who bequeathed it to Gerard; Venetian origin, bronze; head is faithful copy of so-called *Venice portrait*, AD ± 130, Venice, Archaeological museum [ill. *op.cit.* fig. 2]; bust completed and clad in sixteenth-century version of Roman cuirass).

Plate: S 27.

28. *Female Portrait Bust*

Engraving: Anonymous.
327 × 194 mm.
ICONES 28: *Domitia*.

Sculpture: Lost.

Provenance: Andrea Vendramin, Venice (*cat.* II, 47); Gerard Reynst.

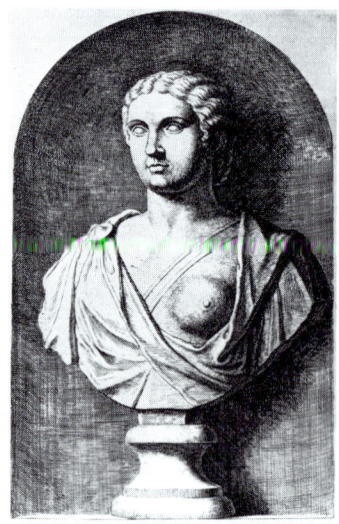

Bibliography: Knorr 1663?
(Fuchs, p. 240: *In Conclavi
secundo... Domitianus.*); Jacobs,
p. 26, note 1 (concordance of
ICONES with *De Sculpturis*); Timmers, no. 163.

Plate: s 28.

29. *Bust of Athena*

Engraving: Anonymous.
325 × 193 mm.
ICONES 29: *Pallas*.

Sculpture: Lost.

Provenance: Andrea Vendramin,
Venice (*cat.* II, 3); Gerard Reynst.

Bibliography: Knorr 1663 (Fuchs,
p. 240: *In Conclavi secundo... Pallas
cum Sphynge in galea.*); Jacobs,
p. 26, note 1 (concordance of
ICONES with *De Sculpturis*);
Timmers, no. 164.

Plate: s 29.

30. *Male Portrait Bust*

Engraving: Anonymous.
326 × 194 mm.
ICONES 30: *Milenus*.

Sculpture: Lost.

Provenance: Andrea Vendramin,
Venice (*cat.* II, 9); Gerard Reynst.

Bibliography: Knorr 1663 (Fuchs,
p. 240: *In Conclavi secundo... Milenus.*); Jacobs, p. 26, note 1 (concordance of ICONES with *De
Sculpturis*); Timmers, no. 165.

Plate: s 30.

31. *Female Portrait Bust*

Engraving: Anonymous.
330 × 199 mm.
ICONES 31: *Livia*.

Sculpture: Rijksmuseum van Oudheden, Leiden (Inv.no. Pb. 130).
White marble. H.: 0,53 m (with
base: 0,66 m). modern.

Provenance: Andrea Vendramin,
Venice (*cat.* II, 56 B); Gerard

Reynst; Jan Six, Amsterdam (sale
Amsterdam, 6 April 1702, no. 11:
Livia, dito, Antiq.); Gerard van
Papenbroek, Amsterdam (*cat.*
no. 130).

Bibliography: Knorr 1663 (Fuchs,
p. 240: *In Conclavi secundo...
Livia.*); Oudendorp, II, no. 33
(reference to Reynst ICONES; as
Julia or *Livia Augusta*); Janssen, I,
no. 147 (*Julia* or *Livia Augusta*;
badly worn surface); Jacobs, p. 26,
note 1 (concordance of ICONES
with *De Sculpturis*) and p. 33, note 3
(ex Papenbroek); Timmers,
no. 166; Brunsting, p. 189, no. 11
(incorrectly identified with
Inv.no. Pb. 121).

Plates: s 31, s 31a.

32. *Female Bust*

Engraving: Anonymous.
329 × 196 mm.
ICONES 32: *Cleopatra*.

s 29
s 30

S 31a
S 32a

S 31
S 32
S 33

Sculpture: Rijksmuseum van Oudheden, Leiden (Inv.no. Pb. 132). White marble. Bust with base: 0,65 m. Head: 0,27 m. Nose missing; veil separate, originally attached to head with stucco; bust modern.

Provenance: Andrea Vendramin, Venice (*cat.* II, 56 A); Gerard Reynst; Gosuin Uilenbroek, Amsterdam; Gerard van Papenbroek, Amsterdam (*cat.* no. 132).

Bibliography: Knorr 1663 (Fuchs, p. 240: *In Conclavi secundo... Cleopatra.*); Sigebert Havercamp, *Museum Uilenbroekianum*, Amsterdam, 1741, p. 287, no. 2 (reference to Reynst ICONES; as *Cleopatra*); Oudendorp, II, no. 44 (reference to Reynst ICONES; as *Berenice*); Jacobs, p. 26, note 1 (concordance of ICONES with *De Sculpturis*) and p. 33, note 3 (ex Uilenbroek and ex Papenbroek); Timmers, no. 167.

Plates: S 32, S 32a.

33. *Male Portrait Bust*

Engraving: Anonymous.
329 × 198 mm.
ICONES 33: *Trajanus*.

Sculpture: Lost.

Provenance: Andrea Vendramin, Venice (*cat.* II, 49); Gerard Reynst; Jan Six, Amsterdam (sale Amsterdam, 6 April 1702, no. 5: *Trajanus, zeer fraay, Antiq. Borstbeeld*).

Bibliography: Knorr 1663 (Fuchs, p. 240: *In Conclavi secundo... Trajanus.*); Jacobs, p. 26, note 1 (concordance of ICONES with *De Sculpturis*); Timmers, no. 168; Brunsting, p. 189, no. 5.

Plate: S 33.

34. *Bust of Gordian III*

Engraving: Anonymous.
325 × 193 mm.
ICONES 34: *Gordianus*.

Sculpture: Skulpturensammlung, Dresden (Inv.no. *Herrmann*, 409).

Marble. H. 0,595 m.
Bust, neck, chin, nose, eyebrows, front and part of hair restored.

Provenance: Andrea Vendramin, Venice (*cat.* II, 13); Gerard Reynst; Great Elector of Brandenburg; by exchange to Dresden in 1724/26.

Bibliography: Knorr 1663 (Fuchs, p. 240 *In Conclavi secundo... Gordianus.*); L. Beger, *Thesauri Regii et Eelectorialis Brandenburgici...* III, Coloniae Marchicae (1696), pp. 351–52 (ill.; as *Gordianus Pius*); Le Plat, pl. 157, 1; Jacobs, p. 26, note 1 (concordance of ICONES with *De Sculpturis*) and p. 35 (provenances); Timmers, no. 169.

Plates: S 34, S 34a.

35. *Female Portrait Bust*

Engraving: Anonymous.
324 × 192 mm.
ICONES 35: *Helena*.

Sculpture: Lost.

S 34
S 34a

Provenance: Andrea Vendramin, Venice (*cat.* II, 7); Gerard Reynst.

Bibliography: Knorr 1663 (Fuchs, p. 240: *In Conclavi secundo... Helena.*); Jacobs, p. 26, note 1 (concordance of ICONES with *De Sculpturis*); Timmers, no. 170.

Plate: s 35.

36. *Female Portrait Bust*

Engraving: Anonymous.
325 × 192 mm.
ICONES 36: *Octavia*.

Sculpture: Lost.

Provenance: Andrea Vendramin, Venice (*cat.* II, 50); Gerard Reynst.

Bibliography: Knorr 1663 (Fuchs, p. 240: *In Conclavi secundo... Octavia aliter.*); Jacobs, p. 26, note 1 (concordance of ICONES with *De Sculpturis*); Timmers, no. 171.

Plate: s 36.

37. *Male Portrait Bust*

Engraving: Anonymous.
326 × 192 mm.
ICONES 37: *Hadrianus*.

Sculpture: Lost.

Provenance: Andrea Vendramin, Venice (*cat.* II, 23); Gerard Reynst; possibly Jan Six, Amsterdam (sale Amsterdam, 6 April 1702, no. 1); possibly Nicolaas Antoni Flinck, Rotterdam (see below).

Bibliography: Knorr 1663 (Fuchs, p. 240: *In Conclavi secundo... Hadrianus.*); Uffenbach (1710, p. 316: *Ferner waren allhier eine Buste von ... Hadriano*); Jacobs, p. 26, note 1 (concordance of ICONES with *De Sculpturis*); Timmers, no. 172; Brunsting, p. 189, under no. 1 (possibly in Six collection).

Plate: s 37.

38. *Male Portrait Bust*

Engraving: Anonymous.
325 × 194 mm.
ICONES 38: *Caracalla*.

Sculpture: Lost.

Provenance: Andrea Vendramin, Venice (*cat.* II, 60); Gerard Reynst.

Bibliography: Knorr 1663 (Fuchs, p. 240: *In Conclavi secundo ... Caracalla.*); Jacobs, p. 26, note 1 (concordance of ICONES with *De Sculpturis*); Timmers, no. 173.

Plate: s 38.

39. *Female Bust*

Engraving: Anonymous.
325 × 193 mm.
ICONES 39: *Terentia*.

Sculpture: Lost.

Provenance: Andrea Vendramin, Venice (*cat.* II, 10); Gerard Reynst; Jan Six, Amsterdam (sale Amsterdam, 6 April 1702, no. 7: *Terentia,*

s 35
s 36
s 37

Borstbeeld, dito); Johan de Vries, Amsterdam (sale Amsterdam, 13 October 1738, no. 5: *Terentia*).

Bibliography: Knorr 1663 (Fuchs, p. 240: *In Conclavi secundo... Terentia.*); Jacobs, p. 26, note 1 (concordance of ICONES with *De Sculpturis*); Timmers, no. 174; Brunsting, p. 189, no. 7.

Plate: s 39.

40. *Female Portrait Bust*

Engraving: Anonymous.
326 × 191 mm.
ICONES 40: *Iulia Mammea*.

Sculpture: Lost.

Provenance: Gerard Reynst; Jan Six, Amsterdam (sale Amsterdam, 6 April 1702, no. 4: *Julia Mammea. Borstbeeld, Antiq.*); Johan de Vries, Amsterdam (sale Amsterdam, 13 October 1738, no. 4: *Julia Mammea*).

Bibliography: Knorr 1663 (Fuchs, p. 240: *In Conclavi secundo... Julia*

Mammea.); Timmers, no. 175; Brunsting, p. 189, no. 4.

Plate: s 40.

41. *Female Bust*

Engraving: Anonymous.
326 × 193 mm.
ICONES 41: *Agrippina*.

Sculpture: No longer traceable.

Provenance: Andrea Vendramin, Venice (*cat.* II, 8); Gerard Reynst; Gerard van Papenbroek, Amsterdam; Jeronimo de Bosch, Amsterdam (donated to Academy in 1744).

Bibliography: Knorr 1663 (Fuchs, pp. 240–41: *In Conclavi secundo... Agrippina minor.*); Oudendorp, II, no. 14 (reference to Reynst ICONES; bought at Papenbroek sale by De Bosch and donated to Academy; not antique); Jacobs, p. 26, note 1 (concordance of ICONES with *De Sculpturis*) and p. 33, note 3 (ex Papenbroek); Timmers, no. 176;

I.Q. van Regteren Altena and P.J.J. van Thiel, *De portret-galerij van de Universiteit van Amsterdam en haar stichter Gerard van Papenbroeck 1673–1743*, Amsterdam, 1964, p. 51 (reference to the resolution of the curators of 14 October 1744, where De Bosch's purchase and subsequent gift is recorded with his stipulation that the bust be included in the catalogue of sculptures from the Papenbroek bequest).

Plate: s 41.

The bust Pb. 102 that has been associated with Oudendorp's description is not close enough to the sculpture represented in this engraving to positively identify it with this so-called *Agrippina*, and no alternate bust has been found in Leiden.

42. *Bust of Jupiter*

Engraving: Anonymous.
328 × 198 mm.
ICONES 42: *Iupiter Ammon*.

s 38
s 39
s 40

s 43a Sculpture: Lost.

>Provenance: Andrea Vendramin, Venice (*cat*. II, 30: *Giove*); Gerard Reynst.
>
>Bibliography: Jacobs, p. 26 and note 1 (concordance of ICONES with *De Sculpturis*); Timmers, no. 177.
>
>Plate: s 42.

43. *Bust of a Youth*

>Engraving: Anonymous.
>325 × 195 mm.
>ICONES 43: *Aristea*.
>
>Sculpture: Staatliche Museen, East Berlin (Inv.no. Sk. 547). White, coarse marble. H. 0,235 m. Restored are nose, chin including left lower cheek, mouth, large section of left eye with eyebrow and eyelid, corresponding, smaller section of right eye, left earlobe, large piece of the part in the hair, left side of strands of hair at left. Face was cleaned. Bust modern.
>
>Provenance: Andrea Vendramin,

s 41
s 42
s 43

Venice (*cat.* II, 11); Gerard Reynst; Great Elector of Brandenburg.

Bibliography: Knorr 1663 (Fuchs, p. 241: *In Conclavi secundo... Aristaeus*); E. Gerhard, *Berlin's antike Bildwerke*, I, Berlin, 1836, p. 63, no. 67 d; (Alexander Conze), *Königliche Museen zu Berlin, Beschreibung der antiken Skulpturen...*, Berlin, 1891, pp. 213–14, no. 547; Jacobs, p. 26, note 1 (concordance of ICONES with *De Sculpturis*); C. Blümel, *Staatliche Museen zu Berlin, Römische Kopien griechischer Skulpturen des fünften Jahrhunderts v.Chr.*, Berlin, 1931, p. 14, no. K 141 (ill.; formerly Vendramin and Reynst collections, probably bought by the Great Elector in Holland; formerly joined with female bust; one of several replicas after argivian original in bronze of ca. 460 BC, representing Orpheus); Timmers, no. 178.

Plates: s 43, s 43a.

44. *Bust of Athena*

Engraving: Anonymous. 325 × 193 mm. ICONES 44: *Pallas*.

Sculpture: Lost.

Provenance: Gerard Reynst.

Bibliography: Knorr 1663 (Fuchs, p. 241: *In Conclavi secundo ... Pallas cum triplici sphynge in galea*); Timmers, no. 179.

Plate: s 44.

45. *Portrait Bust of Geta*

Engraving: Anonymous. 325 × 195 mm. ICONES 45: *Geta*.

Sculpture: Lost.

Provenance: Andrea Vendramin, Venice (*cat.* II, 16); Gerard Reynst; Nicolaas Antoni Flinck, Rotterdam (see below).

Bibliography: Knorr 1663 (Fuchs, p. 241: *In Conclavi secundo... Geta.*); Uffenbach (1710, p. 316: *Ferner waren allhier eine Buste von ...Geta...*); Jacobs, p. 26, note 1 (concordance of ICONES with *De Sculpturis*); Timmers, no. 180.

Plate: s 45.

Possibly identical with the bust seen by Uffenbach in Flinck's collection in 1710.

46. *Female Portrait Bust*

Engraving: Anonymous. 328 × 196 mm. ICONES 46: *Flavia*.

Sculpture: Lost.

Provenance: Andrea Vendramin, Venice (cat. II, 65 B); Gerard Reynst.

Bibliography: Knorr 1663 (Fuchs, p. 241: *In Conclavi secundo ... Flavia.*); Jacobs, p. 26, note 1 (concordance of ICONES with *De Sculpturis*); Timmers, no. 181.

Plate: s 46.

s 44
s 45
s 46

47. Bust of a Faun

Engraving: Anonymous.
327 × 195 mm.
ICONES 47: *Faunus*.

Sculpture: Lost.

Provenance: Andrea Vendramin, Venice (*cat.* II, 26); Gerard Reynst.

Bibliography: Knorr 1663 (Fuchs, p. 241: *In Conclavi secundo … Faunus cum pelle caprina ridens*); Jacobs, p. 26, note 1 (concordance of ICONES with *De Sculpturis*); Timmers, no. 182.

Plate: s 47.

48. Male Portrait Bust

Engraving: Anonymous.
323 × 192 mm.
ICONES 48: *Antoninus*.

Sculpture: Lost.

Provenance: Andrea Vendramin, Venice (*cat.* II, 37); Gerard Reynst; Gerard van Papenbroek, Amsterdam.

Bibliography: Knorr 1663 (Fuchs, p. 241: *In Conclavi secundo… Antoninus.*); Oudendorp, II, no. 27 (reference to Reynst ICONES and Vendramin collection; sometimes identified as *Caracalla*); Janssen, I, no. 131 (*M. Aurel. Antonin. Caracalla*); Jacobs, p. 26, note 1 (concordance of ICONES with *De Sculpturis*) and p. 33, note 3 (ex Papenbroek); Timmers, no. 183.

Plate: s 48.

The identification with the bust Inv.no. Pb. 115 does not convince, and no other bust that is closer to the representation in the ICONES is extant at the Rijksmuseum van Oudheden, Leiden.

49. Portrait Bust of Hadrian

Engraving: Anonymous.
326 × 195 mm.
ICONES 49: *Hadrianus*.

Sculpture: Rijksmuseum van Oudheden, Leiden (Inv.no. Pb. 123). White marble. H. 0,18 m.

Bust modern, still extant; hair was partly added in stucco, now removed; nose missing.

Provenance: Andrea Vendramin, Venice (*cat.* II, 20); Gerard Reynst; possibly Jan Six, Amsterdam (sale Amsterdam, 6 April 1702, no. 1: *Hadrianus, een schoen Borstbeeld, Antiq. op een Pedestal*); possibly Nicolaas Antoni Flinck,

s 47
s 48
s 49

Rotterdam (see below); Gerard van Papenbroek, Amsterdam (*cat.* no. 123).

Bibliography: Knorr 1663 (Fuchs, p. 241: *In Conclavi secundo ... Hadrianus aliter.*); Uffenbach (1710, p. 316: *Ferner waren allhier eine Buste von ... Hadriano*); Oudendrop, II, no. 35 (reference to Reynst ICONES and Vendramin collection; *Hadrian*); Janssen, I, no. 129 (*Hadrian*; only head antique); Jacobs, p. 26, note 1 (concordance of ICONES with *De Sculpturis*) and p. 33, note 3 (ex Papenbroek); Timmers, no. 184; M. Wegner, *Hadrian*, Berlin, 1956, pp. 30 and 100 (unidentifiable with existing types; antique?); Brunsting, p. 189, no. 1 (possibly formerly with Six).

Plates: s 49, s 49a.

50. *Bust of Cupid*

Engraving: Anonymous.
328 × 193 mm.
ICONES 50: *Cupido*.

Sculpture: Lost.

Provenance: Andrea Vendramin, Venice (*cat.* II, 69 A); Gerard Reynst.

Bibliography: Knorr 1663 (Fuchs, p. 241: *In Conclavi secundo ... Cupido alius.*); Jacobs, p. 26, note 1 (concordance of ICONES with *De Sculpturis*); Timmers, no. 185.

Plate: s 50.

51. *Bust of a Faun*

Engraving: Anonymous.
329 × 198 mm.
ICONES 51: *Faunus*.

Sculpture: Lost.

Provenance: Andrea Vendramin, Venice (*cat.* II, 14); Gerard Reynst.

Bibliography: Knorr 1663 (Fuchs, p. 241: *In Conclavi secundo ... Faunus alius.*); Jacobs, p. 26, note 1 (concordance of ICONES with *De Sculpturis*); Timmers, no. 186.

Plate: s 51.

52. *Portrait Bust of a Man with Phrygian Cap*

Engraving: Anonymous.
327 × 193 mm.
ICONES 52: *Cijrus*.

Sculpture: Lost.

Provenance: Andrea Vendramin, Venice (*cat.* II, 64); Gerard Reynst.

Bibliography: Knorr 1663 (Fuchs, p. 241: *In Conclavi secundo ... Cyrus tiara tectus.*); Jacobs, p. 26, note 1 (concordance of ICONES with *De Sculpturis*); Timmers, no. 187.

Plate: s 52.

53. *Portrait Bust of a Young Woman*

Engraving: Anonymous.
327 × 193 mm.
ICONES 53: *Flavia*.

Sculpture: Staatliche Museen, East Berlin (Inv.no. Sk. 438). White marble. H. 0,37 m. Bust, nose and left ear restored. Face and hair reworked.

s 50
s 51
s 52

Provenance: Andrea Vendramin, Venice (*cat.* II, 51); Gerard Reynst; Great Elector of Brandenburg.

Bibliography: Knorr 1663 (Fuchs, p. 241: *In Conclavi secundo ... Flavia alia.*); E. Gerhard, *Berlin's antike Bildwerke*, I, Berlin, 1836, p. 104, no. 181 (unknown provenance; as *Plotina*); (Alexander Conze), *Königliche Museen zu Berlin, Beschreibung der antiken Skulpturen...*, Berlin, 1891, p. 171, no. 438 (Royal collection; Roman woman, time of Hadrian); Jacobs, p. 26, note 1 (condordance of ICONES with *De Sculpturis*); C. Blümel, *Staatliche Museen zu Berlin, Römische Bildnisse*, Berlin, 1933, p. 17, no. R 39 (ill.; reference to Gerard Reynst collection; first decades of 2nd century AD); Timmers, no. 188.

Plates: S 53, S 53a.

54. *Portrait Bust of a Youth (Commodus ?)*

Engraving: Anonymous.
322 × 194 mm.
ICONES 54: *Lucilla*.

Sculpture: Lost.

Provenance: Gerard Reynst; Gosuin Uilenbroek, Amsterdam.

Bibliography: Knorr 1663 (Fuchs, p. 241: *In Conclavi secundo ... Lucilla.*); Sigebert Havercamp, *Museum Uilenbroekianum*, Amsterdam, 1741, p. 288, no. 8 (with incorrect reference to ICONES 9); Jacobs, p. 33, note 3 (incorrect concordance of ICONES with *De Sculpturis* and cat. *Museum Uilenbroekianum*); Timmers, no. 189.

Plate: S 54.

55. *Portrait Head of Faustina*

Engraving: Anonymous.
328 × 198 mm.
ICONES 55: *Faustina*.

Sculpture: No longer traceable.

Provenance: Andrea Vendramin, Venice (*cat.* II, 18); Gerard Reynst; Jan Six, Amsterdam (sale Amsterdam, 6 April 1702, no. 10: *Faustina, schoone Borstbeeld, Antiq.*); Gerard van Papenbroek, Amsterdam.

Bibliography: Knorr 1663 (Fuchs, p. 241: *In Conclavi secundo...*

S 53
S 54
S 55

Faustina.); Oudendorp, II, no. 42 (reference to Reynst ICONES and Vendramin collection; *Faustina*); Jacobs, p. 26, note 1 (concordance of ICONES with *De Sculpturis*); Timmers, no. 190; Brunsting, p. 189, no. 10 (incorrectly identified with Inv.no. Pb. 130 which is depicted in ICONES 31).

Plate: s 55.

Although described by Oudendorp, the bust can no longer be identified among the sculptures in Leiden. Oudendorp seems to have confused this portrait head with the bust of the so-called *Livia* (Papenbroek 130; ICONES 31).

56. *Portrait Bust of a Youth (Alexander the Great?)*

Engraving: Anonymous.
327 × 198 mm.
ICONES 56: *Silvius Postumius*.

Sculpture: Lost.

Provenance: Andrea Vendramin,

Venice (*cat.* II, 46); Gerard Reynst.

Bibliography: Knorr 1663 (Fuchs, p. 241: *In tertio conclavi… Sylvius, Posthumus*.); Jacobs, p. 26, note 1 (concordance of ICONES with *De Sculpturis*); Timmers, no. 191.

Plate: s 56.

57. *Portrait Bust of a Young Boy*

Engraving: Anonymous.
326 × 198 mm.
ICONES 57.

Sculpture: Lost.

Provenance: Gerard Reynst.

Bibliography: Timmers, no. 192.

Plate: s 57.

58. *Portrait Bust of a Young Boy*

Engraving: Anonymous.
326 × 198 mm.
ICONES 58: *Germanicus*.

Sculpture: Lost.

Provenance: Andrea Vendramin, Venice (*cat.* II, 45); Gerard Reynst.

Bibliography: Jacobs, p. 26, note 1 (concordance of ICONES with *De Sculpturis*); Timmers, no. 193.

Plate: s 58.

During his visit to Holland ca. 1717–18, J.R. Richardson saw a bust of a *Young Germanicus* in the collection of Lambert ten Kate in Amsterdam which may possibly refer to this sculpture. (see *An Account of the Statues, Bas-reliefs, Drawings and Pictures in Italy, France…*, London, 1722, p. 4; I would like to thank Professor Van Gelder for this reference).

59. *Male Portrait Bust*

Engraving: Anonymous.
327 × 198 mm.
ICONES 59: *Papirius*.

Sculpture: Lost.

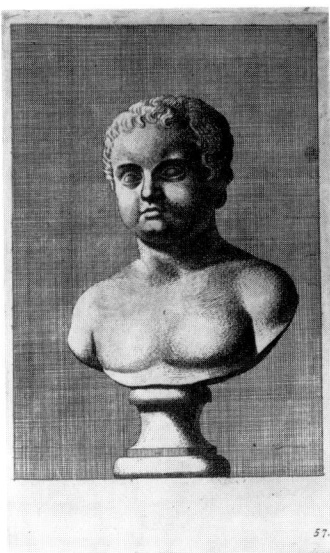

s 56
s 57
s 58

Provenance: Andrea Vendramin, Venice (*cat.* II, 36); Gerard Reynst.

Bibliography: Knorr 1663 (Fuchs, p. 241: *In tertio conclavi... Papyrius.*); Jacobs, p. 26, note 1 (concordance of ICONES with *De Sculpturis*); Timmers, no. 194.

Plate: s 59.

60. *Female Portrait Bust*

Engraving: Anonymous.
328 × 196 mm.
ICONES 60: *Antonia Major*.

Sculpture: Lost.

Provenance: Andrea Vendramin, Venice (*cat.* II, 39); Gerard Reynst.

Bibliography: Knorr 1663 (Fuchs, p. 241: *In tertio conclavi... Antonia major.*); Jacobs, p. 26, note 1 (concordance of ICONES with *De Sculpturis*); Timmers, no. 195.

Plate: s 60.

61. *Male Bust*

Engraving: Anonymous.
326 × 195 mm.
ICONES 61: *Hadrianus*.

Sculpture: Lost.

Provenance: Andrea Vendramin, Venice (*cat.* II, 2); Gerard Reynst; possibly Jan Six, Amsterdam (sale Amsterdam, 6 April 1702, no. 1: *Hadrianus, een schoen Borstbeeld, Antiq., op een Pedestal*); possibly Nicolaas Antoni Flinck, Rotterdam (see below).

Bibliography: Knorr 1663 (Fuchs, p. 241: *In tertio conclavi... Hadrianus denuo.*); Uffenbach (1710, p. 316: *Ferner waren allhier eine Buste von ... Hadriano*); Jacobs, p. 26, note 1 (concordance of ICONES with *De Sculpturis*); Timmers, no. 196; Brunsting, p. 189, under no. 1 (possibly in Six collection).

Plate: s 61.

62. *Female Portrait Bust*

Engraving: Anonymous.
326 × 196 mm.
ICONES 62: *Claudia*.

Sculpture: Lost.

Provenance: Andrea Vendramin, Venice (*cat.* II, 66 B); Gerard Reynst.

Bibliography: Knorr 1663 (Fuchs, p. 241: *In tertio conclavi... Claudia*); Jacobs, p. 26, note 1 (concordance of ICONES with *De Sculpturis*); Timmers, no. 197.

Plate: s 62.

63. *Bust of Youthful Pan*

Engraving: Anonymous.
327 × 197 mm.
ICONES 63: *Bachus*.

Sculpture: Rijksmuseum van Oudheden, Leiden (Inv.no. Pb. 112). White marble. H. 0,64 m.
Bust, lower part of head with chin, lips, ears, part of flowers modern.

Provenance: Gerard Reynst; Jan

s 59
s 60
s 61

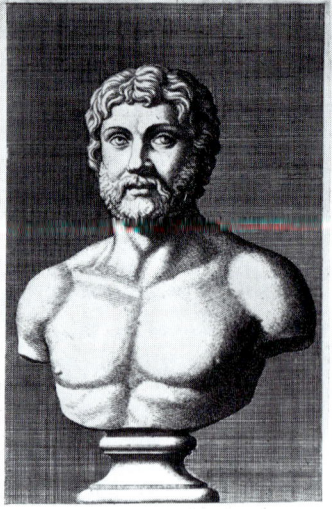

Six, Amsterdam (sale Amsterdam, 6 April 1702, no. 6: *Jonge Bachus dito, Borststuk*); Gerard van Papenbroek, Amsterdam (*cat.* no. 112).

Bibliography: Knorr 1663 (Fuchs, p. 241: *In tertio conclavi... Bacchus.*); Oudendorp, II, no. 24 (reference to Reynst ICONES; youthful Pan); Janssen, I, no. 111 (young Pan; only upper part of face until lower lip antique); Timmers, no. 198; Brunsting, p. 189, no. 6.

Plates: s 63, s 63a.

64. *Portrait Bust of Severus*

Engraving: Anonymous.
326 × 191 mm.
ICONES 64: *Severus*.

Sculpture: Lost.

Provenance: Gerard Reynst.

Bibliography: Knorr 1663 (Fuchs, p. 241: *In tertio conclavi... Severus, pectore tenus.*); Timmers, no. 199.

Plate: s 64.

65. *Herm of Priap* (front view)

Engraving: Anonymous.
328 × 199 mm.
ICONES 65: *Priapus* (front).

Sculpture: Skulpturensammlung, Dresden (Inv.no. *Inventar* 1810, no. 439).
Marble. H. 0,465 m.

Provenance:
Gerard Reynst; Great Elector of Brandenburg; by exchange to Dresden in 1724/26.

Bibliography: Knorr 1663 (Fuchs, p. 241: *In tertio conclavi... Priapus inguine tenus*); Heimbach, *Cimeliarchium Brandenburgicum...*, 1672 (ms. Berlin), p. 5; L. Beger, *Thesauri Regii et Eelectorialis Brandenburgici...*, III, Coloniae Marchicae (1696), pp. 261–64 (ill.); Le Plat, pl. 154, 3/4; Reinach, II, p. 75, no. 7 (ill.); Jacobs, p. 34 (traces herm to Berlin and Dresden); Timmers, nos. 152–53; H. Ladendorf, *Antikenstudium und Antikenkopie* (*Abhandlungen der sächsischen Akademie der Wissenschaften zu Leipzig, Philologisch-historische Klasse*), 46, Berlin, 1958, p. 194.

Plates: s 65, s 65a.

Herm of Priap (back view)

Engraving: Anonymous.
326 × 193 mm.
ICONES 66: *Priapus* (back).

s 62
s 63
s 64

202 SCULPTURES IN THE COLLECTION OF GERARD REYNST

s 65a

s 65
s 65b
s 66

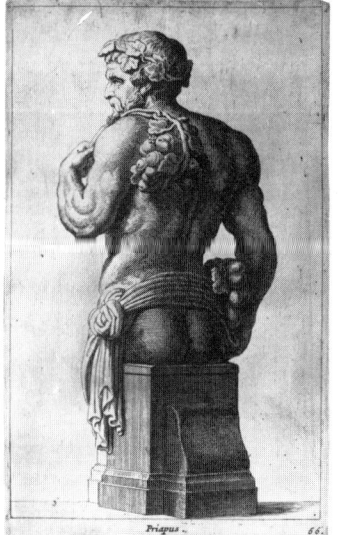

Sculpture: See ICONES 65.

Plate: s 65b.

66. *Male Portrait Bust*

Engraving: Anonymous.
325 × 194 mm.
ICONES 67.

Sculpture: Lost.

Provenance: Gerard Reynst.

Bibliography: Timmers, no. 200.

Plate: s 66.

67. *Male Portrait Bust*

Engraving: Anonymous.
326 × 194 mm.
ICONES 68.

Sculpture: Lost.

Provenance: Gerard Reynst.

Bibliography: Timmers, no. 201.

Plate: s 67.

68. *Female Portrait Bust*

Engraving: Anonymous.
327 × 192 mm.
ICONES 69.

Sculpture: Lost.

Provenance: Gerard Reynst.

Bibliography: Timmers, no. 202.

Plate: s 68.

69. *Portrait Bust of Septimius Severus*

Engraving: Anonymous.
325 × 193 mm.
ICONES 70.

Sculpture: Lost.

Provenance: Gerard Reynst.

Bibliography: Timmers, no. 203.

Plate: s 69.

70. *Female Portrait Bust*

Engraving: Anonymous.
326 × 195 mm.
ICONES 71.

Sculpture: Rijksmuseum van Oudheden, Leiden (Inv.no. Pb. 105).
Head in white, bust in red marble.
Head: 0,34 m. Bust 0,24 m.
Bust and nose modern, round base lost.

Provenance: Gerard Reynst;
Robert de Neufville, Leiden;
Jan de Witt Jr. (sale Amsterdam, 1701, no. 127); Gerard van Papenbroek, Amsterdam (*cat.* no. 105).

Bibliography: Oudendorp, II, no. 17 (reference to Reynst ICONES and later provenances; as *Faustina*); Janssen, I, no. 149 (*Faustina*; nose modern; crown damaged); Frederik Poulsen, in *Kunstmuseets Aarsskrift*, XIII-XV, 1926–28, p. 24, figs. 32–33 (4th century AD); Timmers, no. 204; *Rijksmuseum van Oudheden te Leiden*, Gids voor de verzameling van griekse en romeinse beeldhouwwerken, 's-Gravenhage, 1966, p. 21, fig. 18 (early 4th century, probably *Helena*, mother of Constantine the Great); Brunsting, p. 189, under no. 10.

s 67
s 68
s 69

S 70a

(bought by Papenbroek at De Witt auction; not to be identified with bust of *Faustina*, listed in Six sale).

Plates: S 70, 70 a.

71. *Portrait Bust of a Man Wearing Priestly Crown*

Engraving: Anonymous.
328 × 197 mm.
ICONES 72.

Sculpture: Skulpturensammlung, Dresden (Inv.no. *Herrmann*, no. 411).
Marble. H.: 0,35 m.
Bust modern; reworked on neck and crown.

Provenance: Gerard Reynst; Great Elector of Brandenburg; by exchange to Dresden in 1724/26.

Bibliography: Le Plat, pl. 162, 5; *Verzeichniss der alten und neuen Bildwerke und übrigen Alterthümer in den Sälen der Kgl. Antikensammlung zu Dresden*, Dresden, 1836, pp. 43–44, no. 147 (King Arsaces XX); Jacobs, p. 34, note 4 (via Berlin to Dresden); Timmers, no. 205.

Plates: S 71, S 71a.

72. *Male Portrait Bust*

Engraving: Anonymous.
327 × 193 mm.
ICONES 73.

Sculpture: Lost.

Provenance: Gerard Reynst.

Bibliography: Timmers, no. 206.

Plate: S 72.

73. *Male Portrait Bust*

Engraving: Anonymous.
328 × 196 mm.
ICONES 74.

Sculpture: Lost.

Provenance: Gerard Reynst.

Bibliography: Timmers, no. 207.

Plate: S 73.

74. *Male Bust*

Engraving: Anonymous.
327 × 196 mm.
ICONES 75.

Sculpture: Lost.

Provenance: Gerard Reynst.

Bibliography: Timmers, no. 208.

Plate: S 74.

S 70
S 71
S 72

S 71a

S 73
S 74
S 75

75. *Male Bust*

Engraving: Anonymous.
326 × 193 mm.
ICONES 76.

Sculpture: Lost.

Provenance: Gerard Reynst.

Bibliography: Timmers, no. 209.

Plate: S 75.

76. *Male Bust*

Engraving: Anonymous.
326 × 192 mm.
ICONES 77.

Sculpture: Lost.

Provenance: Gerard Reynst.

Bibliography: Timmers, no. 210.

Plate: S 76.

77. *Portrait Bust of a Philosopher*

Engraving: Anonymous.
326 × 194 mm.
ICONES 78.

Sculpture: Lost.

Provenance: Gerard Reynst.

Bibliography: Timmers, no. 211.

Plate: S 77.

78. *Ideal Portrait Bust of a Poet (?)*

Engraving: Anonymous.
327 × 198 mm.
ICONES 79.

Sculpture: Lost.

Provenance: Andrea Vendramin, Venice (*cat.* II, 28); Gerard Reynst.

Bibliography: Jacobs, p. 26, note 1 (concordance of ICONES with *De Sculpturis*); Timmers, no. 212.

Plate: S 78.

79. *Female Bust*

Engraving: Anonymous.
326 × 195 mm.
ICONES 80.

Sculpture: Lost.

Provenance: Gerard Reynst.

Bibliography: Timmers, no. 213.

Plate: S 79.

80. *Laureate Male Bust*

Engraving: Anonymous.
328 × 196 mm.
ICONES 81.

Sculpture: Lost.

Provenance: Gerard Reynst.

Bibliography: Timmers, no. 214.

Plate: S 80.

81. *Laureate Male Bust*

Engraving: Anonymous.
325 × 192 mm.
ICONES 82.

Sculpture: Lost.

Provenance: Gerard Reynst.

Bibliography: Timmers, no. 215.

Plate: S 81.

S 76
S 77
S 78

SCULPTURES IN THE COLLECTION OF GERARD REYNST

S 79
S 80
S 81

S 82
S 83

208 SCULPTURES IN THE COLLECTION OF GERARD REYNST

s 85a 82. *Female Portrait Bust*

Engraving: Anonymous.
326 × 198 mm.
ICONES 83.

Sculpture: Lost.

Provenance: Gerard Reynst.

Bibliography: Timmers, no. 216.

Plate: s 82.

Similar to portraits of the young *Julia Domna* (married ca. 185 AD to Septimius Severus); compare marble head in the Musée St. Raymond, Toulouse (E. Espérandieu, *Recueil général des bas-reliefs ... de la Gaule romaine*, Paris, 1907 ff., II, pp. 77–78, no. 979).

83. *Female Bust as Oratress* (?)

Engraving: Anonymous.
327 × 194 mm.
ICONES 84.

Sculpture: Lost.

Provenance: Gerard Reynst.

Bibliography: Timmers, no. 217.

Plate: s 83.

84. *Male Portrait Bust*

Engraving: Anonymous.
326 × 193 mm.
ICONES 85.

Sculpture: Lost.

Provenance: Gerard Reynst; Gosuin Uilenbroek, Amsterdam.

Bibliography: Sigebert Havercamp, *Museum Uilenbroekianum*, Amsterdam, 1741, p. 288, no. 7 (reference to Reynst collection; as *Hadrian*); Timmers, no. 218.

Plate: s 84.

85. *Male Portrait Bust*

Engraving: Anonymous.
327 × 196 mm.
ICONES 86.

Sculpture: Rijksmuseum van Oudheden, Leiden (Inv.no. Pb. 109). White marble. Head: 0,25 m.; Bust: 0,17 m. Nose, ears, part of beard, bust modern.

Provenance: Gerard Reynst; Gerard van Papenbroek, Amsterdam (*cat.* no. 109).

Bibliography: Oudendorp, II, no. 21 (reference to Reynst ICONES; as *T. Antoninus Pius*); Janssen, I,

s 84
s 85
s 86

SCULPTURES IN THE COLLECTION OF GERARD REYNST

no. 132 (probably *Alexander Severus*; nose, ears and part of beard modern); Timmers, no. 219.

Plates: s 85, s 85a.

86. *Female Bust as Oratress* (?)

Engraving: Anonymous.
327 × 192 mm.
ICONES 87.

Sculpture: Lost.

Provenance: Gerard Reynst.

Bibliography: Timmers, no. 220.

Plate: s 86.

87. *Female Bust*

Engraving: Anonymous.
329 × 196 mm.
ICONES 88.

Sculpture: Lost.

Provenance: Gerard Reynst.

Bibliography: Timmers, no. 221.

Plate: s 87.

88. *Bust of Julius Caesar* (?)

Engraving: Anonymous.
326 × 197 mm.
ICONES 89.

Sculpture: Rijksmuseum van Oudheden, Leiden (Inv.no. Z.⁰⁷/9¹; old number B 54).
White marble. H. 0,35 m. Bust modern.

Provenance: Gerard Reynst; possibly Jan Six, Amsterdam (sale Amsterdam, 6 April 1702, no. 2 or no. 15: *Julius Caesar, dito Borststuk, Antiq.* and *Julius Caesar dito*); possibly Gosuin Uilenbroek, Amsterdam (see under Bibliography).

Bibliography: Sigebert Havercamp, *Museum Uilenbroekianum*, Amsterdam, 1741, p. 288, no. 5 (*Julii Caesaris nudum caput, pectore tenus, in pectore caput alatum, ex Marmore Graeco. Inter cones D. Reinst num 92*, which might refer to this bust); Brunsting, p. 189, no. 15 (possibly ICONES 89).

Plates: s 88, s 88a.

89. *Male Portrait Bust*

Engraving: Anonymous.
325 × 191 mm.
ICONES 90.

s 88a

s 87
s 88

Sculpture: No longer traceable.

Provenance: Gerard Reynst; Gerard van Papenbroek, Amsterdam.

Bibliography: Oudendorp, II, no. 41 (reference to Reynst ICONES); Janssen, I, no. 139; Timmers, no. 213.

Plate: s 89.

The bust cannot be identified among the sculptures in the Rijksmuseum van Oudheden, Leiden. The two heads (both Inv.no. Pb. 129?) associated with Oudendorp's description are unconvincing.

90. *Bust of a Youth (Caracalla ?)*

Engraving: Anonymous.
326 × 192 mm.
ICONES 91.

Sculpture: Lost.

Provenance: Gerard Reynst.

Bibliography: Timmers, no. 224.

Plate: s 90.

91. *Female Bust*

Engraving: Anonymous.
325 × 195 mm.
ICONES 92.

Sculpture: Lost.

Provenance: Andrea Vendramin, Venice (*cat.* II, 22); Gerard Reynst; Gosuin Uilenbroek, Amsterdam.

Bibliography: Sigebert Havercamp, *Museum Uilenbroekianum*, Amsterdam, 1741, pp. 287-8, no. 4 (with reference to Reynst collection; as *Bacchae*); Jacobs, p. 26, note I (concordance of ICONES with *De Sculpturis*) and p. 33, note 3 (Uilenbroek collection); Timmers, no. 225.

Plate: s 91.

92. *Half-figure Statue of a Faun*

Engraving: Anonymous.
324 × 193 mm.
ICONES 93.

Sculpture: Lost.

Provenance: Andrea Vendramin, Venice (*cat.* I, 21); Gerard Reynst; possibly Jan Six, Amsterdam (sale Amsterdam, 6 April 1702, no. 25: *Een schoon Terminus*).

Bibliography: Clarac, III, p. CCLXVI (with Gerard Reynst); Jacobs, p. 26, note I (concordance of ICONES with *De Sculpturis*); Timmers, no. 150; Brunsting, p. 189, no. 25 (possibly formerly in Jan Six collection).

Plate: s 92.

93. *Bust of a Poet*

Engraving: Anonymous.
329 × 196 mm.
ICONES 94.

Sculpture: Lost.

Provenance: Gerard Reynst.

Bibliography: Timmers, no. 226.

Plate: s 93.

s 89
s 90
s 91

SCULPTURES IN THE COLLECTION OF GERARD REYNST

94. *Herm of a Bearded God*

Engraving: Anonymous.
326 × 192 mm.
ICONES 95.

Sculpture: Skulpturensammlung, Dresden (Inv.no. *Herrmann*, no. 69).
Marble. H. 0,45 m.

Provenance: Gerard Reynst; Great Elector of Brandenburg; by exchange to Dresden in 1724/26.

Bibliography: L. Beger, *Thesauri Regii et Electorialis Brandenburgici ...*, III, Coloniae Marchicae (1696), pp. 322–23 (ill.; as *Plato*); Le Plat, pl. 155, 3; Oudendorp, II, no. 30 (erroneously identifies *Bacchus-Indicus*, Papenbroek, *cat*.no. 118, with this bust; see cat. no. 114); Timmers, no. 227 (as *Aesculapius*).

Plates: S 94, S 94a.

95. *Bust of Apollo (?) or Diana (?)*

Engraving: Anonymous.
328 × 192 mm.

S 94a

S 92
S 93
S 94

212 SCULPTURES IN THE COLLECTION OF GERARD REYNST

S 96a
S 98a

S 95
S 96
S 97
S 98

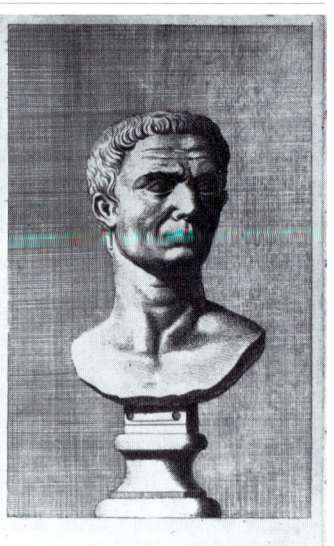

ICONES 96.

Sculpture: Rijksmuseum van Oudheden, Leiden (Inv.no. Pb. 104). White marble. H. 0,19 m. Bust removed and no longer traceable.

Provenance: Gerard Reynst; Nicolaes Witsen, Amsterdam; Gerard van Papenbroek, Amsterdam (*cat*. no. 104).

Bibliography: Oudendorp, II. no. 16 (reference to Reynst ICONES; from Witsen; female head); Janssen, I, no. 117 (*Diana*); Timmers, no. 229 (*Apollo*).

Plates: s 95, s 95a.

96. *Female Head to Left* (top)

Engraving: Anonymous.
327 × 191 mm.
ICONES 97.

Sculpture: Rijksmuseum van Oudheden, Leiden (Inv.no. Pb. 114 [head]).
White marble. Head: 0,09 m. (Added to statuette of *Hygieia*, formerly in Vendramin collection; see cat. no. 113).

Provenance: Andrea Vendramin, Venice; Gerard Reynst; Gerard van Papenbroek, Amsterdam (*cat*. no. 114).

Bibliography: See under *Hygieia*, cat. no. 113.

Plates: s 96, s 96a.

97. *Male Head to Right* (bottom) s 95a

Sculpture: Lost.

Provenance: Gerard Reynst.

Bibliography: Timmers, no. 228.

Plate: s 97.

98. *Male Portrait Head*

Engraving: Anonymous.
327 × 193 mm.
ICONES 98.

Sculpture: Rijksmuseum van Oudheden, Leiden (Inv.no. Pb. 137). White marble. Head: 0,40 m. Modern, nose restored; base lost.

Provenance: Gerard Reynst; Nicolaes Witsen, Amsterdam; Gerard van Papenbroek, Amsterdam (*cat*. no. 137).

Bibliography: Oudendorp, II, no. 49 (reference to Reynst ICONES; from Witsen; *Germanicus*); Janssen, I, no. 127; Timmers, no. 230.

Plates: s 98, s 98a.

S 106a
S 108a

reproduced in the manuscript inventory at Windsor Castle, *Busts & Statues in Whitehall Gardens* (Inv.no. 8922, top right; Pl. S 106a). No longer traceable in the English royal collections.

107. *Male Portrait Bust*

Engraving: Anonymous.
324 × 194 mm.
ICONES I: *Commodus*.

Sculpture: Lost.

Provenance: Andrea Vendramin, Venice (*cat.* II, 27); Gerard Reynst; Charles II of England.

Bibliography: Knorr 1663 (Fuchs, p. 241: *Atque hinc ablatae sunt statuae…Commodi…Angliae Regi oblatae*); Jacobs, p. 26, note 1 (concordance of ICONES with *De Sculpturis*); Timmers, no. 235.

Plate: S 107.

108. *Portrait Bust of Faustina the Younger*

Engraving: Anonymous.
324 × 194 mm.
ICONES K: *Faustina*.

Sculpture: Hampton Court Palace.
Marble.

Provenance: Gerard Reynst; Charles II of England.

Provenance: Andrea Vendramin, Venice (*cat.* II, 29); Gerard Reynst; Charles II of England.

Bibliography: Jacobs, p. 26, note 1 (concordance of ICONES with *De Sculpturis*); Timmers, no. 234.

Plates: S 106, S 106a.

Probably identical with the bust

S 105
S 106
S 107

Bibliography: Knorr 1663 (Fuchs, p. 241: *Atque hinc ablatae sunt statuae...Faustinae... Angliae Regi oblatae*); Timmers, no. 236.

Plates: s 108, s 108a, b.

This is the only sculpture from the 'Dutch Gift' to Charles II in 1660 that has been identified. It apparently was placed in the Whitehall Gardens and thus escaped the fire in 1696. The bust was reproduced in the manuscript inventory *Busts & Statues in Whitehall Gardens*, kept at Windsor Gardens, kept at 8923, bottom left; Pl s 108a).

109. *Male Portrait Bust*

Engraving: Anonymous.
324 × 193 mm.
ICONES L: *Scipio Africanus*.

Sculpture: Lost.

Provenance: Gerard Reynst; Charles II of England.

Bibliography: Knorr 1663 (Fuchs, p. 241: *Atque hinc ablatae sunt statuae...Scipionis Africani...Angliae Regi oblatae*); Timmers, no. 237.

Plate: s 109.

110. *Male Portrait Bust*

Engraving: Anonymous.
322 × 193 mm.

ICONES M: *M. Brutus*.   s 108b

Sculpture: Lost.

Provenance: Andrea Vendramin, Venice (*cat.* II, 15); Gerard Reynst; Charles II of England.

Bibliography: Knorr 1663 (Fuchs, p. 241: *Atque hinc ablatae sunt statuae...M. Bruti...Angliae Regi oblatae*); Jacobs, p. 26, note 1 (concordance of ICONES with *De Sculpturis*); Timmers, no. 238.

Plate: s 110.

s 108
s 109
s 110

SCULPTURES ILLUSTRATED IN THE VENDRAMIN CATALOGUE 'DE SCULPTURIS' OR WITH AN OLD PROVENANCE FROM THE VENDRAMIN COLLECTION, BUT NOT ENGRAVED FOR THE ICONES

111. *Statuette of a Youth*
Not in ICONES.

Sculpture: Staatliche Museen, East Berlin (Inv.no. Sk. 521). White marble, H. 0,91 m. Restored (according to Conze, *loc.cit.*, probably already during 16th century) are head, section of arms holding dish and fruit, legs from middle of thighs down; support and plinth.

Provenance: Andrea Vendramin, Venice (*cat.* 1, 18: *Bacco*); Gerard Reynst; Great Elector of Brandenburg.

Bibliography: L. Beger, *Thesauri Regii et Electorialis Brandenburgici...*, III, Coloniae Marchicae (1696), pp. 286–88 (ill.; as *Triptolemus*); E. Gerhard, *Berlin's antike Bildwerke*, I, Berlin, 1836, p. 51, no. 49, d; (Alexander Conze), *Königliche Museen zu Berlin, Beschreibung der antiken Skulpturen ...*, Berlin, 1891, p. 204, no. 521 (insignificant work); Reinach, II, p. 590, no. 6 (ill.);

Jacobs, p. 35 (from Vendramin and Reynst collections; to Berlin via Uylenburgh).

Plate: S 111.

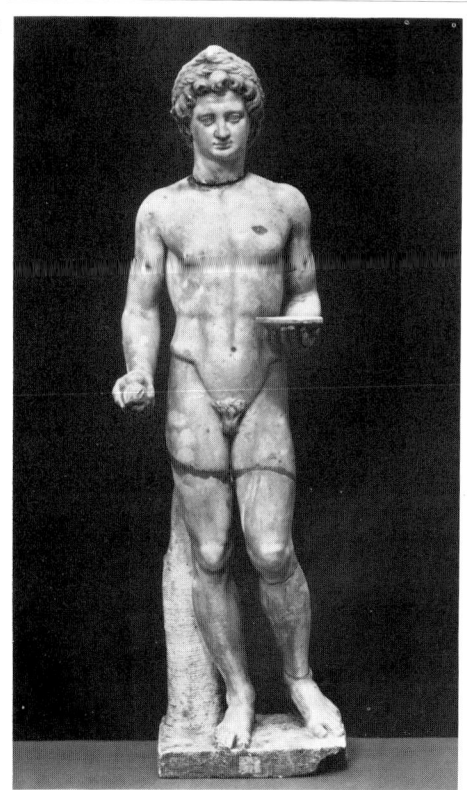
S 111

SCULPTURES IN THE RIJKSMUSEUM VAN OUDHEDEN, LEIDEN, FORMERLY IN THE COLLECTION OF ANDREA VENDRAMIN (BASED ON OUDENDORP)

112. *Statuette of Venus with Cupid*

Not in ICONES.

Sculpture: Rijksmuseum van Oudheden, Leiden (Inv.no. Pb. 101). White, fine grain marble. Torso: 0,34 m. Head: 0,11 m. Both hands, wrists, right leg from knee, left leg from middle of thigh, and both feet missing. Legs of Venus and Cupid modern (0,26 m). Head perhaps modern and from different statue.

Provenance: Andrea Vendramin, Venice; Gerard Reynst; Gerard van Papenbroek, Amsterdam (*cat.* no. 101).

Bibliography: Oudendorp, II, no. 13 (formerly in Vendramin collection); Brants, no. 35, pl. XVII, 35 (motif of Capitoline Venus; poor work).

Plate: S 112.

113. *Statuette of Hygieia, Sitting on a Throne*

Statuette not in ICONES, only head of *Hygieia* (see ICONES 97; cat. no. 96).

Sculpture: Rijksmuseum van Oudheden, Leiden (Inv.no. Pb. 114). White, fine grain marble. H.: 0,39 m. Both forearms, lower part of cornucopia, greater part of serpent missing; chair hollowed at back. Head does not belong.

Provenance: Andrea Vendramin, Venice; Gerard Reynst; Gerard van Papenbroek, Amsterdam (*cat.* no. 114).

Bibliography: Oudendorp, II, no. 26 (formerly in Vendramin collection); Janssen, I, no. 70 (*Hygieia*); Brants, no. 43, pl. XVIII, 43 (probably not antique).

Plate: S 96a.

114. *Herm of Bacchus-Indicus*

Not in ICONES.

Sculpture: Rijksmuseum van Oudheden, Leiden (Inv.no. Pb. 118). White marble. Head: 0,14 m. Nose damaged; back of herm missing; bust modern.

Provenance: Andrea Vendramin, Venice; Gerard Reynst; Gerard van Papenbroek, Amsterdam (*cat.* no. 118).

Bibliography: Oudendorp, II, no. 30 (incorrect reference to Reynst ICONES 95 [for the latter see cat. no. 94]; formerly in Vendramin collection); Janssen, I, no. 107 (nose damaged; back missing; bust modern).

Plate: S 114.

115. *Standing Figure of a Draped Woman*

Not in ICONES.
Sculpture: Rijksmuseum van Oudheden, Leiden (Inv.no. Pb. 125).

S 112
S 114

Yellowish-gray marble. H.: 0,30 m.
Upper part of head and mantle
missing. Back unfinished.

Provenance: Andrea Vendramin,
Venice; Gerard Reynst; Gerard
van Papenbroek, Amsterdam
(*cat.* no. 125).

Bibliography: Oudendorp, II,
nos. 37/38 (formerly in Vendramin
collection); Janssen, I, no. 85
(*Vestal*); Brants, no. 41, pl. XVIII, 41
(probably from Asia minor;
attitude of so-called *Pudicitia*).

Plate: S 115.

116. *Large Right Foot*

Not in ICONES

Sculpture: Rijksmuseum van Oud-
heden, Leiden (Inv. no. Pb. 144).
White marble, H.: 0,35 m.;
L.: 0,55 m.

Provenance: Andrea Vendramin,
Venice; Gerard Reynst; Gerard van
Papenbroek, Amsterdam
(*cat.* no. 144).

Bibliography: Knorr 1663 (Fuchs,
p. 242: *Membra quaedam humana
marmorea, inter caetera pes colossae
statuae, arte multa insignis*); Oudendorp, II, no. 56 (formerly in
Vendramin collection); Janssen, I,
no. 197.

Plate: S 116.

S 115
S 116

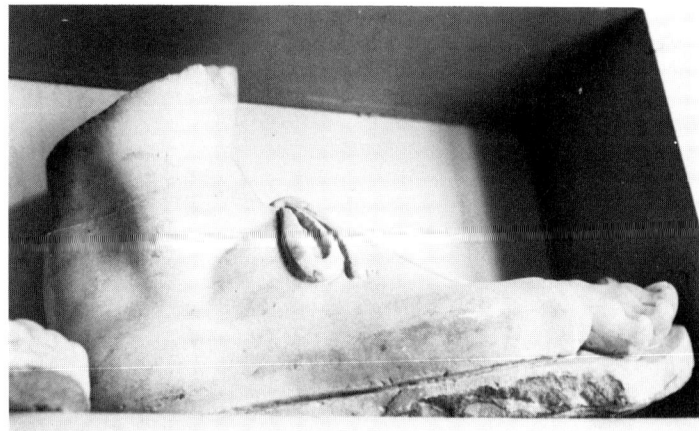

TOMB MONUMENTS FORMERLY IN THE COLLECTION OF GERARD REYNST, REPRODUCED IN THE VENDRAMIN CATALOGUE 'DE ANTIQUORUM TUMULIS', VENICE, 1627*

117. *Funerary Chest*

Rijksmuseum van Oudheden, Leiden (Inv.no. Pb. 32, a,b). Marble. H. 0,315 m.; b.: 0,44 m.

Provenance: Andrea Vendramin, Venice (*cat.*, fol. 2); Gerard Reynst; Nicolaes Witsen, Amsterdam; Gerard van Papenbroek, Amsterdam (*cat.* no. 32 a,b).

Bibliography: Oudendorp, I, no. 32.

Plates: S 117, S 117a.

118. *Tomb Monument*
Rijksmuseum van Oudheden, Leiden (Inv.no. Pb. 79). Marble. H. 0,215 m.; b.: 0,25 m.

Provenance: Andrea Vendramin, Venice (*cat.*, fol. 6); Gerard Reynst; Gerard van Papenbroek, Amsterdam (*cat.* no. 79; found in a canal in Amsterdam).

Bibliography: Oudendorp, I, no. 79; Janssen, I, no. 315.

Plates: S 118, S 118a.

119. *Tomb Relief of Two Couples*

Centraal Museum, Utrecht.

Roman, Northitalian, 2nd quarter of 1st century AD.
Limestone. H.: 1,23 m.; b.: 0,59m.

Inscription: MINGONIUSNI/MARCEL-LUSIH/SIBILIAGISIACAE/EFLUCI-LIAEUXSO (Marcus Ingonius Marcellus, son of Marcus and Sibilia Agisiaca Lucilia, daughter of Lucius).

Provenance: Andrea Vendramin, Venice (*cat.*, fol. 9); Gerard Reynst.

Exhibition: *Klassieke kunst uit particulier bezit*, Leiden (Rijksmuseum van Oudheden), 1975, no. 25 (references to Vendramin and Reynst collections).

Bibliography: *Centraal Museum Utrecht, Mededelingen*, 12, 1975, ill. p. 15 (reference to Reynst).

Plates: S 119, S 119a.

120. *Tomb Relief*

Rijksmuseum van Oudheden, Leiden (Inv. no. Pb. 53). Marble. H.: 0,585 m.; b.: 0,295 m.

Inscription:
ΖΗΝΟΔΟΤΟΣ ΜΕΝΕΣΤΡΑΤΟΥ.
ΜΕΝΕΣΤΡΑΤΟΣ ΖΗΝΟΔΟΤΟΥ.
ΔΗΜΗΤΡΙΑ ΦΙΛΟΞΕΝΟΥ.

Provenance: Andrea Vendramin, Venice (*cat.*, fol. 17); Gerard Reynst; Jan Six, Amsterdam; Gosuin Uilenbroek, Amsterdam (bought at Six auction; presented to Papenbroek; see below); Gerard van Papenbroek, Amsterdam (*cat.* no. 53).

Bibliography: Jacobi Tollii, *Epistolae Itinerariae...*, Amsterdam, 1700, p. 4 (incorrectly associated with another relief); Oudendorp, I, no. 53; Janssen, I, no. 256; Brunsting, p. 189, under no. 41 (3 reliefs).

Plates: S 120, S 120a.

This tomb relief is also mentioned in the Papenbroek manuscript, Cod. 17 at the University Library in Leiden, based on a passage copied from a manuscript written by Gosuin Uilenbroek (p. LII: EX autographo GOSVINI VILEN-BROECK. pag. 5, 6, 7 and 8: ... [with respect to Tollius] de inscriptie ...op pag. 4... behoort tot een ander Marmer met een geheel ander Afbeeltsel, 't welk de Eigenaar dezes te gelijk met het voorige met de Griekscbe Veersen uit de Auctie van welgemelden Heer Burgermeester Six gekocht en veele jaaren bezeeten heeft; doch onlanx aan den weledelen Achtbaaren Heere Gerard van Papenbroeck out President Schepen deezer Stadt enz. zijnen zeer geeerden en hoog geachten vriendt heeft geoffereert om te dienen tot een weergade van een ander in zijn weledelheits uitstekent en voortreffelijk Cabinet van Antijke Beelden, Busten Marmers, inscriptien, enz. die geplaatst zijn in een Gallerij op den Huize te Papenburg bij Velzen, en aldaar te vinden).

121. *Tomb Relief*

Rijksmuseum van Oudheden, Leiden (Inv.no. Pb. 46). Marble. H.: 0,54 m.; b.: 0,618 m.

Provenance: Andrea Vendramin, Venice (*cat.*, fol. 18); Gerard Reynst; Jan Six, Amsterdam; Gerard van Papenbroek, Amsterdam (*cat.* no. 46).

Bibliography: Oudendorp, I, no. 46; Janssen, I, no. 273; Brunsting, p. 189, no. 41 (3 reliefs).

Plates: S 121, S 121a.

122. *Funerary Chest*

Lowther Castle, England.

Provenance: Andrea Vendramin, Venice (*cat.*, fol. 24); Gerard Reynst; Gosuin Uilenbroek, Amsterdam (*Museum Uilenbroekianum*, p. 290, no. 27); Philippe d'Orville.

Bibliography: A. Michaelis, *Ancient Marbles in Great Britain*, Cambridge, 1882, p. 495, no. 53; Jacobs 1925, p. 33, note 3.

---

* Listed in order found in manuscript

222 TOMB MONUMENTS

S 117
S 118

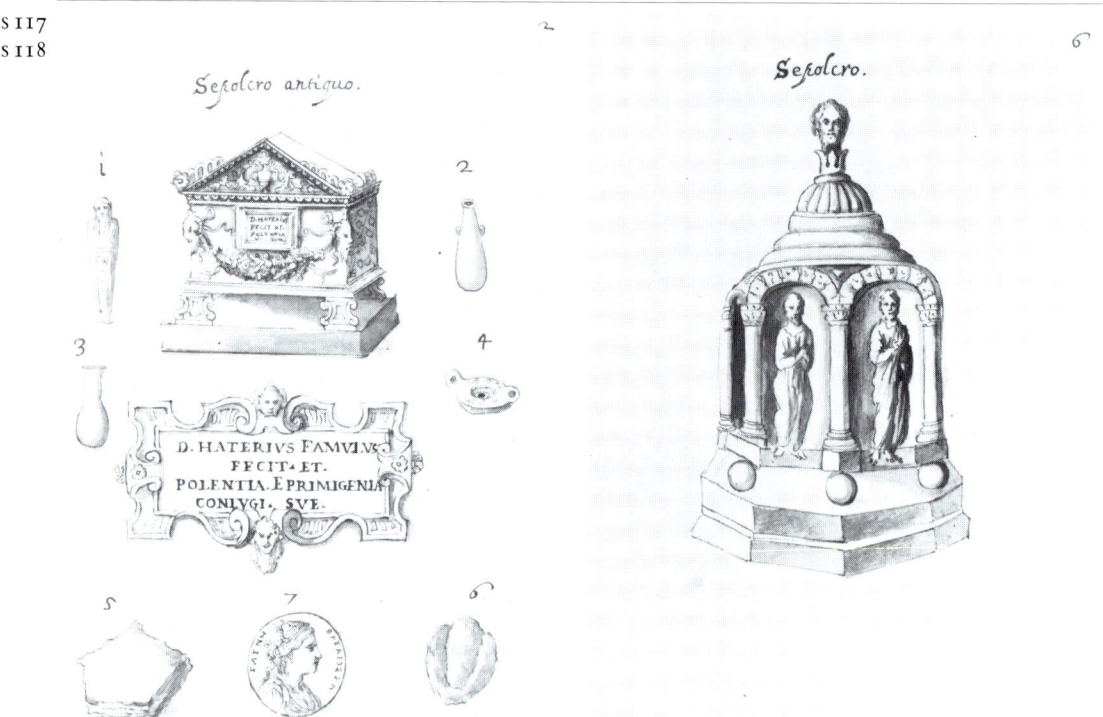

S 117a

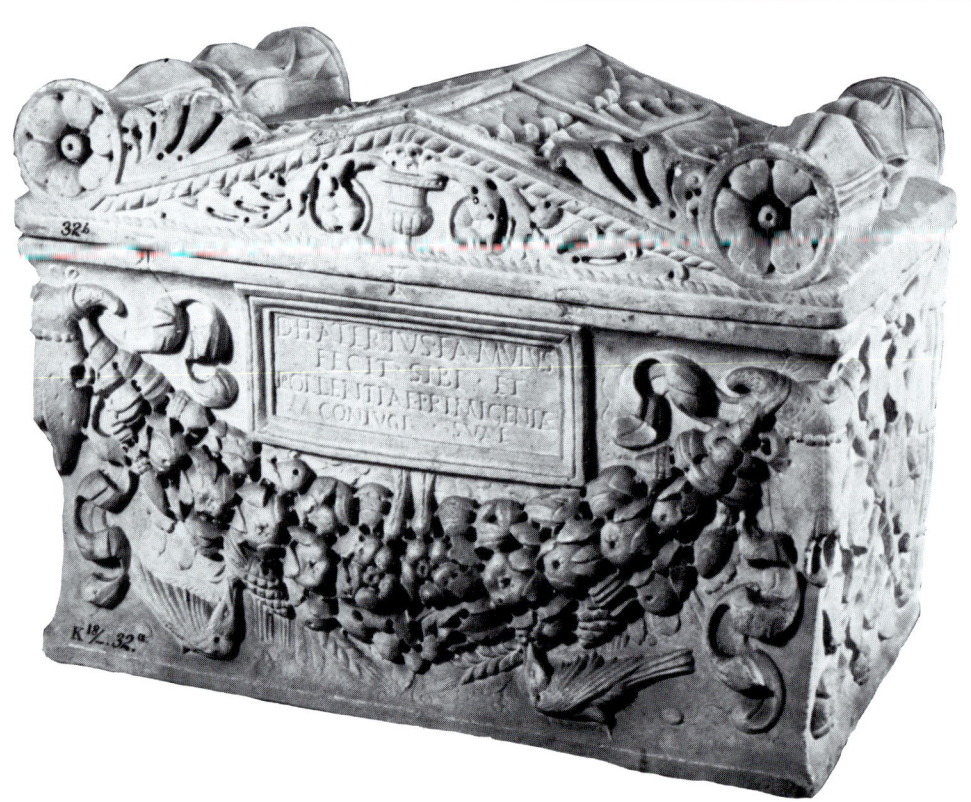

TOMB MONUMENTS

S118a

S119
S119a

S 120

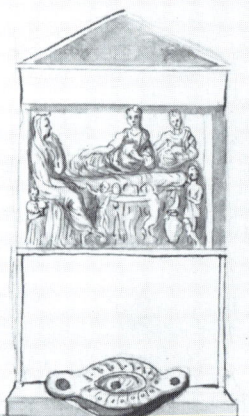

Altro sepolchro di marmo, di mezo rilievo, co'l defonto sedente sopra il lettisterno.

S 121

Memoria, sive sepolchro di Esculapio, di marmo di mezo rilievo; il quale sedente sopra il lettisterno pare che dichiari la medicina alli Ascoltanti, e si vedono in detto rilievo, nella parte superiore alcuni Hieroglifici, cioè, il globbo, alcune figure, et una testa di cavallo.

S 120a

226 TOMB MONUMENTS

S 121a

APPENDIX

DUTCH AND FLEMISH PAINTINGS AND ENGRAVINGS ASSOCIATED INCORRECTLY WITH THE COLLECTION OF GERARD REYNST*

A1. Abraham Bloemaert

St. John the Baptist Preaching

Engraving: Jeremias Falck, *1661* (Heinecken, p. 84, no. 4, should be included in supplement of CAELATURAE instead of engraving after Bassano's *Entombment* [see cat. no. 5]; Stimmel, under no. 39, repeats Heinecken; Hagen, no. 82, with reference to Reynst collection; Wussin, p. 274, under no. 39 and p. 276, under no. 18, repeats Heinecken; Block, no. 14, with reference to Reynst collection; Wurzbach, Jeremias Falck, no. 5, as part of Reynst collection; Hollstein, VI, Jeremias Falck, no. 14). 468 × 595 mm. ($18\frac{7}{16} \times 23\frac{7}{16}$ in.).

Painting: Not identified.

Plate: A 1.

The engraving is included infrequently in the CAELATURAE. Its date *1661* would indicate that Falck continued to work for Reynst's widow after Gerard's death on June 29, 1658. This is unlikely for Falck's letter deploring Reynst's death was written already in Hamburg, on December 10/20, 1658.

The engraving was first associated with the Reynst collection by Heinecken in 1771 in his *Idée générale* and subsequent authors have repeated this information. Unless further evidence is found, Bloemaert's painting most likely was not part of the Reynst collection. Although a number of different pictures of this subject by Abraham Bloemaert are known, the prototype for Falck's engraving remains to be identified.

A2. Cornelis Cornelisz van Haarlem

The Duet

Engraving: Jeremias Falck (Heinecken, p. 84, no. 5, added to CAELATURAE in Dresden, after painting in Reynst collection, attributed to Jan Liss; Stimmel, no. 38, perhaps after Jan Liss, added to CAELATURAE by Winckler; Wussin, p. 274, no. 38 and p. 275, no. 12, after Jan Liss, added by Winckler; Block, no. 157, after Cornelis van Haarlem with reference to Reynst collection; Wurzbach, Jeremias Falck, no. 12, as after Jan Liss; Hollstein, VI, Jeremias Falck, no. 157, after Jan Liss or Cornelis van Haarlem, and XI, after Jan Liss, p. 148, no. 7). 325 × 254 mm. ($12\frac{13}{16} \times$ 10 in.).

Painting: Kunstsammlungen der Georg-August-Universität, Göttingen.
Oil on canvas, 89 × 74 cm.

Provenance: Zschorn collection.

Bibliography: [W. Stechow], *Katalog der Gemäldesammlung der Universität Göttingen*, Göttingen, 1926, no. 36 (copy after Falck's engraving; the original was ca.1656 in Reynst collection); K. Steinbart, *Johann Liss*, Berlin, 1940, p. 172 (wrongly attributed to Liss; engraved for Reynst by Falck

A1

* Listed alphabetically according to artists

PAINTINGS AND ENGRAVINGS ASSOCIATED INCORRECTLY WITH THE COLLECTION OF GERARD REYNST

A 2

A 2a

between 1655 and 1661; painting Göttingen is copy after engraving).

Plates: A 2, A 2a.

Falck's engraving was only added by Winckler to his copy of the CAELATURAE and does not figure in the various other examples. Thus it is highly unlikely that the painting ever was in the Reynst collection. The painting in Göttingen, thought to be a copy (Stechow, 1926; Steinbart, 1940), is now accepted as the original by both Stechow (letter of February 2, 1970) and P.J.J. van Thiel (letter of February 15, 1967).

A 3. Anthony van Dyck (after)

Christ on the Cross

Engraving: Jeremias Falck (Hagen, no. 85, with reference to Reynst collection, as *Christ Carrying the Cross* ; Wussin, p. 275, no. 8, as *Christ Carrying the Cross* ; Block, no. 21, with reference to Reynst collection; not *Christ Carrying the Cross*, but the *Large Crucifixion*, based on altar of St. Michael, Ghent, of ca. 1630; Wurzbach, Jeremias Falck, no. 8, with reference to Reynst collection, as *Christ Carrying the Cross* ; not listed under Falck after Van Dyck; Hollstein, VI, Jeremias Falck, no. 21, as *Christ Carrying the Cross* and VI, after A. van Dyck, Jeremias Falck, no. 204).
600 × 440 mm. (23 $\frac{5}{8}$ × 17 $\frac{3}{8}$ in.).

Painting: Lost.

The engraving by Falck after Van Dyck, listed by Block with the reference that the painting was in the Reynst collection, represents *Christ on the Cross* not *Christ Carrying the Cross*, as was first stated by Hagen.

This very rare print reproduces (based on Block's description) Van Dyck's painting of ca. 1630, commissioned for the main altar of the St. Michael church in Ghent, where it still is. Block's statement, therefore that the painting was in the Reynst collection is erroneous. Falck's print never was included in the CAELATURAE consulted and its association with Reynst thus should be rejected.

A 4. Justus van Egmont

The Virgin as Queen of Heaven

Engraving: Jeremias Falck (Wussin, p. 275, no. 10, as part of Reynst collection according to Hagen and Winckler see below; Block, no. 11, without reference to Reynst collection; Wurzbach, Jeremias Falck, no. 10, as part of Reynst collection; Hollstein, VI, Jeremias Falck, no. 11).
356 × 271 mm. (14 × 10 $\frac{11}{16}$ in.).

Painting: Presumably lost.

The print was only associated with the Reynst collection by Hagen (1848, no. 87), but the reference was not incorporated by Block. Falck's engraving never is included in the CAELATURAE and thus Justus van Egmont's painting most probably was not in the Reynst collection.

A 5. Jacob Jordaens

Flora, Silenus, and Zephyrus

Engraving: Schelte Adams Bolswert (Heinecken, p. 84, no. 3, added to CAELATURAE in Dresden, after a painting in Reynst collection; Hecquet, no. 7; Stimmel, no. 37, as added by Winckler; Wussin, p. 274, no. 1, probably same print as the one listed on p. 275, no. 6 as by Jeremias Falck;

Block, p. 18, no. 4, by S.A. Bolswert; Wurzbach, Jeremias Falck, no. 6; Hollstein, III, S.A. Bolswert, no. 289 and IX, after Jacob Jordaens, S.A. Bolswert, no. 7 and Jeremias Falck, no. 15); Michael Jaffé, *Jordaens* (exh.cat.), Ottawa (National Gallery of Canada), 1968–69, p. 246, no. 305 (ill.; as S.A. Bolswert, based on composition of a Jordaens drawing in Berlin [see below]).
402 × 286 mm. (15 $\frac{13}{16}$ × 11 $\frac{1}{4}$ in.).

Painting: Presumably lost.

Bibliography: Max Rooses, *Jacob Jordaens: His Life and Work*, London, 1908, p. 181, reprod. on p. 239 (for a painting corresponding in composition to this print, in 1905 with Mrs. Parmentier, Knokke; according to R.-A. d'Hulst [see below] a weak workshop piece); Roger-A. d'Hulst, *De Tekeningen van Jacob Jordaens*, Brussels, 1956, pp. 190 and 348, cat. no. 71 (for drawing in Berlin, Staatliche Museen, Inv.no. 2822, dated 1639, related in composition to this print, and for various painted versions corresponding to the drawing); Jaffé, *loc.cit.* (mentions one further painted studio version in a private collection in England); R.-A. d'Hulst, *Jordaens Drawings*, London/New York, 1974, p. 240, under no. A 147 (does not mention Reynst collection).

Plate: A 5.

There is no proof of Heinecken's statement that the print was made after a painting formerly in the Reynst collection.

PAINTINGS AND ENGRAVINGS ASSOCIATED INCORRECTLY WITH THE COLLECTION OF GERARD REYNST

A 6. Pieter van Laer

Marauders Setting a House on Fire

Engraving: Cornelis Visscher (Hecquet, no. 21, without reference to Reynst collection; Wussin, no. 164, without reference to Reynst collection; Wurzbach, no. 164; Hollstein, x, after Pieter van Laer, p. 10, no. 13). 358×289 mm. ($14\frac{1}{8} \times 11\frac{3}{8}$ in.).

Painting: Prince of Liechtenstein, Vaduz (Inv.no. 639). Oil on canvas, 38×29 cm. ($15 \times 11\frac{7}{16}$ in.).

Provenance: Sale Amsterdam, 10 April 1743, Isaak Hoogenbergh, no. 58.

Bibliography: Axel Janeck, *Untersuchung über den holländischen Maler Pieter van Laer, genannt Bamboccio*, Diss. Würzburg, 1968, pp. 75–76, cat. A I 6 (full reference to painting and engraving), and pp. 192–195 (Roman period, before 1635).

See the following catalogue number.

A 7. Pieter van Laer

The Resting Herd

Engraving: Cornelis Visscher (Hecquet, no. 22, without reference to Reynst collection; Wussin, no. 165, without reference to Reynst collection; Wurzbach, no. 165; Hollstein, x, after Pieter van Laer, p. 10, no. 14). 357×289 mm. ($14\frac{1}{16} \times 11\frac{3}{8}$ in).

Painting: Presumably lost.

Bibliography: Axel Janeck, *Untersuchung über den holländischen Maler Pieter van Laer, genannt Bamboccio*, Diss. Würzburg, 1968, p. 95, cat. A IV 6 (reference to engraving; painting lost), and pp. 195–196 (Roman period, before 1635); A. Blankert, 'Over Pieter van Laer als dier- en landschapschilder,' *Oud Holland*, LXXXIII, 1968, pp. 118–119 (refers to painting by or after Van Laer, formerly with Dr. Läuffer, Basel, which was prototype for Visscher's engraving; panel, 31×29 cm, as school of Adriaen van de Velde; ill.).

This and the engraving listed under the previous catalogue number are pendants.

None of the old sources mentions the Reynst collection when describing Visscher's engravings. It seems, therefore, most likely that the two prints were simply added to the CAELATURAE in the Cincinnati Public Library, complementing and immediately preceding the three engravings by Visscher after Van Laer's paintings traditionally associated with the Reynst collection (see catalogue numbers 40–42).

A 8. Pieter van Laer

The Horse Stable

Engraving: Cornelis Visscher (Hecquet, no. 33; Wussin, no. 167,

A 5

inclusion in Reynst collection based on opinion of Rudolf Weigel, and p. 274, no. 40, without name of engraver, included because of Weigel's opinion; not listed by Stimmel; Wurzbach, no. 167; Hollstein, X, after Pieter van Laer, p. 10, no. 16).
295 × 390 mm. ($11\frac{5}{8} \times 15\frac{3}{8}$ in.).

Painting: Presumably lost.

Bibliography:
Axel Janeck, *Untersuchung über den holländischen Maler Pieter van Laer*, Diss. Würzburg, 1968, especially pp. 89–90, cat. A IV 3 (lists two paintings of this subject, one exhibited in Madrid [Sala Vilches], 1944[?], cat. no. 7 [ill.], the other one in Munich, Bayerische Staatsgemäldesammlungen, Depot, Inv.no. 7259; Wussin's reference to the Reynst collection not mentioned), and pp. 212–215 (Roman Period, 1635–1636).

Visscher's engraving differs somewhat from the two known painted versions. The inclusion of Van Laer's painting in the Reynst collection, based primarily on a supposition by Weigel and perpetuated by Wussin, remains highly speculative and needs further proof.

A9. Matthias Stomer (attributed to)
Esau Selling His Birthright to Jacob

Engraving: Jeremias Falck, *1663* (Heinecken, p. 84, no. 2, added to CAELATURAE in Dresden, after painting in Reynst collection; F. Basan, *Catalogue d'Estampes des plus grands Maitres Italiens, Flamands et Français … de M. Mariette*, Paris, 1775, p. 9, no. 85, by Falck after Tintoretto, 1663, supposedly to be inserted in Reynst collection; Stimmel, no. 36, as after Tintoretto, 1663, added to CAELATURAE by Winckler; Hagen, no. 88, with reference to Reynst collection, as after Tintoretto; Wussin, p. 274, no. 36, after Tintoretto, added by Winckler to his copy of the CAELATURAE; Block, no. 1, 1663, after Tintoretto, with reference to Reynst collection; Wurzbach, Jeremias Falck, no. 11, for Reynst collection; Hollstein, VI, Jeremias Falck, no. 1, and IX, after Gerard van Honthorst, no. 12, painting in collection Semenoff, Leningrad).
299 × 386 mm. ($11\frac{3}{4} \times 15\frac{1}{4}$ in.).

Painting: Hermitage, Leningrad (Inv.no. 2913).
Oil on canvas, 118 × 164 cm.

Bibliography: P. Semenov, *Etudes sur les peintres des écoles hollandaise, flamande et néerlandaise qu'on trouve dans la collection Semenov et les autres collections publiques et privées de St.-Pétersbourg*, St.-Pétersbourg, 1906, p. 93, no. 222 (formerly in Reynst collection; by Honthorst); Roberto Longhi, 'Ultimi studi sul Caravaggio e la sua cerchia,' *Proporzioni*, I, 1943, p. 60, note 85 (attributes painting formerly in Reynst collection to Stomer); H. Pauwels, 'De Schilder Matthias

A9

Stomer,' *Gentse Bijdragen tot de Kunstgeschiedenis*, XIV, 1953, pp. 155 and 189 (reference to Reynst collection, whereabouts of painting unknown); Hermitage, *Catalogue of Paintings*, Leningrad/Moscow, 1958, II, p. 272, no. 2913; J. Richard Judson, *Gerrit van Honthorst*, The Hague, 1959, p. 146, under cat. no. 4 (states incorrectly that Falck's engraving was made after Stomer's painting of this subject in Berlin, Inv.no.434, 132 × 166 cm, now lost); M. Shcherbacheva, 'Paintings by Mathias Stomer in the Hermitage (in Russian),' *Studies, Hermitage*, XXV, 1964, p. 25 (ill.; with Reynst in Amsterdam, where engraved by Falck; by Stomer, ca. 1640).

Plates: A 9, A 9a.

Heinecken was the first to mention the Reynst collection when describing Falck's engraving of *1663*. According to Stimmel, the print was merely added by Winckler to his copy of the CAELATURAE. It therefore remains uncertain, whether the painting actually ever was part of the Reynst collection. Although traditionally attributed to Tintoretto through the inscriptions on the engraving, the painting is now believed to be by Matthias Stomer. Falck's engraving however was not executed after the painting formerly in Berlin but after the painting in the Hermitage, formerly in the Semenov collection.

The date *1663* would indicate that Falck continued to work for Reynst's widow after Gerard Reynst's death on June 29, 1658. This is highly unlikely since Falck deplores Reynst's passing away in a letter of December 10–12, 1658, written from Hamburg, where he lived until 1667 or 1668 (Block, *op.cit.*, p. 12).

A 9a

FRENCH AND ITALIAN PAINTINGS AND ENGRAVINGS ASSOCIATED INCORRECTLY WITH THE COLLECTION OF GERARD REYNST

A 10. Jacques Stella

The Virgin and Child, St. John and a Lamb

Engraving: Jeremias Falck (Wussin, p. 275, no. 9, as part of Reynst collection according to Hagen and Winckler [see below]; Block, no. 12, with reference to Reynst collection; Wurzbach, Jeremias Falck, no. 9, as part of Reynst collection; Hollstein, VI, Jeremias Falck, no. 12). 364 × 286 mm. ($14\frac{3}{8} \times 11\frac{1}{4}$ in.).

Painting: Presumably lost.

Hagen (1848, no. 86) added this engraving by Falck to the prints after paintings in the Reynst collection. It is never included in the CAELATURAE, however, and thus Stella's painting most likely was not part of the Reynst collection.

A 11. Caravaggio (Michelangelo Merisi da) (attributed to)

Four Cyclops in a Forge

Engraving: Jeremias Falck (Heinecken, p. 84, added to Dresden copy of CAELATURAE; Stimmel, no. 35, as added by Winckler; Wussin, p. 271, no. 1, in Supplement of CAELATURAE in Dresden, and p. 275, no. 3; Block, no. 54, with reference to Reynst collection; Wurzbach, Jeremias Falck, no. 3, as part of Reynst collection; Hollstein, VI, Jeremias Falck, no. 54). 359 × 333 mm. ($14\frac{1}{8} \times 13\frac{1}{8}$ in.).

Painting: Presumably lost.

Bibliography: Carl Justi, *Diego Velasquez und sein Jahrhundert*, Zürich, 1933, p. 306 (with reference to Reynst collection); A. Pigler, *Barockthemen*, II, Budapest, 1956, p. 256 (as attributed to Caravaggio, with reference to Reynst collection).

Plate: A 11.

Among the prints added as a supplement to the Dresden copy of the CAELATURAE, mentioned by Heinecken, and generally not included in the volume. The painting, therefore, was not in the Reynst collection. No such composition is listed among Caravaggio's authentic works either. It appears more likely that the painting reproduced in Falck's engraving was derived from a work by Palma Giovane, since a very similar composition of a forge surrounded by four blacksmiths is incorporated in his signed picture of *Venus and Amor in front of the Forge of Vulcan*, Cassel (Gemäldegalerie, *Katalog*, Cassel, 1958, p. 108, no. 502, ill.).

A 12. Parmigianino (attributed to)

Lady with a Dog and an Orrery

Painting: Hampton Court Palace (No. 174; Inv.no. 553). Oil on canvas, 40 × 30 in.

Provenance: Charles II (Inventory, no. 258: *A woman sitting in a chayre ... a Gloabe by her and a Dogge : by*

A 11

Parmezano); James II (Catalogue, ms., no. 154: *A Dutch Woman, a globe by her, and a book in one hand*); William III (Catalogue, no. 48: *A German Woman with a Globe*); Queen Anne (Catalogue, no. 87: *by Parmegino*).

Exhibition: London 1946-47, no. 221 (as possibly in Reynst collection and 'Dutch Gift').

Bibliography: Law 1898, pp. 65-66, no. 174 (perhaps from Reynst collection and part of 'Dutch Gift'); Baker 1929, pp. 81-82, no. 174 (North Italian school; possibly from Reynst collection and part of 'Dutch Gift'); S.J. Freedberg, *Parmigiano*, Cambridge, 1950, p. 227 (under attributed works; Emilian [?], ca. 1540); Mahon 1950, p. 17, note 6 (points out that reference to Reynst collection was first introduced by Law, *loc.cit.*, and that it is pure speculation); Bernard Berenson, *Italian Pictures of the Renaissance, Central and North Italian Schools*, 1, London, 1968, p. 319 (as copy after Parmigianino?).

There is no reference in Charles II's inventory that this painting was part of the 'Dutch Gift'. The speculative provenance from the Reynst collection, first introduced by Law, should therefore be rejected.

LIST OF ITALIAN PAINTINGS IN THE COLLECTION OF GERARD REYNST

FEDERICO BAROCCI

Woman with a Dog
Cat. 1
Gerard and Jan Reynst (?)
Charles II
'Dutch Gift' 1660
Lost

JACOPO BASSANO

Abraham Leaving Haran
Cat. 2 ; Pl. P 2
Gerard Reynst
CAELATURAE 34
Lost

Annunciation to the Shepherds
Cat. 3 ; Pl. P 3
Gerard Reynst
CAELATURAE 33
Lost

Christ Carrying the Cross
Cat. 4 ; Pls. P 4, P 4a
Gerard Reynst
CAELATURAE 12
Charles II 'Dutch Gift' 1660
Weston Park, Earl of Bradford

Entombment of Christ
(alternate for Tintoretto's
Pietà)
Cat. 5 ; Pl. P 5
Gerard Reynst
CAELATURAE 14
Lost

Virgin and Child and St. John
Cat. 6 ; Pl. P 6
Gerard Reynst
CAELATURAE 11
Presumably lost

BONIFAZIO VERONESE

Virgin and Child
Cat. 7 ; Pls. P 7, P 7a
Gerard Reynst
CAELATURAE 22
Charles II
no reference to 'Dutch Gift'
(1660?) Hampton Court,
Inv. no. 140

PARIS BORDONE

*Portrait of a Man
Holding a Document*
Cat. 8 ; Pl. P 8
Gerard and Jan Reynst (?)
Charles II
'Dutch Gift' 1660
Hampton Court, Inv. no. 52

CARIANI

Reclining Venus
Cat. 9 ; Pl. P 9
Andrea Vendramin,
Venice ;
Gerard Reynst
Charles II
'Dutch Gift' 1660
Hampton Court, Inv. no. 1103

DOMENICO FETTI

Vision of St. Peter
Cat. 10 ; Pl. P 10
Gerard Reynst
CAELATURAE 15
Presumably lost

GIORGIONE (ATTR. TO)

The Concert
Cat. 11 ; Pls. P 11, P 11a
Gerard Reynst
CAELATURAE 27
Charles II
'Dutch Gift' 1660
Hampton Court, Inv. no. 554

GIULIO ROMANO

Isabella d'Este
Cat. 12 ; Pls. P 12, 12a
Gerard Reynst
CAELATURAE 9
Charles II
no reference to 'Dutch Gift'
(1660?)
Hampton Court, Inv. no. 76

GUERCINO

Semiramis
Cat. 13 ; Pls. P 13, 13a
Daniele Ricci, 1624
Gerard Reynst
CAELATURAE 29
Descendants of Charles II :
Dukes of Grafton (1660?)
Boston, Museum of Fine Arts,
Inv. no. 48.1028

JAN LISS

St. Paul in Ecstasy
Cat. 14 ; Pls. P 14, 14a
Gerard Reynst
CAELATURAE 16
Van de Amory, sale 1722 (?)
Berlin, Staatl. Museen,
Inv. no. 1858

The Brothel
Cat. 15 ; Pls. P 15, P 15a
Gerard Reynst
CAELATURAE 30
Seen by Knorr in 1663 (?)
Jan Six, Amsterdam
Pieter Six, Amsterdam
Cassel, Gemäldegalerie,
Inv. no. 187

LORENZO LOTTO

Andrea Odoni
Cat. 16 ; Pls. P 16, P 16a
Andrea Odoni,
Venice ;
Gerard Reynst
CAELATURAE 20
Charles II
no reference to 'Dutch Gift'
(1660?)
Hampton Court, Inv. no. 72

Portrait of a Gentleman
Cat. 17 ; Pls. P 17, P 17a
Gerard Reynst
CAELATURAE 3
Charles II
'Dutch Gift' 1660
Hampton Court,
Inv. no. 486

ITALIAN PAINTINGS IN THE COLLECTION OF GERARD REYNST

LORENZO LOTTO
(ATTR. TO)

Adoration of the Shepherds
Cat. 18 ; Pl. P 18
Jan Reynst
Gerard Reynst
CAELATURAE 20
Lost

MARCO D'OGGIONO
(ATTR. TO)

Christ Child and St. John
Cat. 19 ; Pl. P 19
Gerard and Jan Reynst (?)
Charles II
'Dutch Gift' 1660
Hampton Court,
Inv. no. 391 (?)

JACOPO PALMA
THE YOUNGER

Dance of Naked Children
Cat. 58 ; Pl. P 58
Gerard Reynst
Jan Six, Amsterdam
Jhr. Six van Hillegom,
Amsterdam

PARMIGIANINO

Athena
Cat. 20 ; Pls. P 20, P 20a
Gerard Reynst
CAELATURAE 1
Charles II
'Dutch Gift' 1660
Windsor Castle,
Inv. no. 1138

SCHOOL OF RAPHAEL

Virgin and Child and St. Anne
Cat. 21 ; Pls. P 21A, B
Gerard Reynst
CAELATURAE 10, 18
Presumably lost

SCHOOL OF RAPHAEL (?)

*Christ on a Lamb, the Virgin
and St. Joseph*
Cat. 22
Gerard and Jan Reynst
Charles II
'Dutch Gift' 1660
Not securely identifiable
in royal collections

GUIDO RENI

Susannah and the Elders
Cat. 23 ; Pls. P 23, P 23a
Gerard Reynst
CAELATURAE 28
Seen by Knorr in 1663 (?)
Not securely
identifiable

GUIDO RENI (ATTR. TO)

St. Bartholomew
Cat. 24 ; Pl. P 24
Gerard Reynst
CAELATURAE 6
Seen by Knorr in 1663 (?)
Lost

St. John the Evangelist
Cat. 25 ; Pls. P 25, P 25a
Gerard Reynst
CAELATURAE 17
Seen by Knorr in 1663 (?)
Museum der bildenden Künste,
Leipzig, Inv.no. 193

Allegory of Painting
Cat. 26 ; Pl. P 26
Gerard Reynst
CAELATURAE 7
Perhaps James II
Not securely identifiable
in royal collections

ANDREA SCHIAVONE

Presentation in the Temple
Cat. 27 ; Pl. P 27
Gerard Reynst
CAELATURAE 21
Lost

Christ Before Pilate
Cat. 28 ; Pl. P 28
Jan Reynst
Gerard Reynst
Charles II
'Dutch Gift' 1660
Hampton Court,
Inv.no. 522

Judgment of Midas
Cat. 29 ; Pl. P 29
Jan Reynst
Gerard Reynst
Charles II
'Dutch Gift' 1660
Hampton Court,
Inv.no. 470

BERNARDO STROZZI

The Old Courtesan
Cat. 30 ; Pls. P 30, P 30a
Gerard Reynst
CAELATURAE 8
Moscow, Pushkin Museum
or Bologna, private collection

JACOPO TINTORETTO

Pietà
(alternate for Bassano's
Entombment)
Cat. 31 ; Pl. P 31
Gerard Reynst (?)
(CAELATURAE 14)
Lost

Portrait of a Dominican Friar
Cat. 32 ; Pls. P 32, P 32a
Gerard Reynst
CAELATURAE 4
Charles II
'Dutch Gift' 1660
Hampton Court,
Inv. no. 772

TITIAN

*Portrait of a Man
('Aretino')*
Cat. 33 ; Pl. P 33
Gerard Reynst
CAELATURAE 5
William III
N. A. Flinck, Rotterdam
Lost

Portrait of a Man
('*Sannazaro*')
Cat. 34 ; Pls. P 34, P 34a
Gerard Reynst
CAELATURAE 2
Charles II
'Dutch Gift' 1660
Hampton Court,
Inv. no. 68

TITIAN (ATTR. TO)

*Holy Family with St. John
and St. Elizabeth*
Cat. 35 ; Pl. P 35
Andrea Odoni,
Venice (?)
Gerard Reynst
CAELATURAE 25
Charles II
'Dutch Gift' 1660
Presumably lost

TITIAN (SCHOOL OF)

*Virgin and Child
with Tobias and Angel*
Cat. 36 ; Pls. P 36, P 36a
Jan Reynst
Gerard Reynst
CAELATURAE 23
Charles II
'Dutch Gift' 1660
Hampton Court,
Inv. no. 465

PAOLO VERONESE

Resurrection of Christ
Cat. 37 ; Pl. P 37
Jan Reynst
Gerard Reynst
CAELATURAE 13
Seen by Knorr in 1663 (?)
Van de Amory, 1722 (?)
Lost

Marriage of St. Catherine
Cat. 38 ; Pls. P 38, P 38a
Jan Reynst
Gerard Reynst
CAELATURAE 24
Charles II
'Dutch Gift' 1660
Hampton Court,
Inv. no. 96

UNIDENTIFIED ARTIST

Christ and the Virgin
Cat. 39
Gerard and Jan Reynst (?)
Charles II
'Dutch Gift' 1660
Lost

LIST OF DUTCH PAINTINGS
IN THE COLLECTION OF GERARD REYNST

PIETER VAN LAER

The Ambush
Cat. 40 ; Pls. P 40, P 40a
Gerard Reynst
CAELATURAE 19
Seen by Knorr in 1663 (?)
Banca Sannitica,
Naples

The Large Limekiln
Cat. 41 ; Pl. P 41
Gerard Reynst
CAELATURAE 31
Perhaps
Joachim von Sandrart
Formerly Prince of Liechtenstein
(lost since World War II)

Shot with the Pistol
Cat. 42 ; Pls. P 42, P 2a
Gerard Reynst
CAELATURAE 32
Leningrad, Hermitage,
Inv. no. 6931

LIST OF PAINTINGS IN THE COLLECTION OF JAN REYNST

GENTILE BELLINI

Madonna and Saints
RIDOLFI, I, p. 62
Cat. 43
Lost

BONIFAZIO VERONESE

Adoration of the Magi
RIDOLFI, I, p. 289
Cat. 44
Renieri (1646?)
Martinioni, p. 378
(Not listed in Renieri sale)
Lost

LORENZO LOTTO
(ATTR. TO)

Adoration of the Shepherds
RIDOLFI, I, p. 145
Cat. 18; Pl. P 18
Amsterdam by 1646
Gerard Reynst
CAELATURAE 20
Lost

HANS ROTTENHAMMER

Madonna
Adoring the Child, Angels
RIDOLFI, II, p. 85
Cat. 45
Lost

ANDREA SCHIAVONE

Christ
Before Pilate
RIDOLFI, I, pp. 250–51
Cat. 28; Pl. P 28
Gerard Reynst
(not in CAELATURAE)
Charles II
'Dutch Gift' 1660
Hampton Court,
Inv. no. 522

Christ
Presented to the People
RIDOLFI, I, p. 251
Cat. 46
Lost

Madonna and Child,
St. Joseph, St. John
RIDOLFI, I, pp. 250–51
Cat. 47
Renieri (1646?)
Boschini, p. 313
Martinioni, p. 378
Renieri (sale Venice, 4 December 1666, G. 26)
Lost

Battle of Lapiths and Centaurs
RIDOLFI, I, p. 251
Cat. 48
Renieri (1646?)
Martinioni, p. 378
Renieri (sale Venice,
4 December 1666, G. 27)
Lost

Judgment of Midas
RIDOLFI, I, p. 251
Cat. 29; Pl. P 29
Gerard Reynst
(not in CAELATURAE)
Charles II
'Dutch Gift' 1660
Hampton Court,
Inv. no. 470

Perseus and Andromeda
RIDOLFI, I, p. 251
Cat. 49
Lost

Heads of Philosophers
RIDOLFI, I, p. 251
Cat. 50
Lost

LAMBERT SUSTRIS

Death of Pharaoh
RIDOLFI, I, p. 226
Cat. 51
Renieri ? (1646?)
Renieri ? (sale Venice,
4 December 1666, G. 55)
Lost

JACOPO TINTORETTO

Pellegrini Family
RIDOLFI, II, p. 55
Cat. 52
Renieri (1646?)
Martinioni, p. 378
Renieri (sale Venice,
4 December 1666, G. 16)
Presumably lost; last recorded
with Lord Barrymore

TITIAN

St. Francis
RIDOLFI, I, p. 201
Cat. 53
Renieri (1646?)
Martinioni, p. 377
(not listed in Renieri sale, 1666)
Lost

Senator
RIDOLFI, I, p. 201
Cat. 54
Lost

TITIAN (SCHOOL OF)

Virgin and Child with
Tobias and Angel
RIDOLFI, I, p. 201
Cat. 36; Pls. P 36, P 36a
Gerard Reynst
CAELATURAE 23
Charles II
'Dutch Gift' 1660
Hampton Court,
Inv. no. 465

PAOLO VERONESE

Sacrifice of Abraham
RIDOLFI, I, p. 340
Cat. 55
Lost

PAINTINGS IN THE COLLECTION OF JAN REYNST

Good Samaritan
RIDOLFI, I, p. 340
Cat. 56
Presumably lost

Marriage of St. Catherine
RIDOLFI, I, p. 340
Cat. 38; Pls. P 38, P 38a
Amsterdam by 1646
Gerard Reynst
CAELATURAE 24

Charles II
'Dutch Gift' 1660
Hampton Court,
Inv. no. 96

Resurrection of Christ
RIDOLFI, I, p. 340
Cat. 37; Pl. P 37
Gerard Reynst
CAELATURAE 13
Seen by Knorr in 1663?
Van de Amory sale (Amsterdam, 23.6.1722, no. 40)
Lost

2 Portraits of Soranza Family
RIDOLFI, I, p. 340
Cat. 57
Renieri (1646?)
Martinioni, p. 378
(not listed in Renieri sale in 1666)
Lost

LIST OF PAINTINGS INCLUDED IN THE 'DUTCH GIFT'*

Paintings given to Charles II in 1660 and specified as Dutch present in his inventory; engraved for the CAELATURAE*

BASSANO
Christ Carrying the Cross (cat.no. 4; Pl. P 4a)
Charles II, inv.no. 161
Weston Park, Earl of Bradford
CAELATURAE 12

GIORGIONE, ATTRIBUTED TO
The Concert (cat.no. 11; Pl. P 11a)
Charles II, inv.no. 534
Hampton Court, Inv.no. 554
CAELATURAE 27

LOTTO
Portrait of a Gentleman (cat.no. 17; Pl. P 17a)
Charles II, inv.no. 116
Hampton Court, Inv.no. 486
CAELATURAE 3

PARMIGIANINO
Athena (cat.no. 20; Pl. P 20a)
Charles II, inv.no. 315
Windsor Castle, Inv.no. 1138
CAELATURAE 1

TINTORETTO
Portrait of a Dominican Friar (cat.no. 32; Pl. P 32a)
Charles II, inv.no. 103
Hampton Court, Inv.no. 772
CAELATURAE 4

TITIAN
Portrait of a Man (Sannazaro) (cat.no. 32; Pl. P 34a)
Charles II, inv.no. 21
Hampton Court, Inv.no. 68
CAELATURAE 2

TITIAN, ATTRIBUTED TO
Holy Family with St. John (cat.no. 35)
Charles II, inv.no. 166
lost
CAELATURAE 25

TITIAN, SCHOOL OF
Virgin and Child with Tobias (cat.no. 36; Pl. P 36a)
Ridolfi, I, p. 201
Charles II, inv.no. 532
Hampton Court, Inv.no. 465
CAELATURAE 23

VERONESE
Marriage of St. Catherine (cat.no. 38; Pl. P 38a)
Ridolfi, I, p. 340
Charles II, inv.no. 165
Hampton Court, Inv.no. 96
CAELATURAE 24

Paintings given to Charles II in 1660 and specified as Dutch present in his inventory; not engraved for CAELATURAE

BAROCCI
Woman with Dog (cat.no. 1)
Charles II, inv.no. 14
lost

BORDONE
Portrait of a Man (cat.no. 8; Pl. P 8)
Charles II, inv.no. 167
Hampton Court, Inv.no. 52

CARIANI
Reclining Venus (cat.no. 9; Pl. P 9)
Charles II, inv.no. 544
Hampton Court, Inv.no. 1103

D'OGGIONO
Christ Child and St. John (cat.no. 19; Pl. P 19)
Charles II, inv.no. 335
Hampton Court, Inv.no. 391?
not securely identifiable

RAPHAEL, SCHOOL OF
Christ on a Lamb, the Virgin, and St. Joseph (cat.no. 22)
Charles II, inv.no. 390
lost

SCHIAVONE
Christ before Pilate (cat.no. 28; Pl. P 28)
Ridolfi, I, pp. 250–51
Charles II, inv.no. 54
Hampton Court, Inv.no. 522

SCHIAVONE
Judgment of Midas (cat.no. 29; Pl. P 29)
Ridolfi, I, p. 251
Charles II, inv.no. 169
Hampton Court, Inv.no. 470

UNIDENTIFIED ARTIST
Christ and the Virgin (cat.no. 39)
Charles II, between inv.nos. 546–47
lost

---

* Listed alphabetically according to artist

LIST OF PAINTINGS IN THE ENGLISH ROYAL COLLECTIONS FORMERLY WITH REYNST

Paintings in the English Royal Collections that were engraved for the CAELATURAE but that were not listed as Dutch present in the royal inventories; engraved for the CAELATURAE:

BONIFAZIO VERONESE
The Virgin and Child
(cat. no. 7; Pl. P 7a)
Charles II, inv.no. 158
Hampton Court, Inv.no. 140
CAELATURAE 22

GIULIO ROMANO
Isabella d'Este (cat.no. 12; Pl. P 12a)
Charles II, inv. no. 4
Hampton Court, Inv.no. 76
CAELATURAE 9

GUERCINO
Semiramis (cat.no. 13; Pl. P 13a)
Dukes of Grafton
Boston, Museum of Fine Arts,
Inv.no. 48.1028
CAELATURAE 29

LOTTO
Andrea Odoni (cat.no. 16; Pl. P 16a)
Charles II, inv.no. 264
Hampton Court, Inv.no. 72
CAELATURAE 26

RENI, ATTRIBUTED TO
Allegory of Painting (cat.no. 26)
perhaps James II, inv.no. 168
presumably lost
CAELATURAE 7

CONCORDANCE OF THE VENDRAMIN CATALOGUE DE SCULPTURIS WITH THE ICONES

ICONES			DE SCULPTURIS		ICONES		DE SCULPTURIS	
	A	Sabina	I,3	Sabina	17	Partly Draped Female Figure	I,31	Venere
	B	Caracalla	I,2	Antonino Caracalla	18	Draped Female	I,22 B	
	C	Aesculapius	I,16	Esculapio	19	Hercules	II,21	Ercole
	D	Cupid	—		20	J. Caesar	II,19	Cesare
	E	Vesta	II,25	Dea Vesta	(B) 21	Agrippina Major	—	
	F	Cijbele	II,17	Cibele	22	Poppea Sabina	II,33	Poppea Sabina
	G	Tiberius	II,31	Tiberio	23	Augustus	II,5	Ottaviano
	H	Domitianus	II,29	Domitiano	24	Tiberius	—	
	I	Commodus	II,27	Comodo	(B) 25	Calphurnia	II,52	Calfurnia
(H)	K	Faustina	—		26	Octavia	II,38	Otavia
	L	Scipio Africanus	—		27	Vitellius	II,1	Vitellio
	M	M. Brutus	II,15	Marco Bruto	28	Domitia	II,47	Domitia
(L)	1	Consul	—		29	Pallas	II,3	Pallade
(L)	2	Hostilius	I,4	Tullo Hostilio	30	Milenus	II,9	Mileno
(L)	3	Venus	I,44 B		(L) 31	Livia	II,56B	Livia
(L)	4	Flora	I,8	Flora	(L) 32	Cleopatra	II,56A	Cleopatra
(L)	5	Venus	I,12	Venere	33	Trajanus	II,49	Traiano
(L)	6	Abundantia	I,5	Dea Abondanza	(D) 34	Gordianus	II,13	Gordiano
(L)	7	Hercules	I,26	Hercole	35	Helena	II,7	Helena
	8	Julia	I,33	Giulia	36	Octavia	II,50	Ottavia
	9	Adonis	I,19	Adone	37	Hadrianus	II,23	Adriano
(B)	10	Gladiator	I,17	Gladiatore	38	Caracalla	II,60	Antonio Caracala
	11	Apollo	I,13	Apollo	39	Terentia	II,10	Terentia
	12	Cupid	—		40	Julia Mammaea	—	
	13	Hermaphroditus	—		41	Agrippina	II,8	Agrippina
	14	Pallas	—		42	Jupiter Ammon	II,30	Giove
	15	Male Nude with Chlamys	—		(B) 43	Aristea	II,11	Aristea
	16	Draped Female Figure	I,23 A		44	Pallas	—	
					45	Geta	II,16	Geta
					46	Flavia	II,65B	Flavia

B: East Berlin, D: Dresden, H: Hampton Court, L: Leiden

ICONES		DE SCULPTURIS		ICONES		DE SCULPTURIS	
	47 Faunus	II,26	Fauno		67 Male Portrait Bust	—	
	48 Antoninus	II,37	Antonino		68 do	—	
(L)	49 Hadrianus	II,20	Adriano		69 Female Portrait Bust	—	
	50 Cupid	II,69A	Cupido		70	—	
	51 Faunus	II,14	Fauno	(L)	71	—	
	52 Cyrus	II,64	Ciro	(D)	72	—	
(B)	53 Flavia	II,51	Flavia		73–78	—	
	54 Lucilla	—			79	II,28	Marco Aurelio
	55 Faustina	II,18	Faustina		80–85	—	
	56 Silvius Postumius	II,46	Silvio Posthumo	(L)	86	—	
	57 Bust of Young Boy	II,69B	Anterote		87–88	—	
	58 Germanicus	II,45	Germanico	(L)	89	—	
	59 Papirius	II,36	Papirio		90–91	—	
	60 Antonia Major	II,39	Antonia Magior		92	II,22	Fauna
	61 Hadrianus	II,2	Hadriano		93	I,21	Fauna
	62 Claudia	II,66B	Claudia		94	—	
(L)	63 Bachus	—		(D)	95	—	
	64 Severus	—		(L)	96	—	
(D)	65 Priapus	—		(L)	97	—	
	66 Priapus	—		(L)	98	—	

DE SCULPTURIS		ICONES		DE SCULPTURIS		ICONES	
I,1		—			Dea Hebbe (?)	—	
I,2	Aontonino Caracalla	B	Carracalla	I,31	Venere	17	
				I,32		—	
I,3	Sabina	A	Sabina	I,33	Giulia	8	Julia
I,4	Tullo Hostilio	2	Hostilius	I,34	Filida (?)	—	
I,5	Dea Abondanze	6	Abundantia		Venere	—	
I,6	Giunone	—		I,35	Diana	—	
I,7	Dea Eleusina	—		I,36	Esculapio	—	
I,8	Flora	4	Flora	I,37	Elide	—	
I,9	Venere	—			Abondanza	—	
I,10	Venere	—		I,38	Diana	—	
I,11	Marco Aurelio	—			Giove	—	
I,12	Venere	5	Venus	I,39	Atti	—	
I,13	Apollo	11	Apollo	I,40		—	
I,14	Consule	—		I,41	Vergini Vestali	—	
I,15	Consule	—			Pastor	—	
I,16	Esculapio	C	Aesculapius	I,42	Venere	—	
I,17	Gladiatore	10	Gladiator		Venere	—	
I,18	Bacco	—			Cupido	—	
I,19	Adone	9	Adonis	I,43		—	
I,20	Marte (?)	—			Fium. Nillo	—	
I,21	Apollo	—		I,44	Marte	—	
	Fauno	93			Venere	3 (?)	
I,22	Venere	—			Esculapio	—	
		18			Cerere	—	
I,23		16		I,45		—	
		—			Adone		
I,24		—			Dea	—	
I,25	R. d. Galli Senusi (?)	—			Dea	—	
				I,46	Fauni	—	
I,26	Hercole	7	Hercules			—	
I,27	Giove	—				—	
I,28	Cerrere	—		I,47	Torsi	—	
I,29		—				—	
	Venere	—				—	
I,30	Dea Bona	—		I,48		—	

CONCORDANCE OF THE VENDRAMIN CATALOGUE DE SCULPTURIS WITH THE ICONES

DE SCULPTURIS		ICONES		DE SCULPTURIS		ICONES	
II,1	Vitellio	27	Vitellius	II,37	Antonino	48	Antoninus
II,2	Hadriano	61	Hadrianus	II,38	Otavia	26	Octavia
II,3	Pallade	29	Pallas	II,39	Antonia Magior	60	Antonia Major
II,4	Fabio Massimo	—		II,40	Lucilla	—	
II,5	Ottaviano	23	Augustus	II,41	Plotinia	—	
II,6	Pantasilea Regina	—		II,42	Giulia Mamea	—	
II,7	Helena	35	Helena	II,43	Getta	—	
II,8	Agrippina	41	Agrippina	II,44	Galeria	—	
II,9	Mileno	30	Milenus	II,45	Germanico	58	Germanicus
II,10	Terentia	39	Terentia	II,46	SilvioPosthumo	56	Silvius Postumius
II,11	Aristea	43	Aristea	II,47	Domitia	28	Domitia
II,12	Giove	—		II,48	Giunia Claudia	—	
II,13	Gordiano	34	Gordianus	II,49	Traiano	33	Traianus
II,14	Fauno	51	Faunus	II,50	Ottavia	36	Octavia
II,15	Marco Bruto	M	M. Brutus	II,51	Flavia	53	Flavia
II,16	Geta	45	Geta	II,52	Calfurnia	25	Calphurnia
II,17	Cibele	F	Cijbele	II,53	Galeria	—	
II,18	Faustina	55	Faustina	II,54	Scittimio Severo	—	
II,19	Cesare	20	J. Caesar	II,55	—	—	
II,20	Adriano	49	Hadrianus	II,56	Cleopatra	32	Cleopatra
II,21	Ercole	19	Hercules		Livia	31	Livia
II,22	Fauna	92		II,57	Fauna	—	
II,23	Adriano	37	Hadrianus		Fauna	—	
II,24	Marco Marcello	—		II,58	Antino	—	
II,25	Dea Vesta	E	Vesta		Adriano	—	
II,26	Fauno	47	Faunus	II,59	Fauno	—	
II,27	Comodo	I	Commodus		Petronia	—	
II,28	Marco Aurelio	79		II,60	Antonino Caracala	38	Caracalla
II,29	Domitiano	H	Domitianus	II,61	Faustina	—	
II,30	Giove	42	Jupiter Ammon	II,62	—	—	
II,31	Tiberio	G	Tiberius	II,63	Ciceron	—	
II,32	Tiberio Greco	70	(?)	II,64	Ciro	52	Cyrus
II,33	Poppea Sabina	22	Poppea Sabina	II,65	Tullia	—	
II,34	Tulliola	—			Flavia	46	Flavia
II,35	Albia Terentia	—		II,66	Eleusina	—	
II,36	Papirio	59	Papirius		Claudia	62	Claudia
				II,69	Cupido	50	Cupid
					Anterote	57	

Not. B. Baddel, N.A.A. *nr* 957, *dd* 29.3.1642 (N.B. akte met brandschade)
...huysraet ende anders ... ten sterffhuyse van Sr. Jacques Nicket beschreven door mij Benedict Baddel bij den Hove van Hollandt geadmitteerden Openbaer Notaris t' Amsterdam op 't versoeck van den E. Gerard Reijnst tot behoeff van dien 't behooren sal ten overstaen van
Jouffer Margareta Nicquet huysvrouw van Jan van Lier, Jouffer Catharina Reijnst huysvrouw van Samuel Blommart, Jouffer Margareta Reijnst huysvrouw van Adam Bessels ende Jouffer Susanna Moor huysvrouw van Jaspar Charles ende Paulus Hilmans uyt laste ende van weghen de weduwen van Thymen, Jacob ende Frants Jacobssen Hinlopen ende ter presentie van Johannes Hamel ende Joachim Thielmans als getuyghen.

Adij XXIX Martij 1642 in Amsterdam.

In een camer aen de suytsijde achter aen den thuyn.
(Een schilderije) ... personagien met een dubbelde vergulde lijste
Item een ditto mede landtschap ende lijste als boven
Item een ditto grooter representerende Elias in de wildernisse, lijste als boven
Item een ditto landtschap met eenighe personagien, lijste als boven
Item een ditto kleyner, lijste als boven
Item ses trijpe sitkussens
Item eenen spiegel met een ebbenhoute lijste
Item een clavercimbel met eenen voet toecomende Margareta Bessels, per memorie
Item een gebroocken ley
Item ses matstoelen daer onder een cleyne
Item een eeckenhoute taeffel
Item een eeckenhouten schabelletgen
Item twee oude gestreepte armosijnen gordijnen ende valletgen van 't selve
Item een groen schoorsteencleedt met een blauw linnen kleedt daer onder
Item vijff porcelijne copgens
Item twee vogelkouwetiens elck met een canarie voghelkens
Item een groen laeckensch tafelkleet
Item een metalen tafelschelletgen
Item een houwertgen met een swarte sluyer
Item een lanckwerpighe schilderije sonder lijste
Item een iseren draeghhamertgen
Item een doosgen met een vierslagh
Item een bedt met een peulue
Item twee oorcussens met haer sloopen

...
    een deel flesschen met wateren ende conserven
Item een deel glasen ende romers
Item een harthouten tresoortgen met een deurtgen in 't midden met veele opene laetgens
Item in een banck verscheijden rommelingh geleyt ende met mijns notaris cachet toegesegelt
Item een lade in 't voors. tresoortgen met eenighe factueren van bloemen daer Sr. Bessels de sleutels van heeft
Item een lade in 't voors. tresoortgen met eenighe rommelingh

Item een groenen karsayen gordijn voor de vensters met een iseren roede
Item de prins Hendrick in een walvischbeen gegoten

In de groote camer aen de noortzijde mede aen de thuyn.

Item twee conterfeytsels representerende eenen peres ende zijne huysvrouw met ebbenhoute lijsten
Item een schilderije representerende een blompot met een swarte lijste
Item een ditto representerende den philosooph Chilon met een vergulde lijste
Item een ditto representerende een romeyn met een roode platte lijste
Item een ditto representerende een moscovisch meijsgen met een platte swarte lijste
Item een conterfeijtsel van een ouden man van de vrienden met een dubbelde vergulde lijste

...

Item acht porcelijne schotelen
Item ses ditto boterschotels
Item ses ditto coppen
Item een grooten spiegel met een ebbenhouten lijste
Item een grooten eeckenhouten taeffel
Item een turcksch taffelkleet
Item een harthouten staende pars
Item een ditto casgen
Item een kistge geslooten ende toegesegelt met 2 schargien
Item acht roode leere spaensche mansstoelen
Item vier ditto vrouwe stoelen
Item een halff harthouten rondt taefeltyen
Item ses oude tapijte kussens
Item twee blauwe engelsch damaste godijnen met orangie frangien
Item een valletgen met een schoorsteencleet van 't selve
Item een corte gordijn met frangien als boven
Item twee oorcussens met sloopen
Item een bedt met sijnen peulue
Item een groene spaensche deecken
Item een ditto witte
Item twee slaeplaeckens
Item eenen bedstock
Item een behanghsel van blauw ende goutleer Sr. Adam Bessels toecomende per memorie

In 't voors. casgen

Item seven mans hemden
Item dry mans lobben
Item elf beffen met canten
Item sesthien servietten gesayde roosgens
Item thien sevetten schaeckwerck met roosenhoet
Item twee roosenhoet ammelaeckens
Item vijff ammelaeckens schaeck ende roosenhoet

...

Item een ditto gesayde roosgens

Item vier flouwijnen
Item vijff ditto slechte
Item twee ditto langhe handdoecken
Item dry doeckskens om den hals
Item twee dweijlen
Item een isere tange
Item een ditto brandiser
Item twee gordijnen voor de glasen met yseren roeden

    In den camer aen de rechterhandt van de gangh

Item een schilderije representerende een affsettinghe van convoy met een swarte lijste
Item een ditto representerende een boerenkermis met een swarten lijste
Item twee conterfeijtsels van den vader ende moeder van den overledenen Jacques Nicket met een vergulde lijsten
Item een schilderije representerende een landschap met eenige personagien met een vergulde lijste
Item een ditto representerende een landtschap met eenighe personagien ende een boerenhuys met een harthouten lijste
Item een ditto representerende een landtschap mede met eenige personagien ende eenen wintmeulen met een lijste als boven
Item een ditto representerende een landschap met een vergulde lijste
Item een ditto representerende een blompot met een dubbelden swarten lijste
Item een ditto representerende een schipsvaert met een dubbelde vergulde lijste

…

Item een cleyn schilderijtgen representerende een kerck met een swarte lijste
Item een achtkantighen spiegel met een swarten lijste
Item een harthouten taeffel
Item een turcksch tapijtcleet
Item een grooten gesneden notebomen kiste
Item een stroyen matras
Item een bedt met een peulue toebehoorende Hendrickge Jacobs
Item twee porcijne schotels
Item twee ditto grooter
Item ses ditto commetgens
Item een ditto oliepot
Item twee ditto cannen met tooten
Item een ditto vlesch
Item twee ronde silvere soutvaten
Item een ditto mostaertpot
Item een ditto peperbusch
Item twee ditto cleyne beeckertgens
Item twaelff ditto lepels
Item neghen ditto forketten
Item twee ditto eijerlepeltgens
Item een ditto spadeltgen
Item een ditto commetgen
Item een groote silvere vergulde schroeff

Item twee ditto cleyner effen
Item twee porcelijne halve lampetschotelen
Item achthien dubbelde boterschotels
Item dry ditto achtcantighe salaetschotels
Item achthien ditto tailloren
Item twee ditto fijne schaeltgens
Item een ditto cleyner
Item een ditto bierkan

...

Item vijff ditto clapmutsiens
Item seven ditto candeelcopgens
Item dry ditto achtcantighe copgens daervan twee geborsten
Item ses ditto cleyner
Item vijff ditto copgens
Item dry ditto witte
Item dry ditto cleyne sausierkens
Item ses ditto boterschoteltgens
Item twee ditto ronde
Item een cleyn schoteltgen
Item een perlemoer copgen
Item twaelff pimpeltgens
Item een cleyn vlesgen
Item dry cleyne copgens by malcanderen
Item twee cleyne geschilderde copere copgens
Item eenen swarten laeckenschen mantel met baeye opslagen
Item eenen turckschen groffgreynen mantel
Item twee swarte laeckensche broecken
Item twee ditto rockjens daer van een met bont
Item een bourat wambeys
Item dry swarte spaensche stoelen met root laecken met frengien

   In de sijdelcamer in 't voorhuys

Item een oude schilderije representerende een boer spelende op 't verkeertbort met een slechte lijste
Item een ditto representerende een fruytschaeltgen met een dubbelde vergulde lijste

...
... een Joseph ende Maria met een swarte lijste
Item een ditto representerende een landschapien met eenige personagien
Item ses albaster bortgens
Item een gestreept armosijn glasegordijntgen met een isere roede
Item een bedt met sijn peulue
Item een achtkantich root taeffeltgen
Item een stroyen matras

   In 't voorhuys

Een schilderije representerende een landtschap met een herder met schaepgens met een swarte platte lijste
Item een ditto met een veranderinghe, lijste als boven
Item een ditto representerende Susanna met een harthouten lijste

Item een ditto representerende een loopent paert, lijste als boven
Item een ditto langh representerende een winter met een dubbelde vergulde lijste
Item een ditto representerende een woestijne met een swarte lijste
Item een ditto representerende een seehaven met een swarte platte lijste
Item een ditto representerende eenige bedelaers ende boeren met een platte lijste
Item een pascaerte van de noordsche see met de handt geschreven met een harthouten lijste
Item dry schilden lackwerck
Item een uurwerck
Item dry roodleere spaensche mansstoelen

... yseren roede
Item een marmeren taeffel
Item een goutleeren behanghsel mit blau ende gout Hr Bessels toebehoorende per memorie
Item een puthaeck

In den gangh

Een caerte representerende Italien
Item eenen capstock
Item eenen cleerbesem

Op den leeghsten solder

Een vuerenhouten taeffel met twee schraghen
Item een yser verguldt ledekant
Item een braetspit
Item een geschilderde oostersche kist met iser beslaghen
Item een slaepbanck
Item twee bloote swarte vrouwenstoelen
Item een en swarten spaenschen leeren mansstoel
Item een ditto vrouwen
Item een houte ledekant
Item een verkeerbert
Item een alembich of forneys (NB = distelleertoestel voor alchemisten)
Item twee groote mans stoelen
Item een groene papegays kouw

In de oostersche kiste

Een papegaey groenen satijne sprey met sijn behanghsel te weten twee gordijnen ende een valletgen met roode ende geele ende groene frengien

... valletgen ende schoorsteencleet met roode, geele ende groene frengien
Item een geele gesteecken sprey met blauw armosijn gevoert
Item een hollandsche huyck
Item een tapijt bedtcleedt met groene, geel ende isabelle frangien
Item een say schoorsteencleet met groene ende roode frengien
Item een blauw engelsch damaste sprey
Item een gordijn van 't selve
Item een geel satyne taeffelcleet met goude frangien
Item een root satijn valletgen met roode ende geele frangien

Item twee gordijnen van 't selve
Item een groene satijne gordijn met roode ende geele sijde frengien
Item een seledon armosijn valletgen met goude ende roode sijde frengien
Item een groen satijn schoorsteenkleet met roode, geele ende groene frengien
Item een fillemort sijde damast valletgen met violette ende geele frengien met blauw linnen gevoedert
Item een fillemort sijde schoukleet
Item een coleure linnen valletgen met root, geel ende blauwe frangien met root linnen gevoedert

… sprey fillemort (met) blauw linnen gevoedert
Item een ditto blauwe sprey met blauw linnen gevoedert
Item een bedttijck
Item een saye valletgen met groene sijde frangien
Item een gestreept armosijn schouwkleet met gele ende groene frangien ende groen linnen gevoedert
Item een gesteecken cattoenen witte sprey
Item een stuck wit linnen
Item een paer camerijcx doecksche flowijnen met groote binnewercken
Item een paer swarte amosijne cousebanden met kanten
Item een paer ditto sonder canten
Item een witte deecken
Item een wit carsaey beddecleet
Item een blouwel
Item thien cleerstocken
Item twee brandisersvoeten
Item twee halve schaverotten
Item dry vuyrenhoute voetbanckgens

   Op de bovenste solder

… gevoedert
Item een vuyrenhouten kiste met ysere banden, daerinne :
Item vier groote tinne schotelen van 5 pond
Item twee saletschotelen, een groot ende een cleyn
Item ses ditto pond schotelen
Item twee coopere brandisers
Item een coopere tanghe
Item een coopere schuppe
Item een coperen mortier met sijnen stamper
Item twee cleyne coopere candelaers
Item eenen grooten heelen belblaecker
Item een ditto halve
Item een cooperen croon van malcanderen
Item een cooperen handtblaecker
Item een wolle matras
Item twee gladde isere brantisers
Item twee beddekens met haer peuluwen waervan een met een linnen overtrecksel
Item een groene deecken
Item een ditto witte
Item vijfftich sevetten

...
Item vijff mans hemden
Item sesthien handtdoecken
Item dry keucken ammelaeckens
Item dry langhe handtdoecken
Item ses flouwijnen
Item een mout
Item een quartel koy
Item een oostersche kiste met isere platen
Item ses doecken om den hals
Item twintich neusdoecken
Item seven kleermanden soo groot als cleijn
Item eenige borden in de rommelingh
Item eenen matstoel

   Op de vlieringh

Een wan
Item twee turffmanden
Item eenich aerdewerck

   Op de bovenachtercamer

Een schilderij representerende een boerekermis met een vergulde lijste

… met een harthouten lijste
Item vijff gesteecken kussens
Item een rond geschildert taeffeltgen
Item twee goude leeren vrouwe stoelen
Item eenen spieghel met een vergulde lijste
Item twee matstoelen
Item een harthouten banckjen
Item eenen kofferstoel
Item een ronde mant
Item een bedt met sijn peulue
Item twee slaeplaeckens
Item twee oorcussens met haer sloopen met binnewerck
Item een wolle matras
Item een groene deecken
Item een ditto wit
Item twee enghelsch blauw damaste gordijnen met blauwe ende orangie frangien
Item een schoorsteencleedt van 't selve
Item een bedtvalletgen van grauwstoff met zijde frengien
Item ses porcelijne schotelen
Item dry ditto hooghe copgens
Item dry witte gemooghlaekens slaeplijven
Item twee ditto onderbroecken

… (nieuw vertrek)
Een bedt met sijn peulue
Item twee laeckens
Item twee groene deeckens
Item dry oorcussens met sloopen ende binnewercken

Item eenen matstoel

   In een ander camertgen daerbij

Een heele groote oostersche kiste met iser beslaghen
Item eenen ouden hoet
Item een braessemspit
Item eenen ouden swarten laeckenschen mantel met baey gevoedert
Item eenen grauwen laeckenschen mantel met baey gevoedert
Item een root karsayen slaeplijff
Item eenen damasten borstlap
Item een koffertgen met wat specerijen
Item dry backgens
Item vier kuypgens ende een tonnetgen

   In een kasgen op de trap

Twee groote tinne wijnkannen
Item een ditto bierkanne

…

Item vierentwintich ditto tailloren
Item twee copere candelaers
Item een tinne tonnetgen
Item een cleijn copere potgen
Item een tinne beckentgen

   In de keucken

Vijff tinne schotelen soo groot als cleijn
Item vierentwintich tailloren
Item eenen tinnen waterpot
Item een tinne commetgen
Item vier porcelijne dubbelde boterschotelen
Item een ditto halve lampet
Item negen witte majorcksche schotelen
Item ses porcelijne taillortgens
Item ses ditto copgens
Item dry ditto saucierkens
Item achtthien roode straetwercksche schotelen
Item dry witte majorcksche kannetgens
Item twee ditto doorluchtighe schalen
Item dry blecke potdecksels
Item twee ditto blaeckers
Item dry roosters
Item een lepelbert met eenige tinne lepels
Item een dop blecklepel

…

Item een isere elle
Item een ditto schop
Item een ditto henghel
Item een harthouten keerslade
Item een aerde biercanne met een silvere decksel
Item een tanghe

Item een heertiser
Item een brandiser
Item een taeffeltgen
Item eenen matstoel
Item eenen aerden wijnkan met een silvere deckseł
Item ses aerde kannen met decksels
Item een blauwe schoorsteenkleet met geele ende blauwe frengien
Item twee ammelaeckens

   In de gangh

Een mant met een hort
Item een coper pannetgen met een iseren steel
Item twee marcktemmers
Item twee copere vischketels
Item twee isere potten
Item twee tinne stooffcommen

…
Item twee majorcasche commen
Item eenich aerdewerck
Item eenen grooten cooperen vleespot
Item een ditto cleijner
Item twee ditto aeckers
Item twee ditto cleijne pannekens
Item een ditto bedtpanne
Item een aschaecker
Item twee isere keteltgens
Item twee copere taertpannen
Item een spit
Item eeniche rommelingh van keuckenwerck

   Boven op 't comptoir

Een vierenhouten kistge vol schriften ende reeckenboecken met mijns notariscachet toegesegelt
Item een ditto met gedruckte boecken als boven toegesegelt
Item eenen groenen glasen lantaren
Continuatie van inventaris van de pampieren, rentebrieven, schriften, reecken-, ende leesboecken, gout ende silver, gemunt ende ongemunt ende anders alsnoch bevonden ... Keijsersgracht ... over d'Accademie beschreven door mij Benedictus Baddel, keijserlijcken ende bij den Hove van Holland geadmitteerden openbaer notaris t'Amsterdam residerende ten versoecke van de E. Heren Samuel Blommaert ende Adam Bessels, cooplieden deser voors. stede als executeuren van den testamente van den selven Nicquet tot behoeff (soo sijlieden verclaerden) van die gene die daertoe gerechticht sullen wesen ten overstaen van deselve ende ter presentie van Jacob van Wijckersloot ende Nicolaes Anthony als getuighen.

   Een doosgen daerinne een memorie luydende als volght :

van Hendrickge Jacobs
Een dubbelde roosenobel
Item twee Jacobusen

Item een ouden dubbelden ducaet met twee hooffden
Item een dubbelde pistolet
Item vier rijders
Item twee franche croonen
Item vijff ducatiens
Item vier rijcksdaelders

...
  deselve specien ende anders in esse bevonden

Een langh doosghen daerinne twee silvere klatertgens
Een silvere spuytgen, eenen leeren coocker met wasse keersgens
Item een doosghen daer op geschreven stont : Rentebrieven
Item eenen rentenbrief daer op geintituleert Jacques Nicquetti tot lijve van Margareta Bessels, jaerlijckx 75 £
Item eenen ditto luydende Jacques Nicquetti tot lijve van Jacques Bessels L £
Item eenen ditto luydende Jacques Nicquet heffer tot lijve van Margareta Bessels, dochter van Adam Bessels, jaerlijckx xxv £
Item eenen ditto luydende Jacques Nicquet heffer tot lijve van Clara Bessels, dochter van Adam Bessels, jaerlijckx xxv £
Item eenen ditto luydende Jacques Nicquet heffer tot lijve van Jacques Blommaert, soone van Samuel Blommaert, jaerlijcks xxv £
Item eenen ditto luydende Hillegont Marcelis tot haren lijve, jaerlijcx xii £ x st.
Item een geschrifte geintituleert 1639 Amsterdam cassa voor Susanna Moor

...
Item een ditto luydende bilance uyt de boecken van Susanna Moor tot 29 november 1639, cassa als boven
Item een geschrift geintituleert jacht vrij
Item een ditto geintituleert Sr Jacques Nicquet
Item een ditto geintituleert acte van de vrije jacht in de heerlijckheijt van Berghen
Item eenen brieff van Abraham van Neck, secretaris, aen Jacques Nicquet
Item een ditto van Chr. Huyghens aen Jacques Nicquet
Item een octroy van prints Mauritius aen Jacques Nicquet wegens de jacht
Item een copye autentick van dien door Notaris Bruyningh geauthauriseert
Item een copye van een brieff aen den secretaris Huygen, belanghende de jacht
Item een brieff daerop geschreven staet Lenart de Vocht van twaelff may daer den houtmeester, de heere van Groenevelt, mijne jacht mede confirmeert
Item een octroy om van sijn goederen te moghen disponeren ; in dato Utrecht 13 april 1632

... bescheijden ende pampieren
Item eenen rentebrieff ten lijven van Jacques Nicquet van 50 £ jaerlijcx

  In een wit sacksgen

negentwintich rijcxdaelders
Item tweeentwintich ducatoenen

   In een cleijn geteerlinct sackgen

dry heele rijders
Item ses ditto halve
Item elff dubbelde ducaten
Item vier ditto enckelde
Item een angelot
Item een halve nieuwe Jacobus
Item een pampier luydende : Extract van de renten ontfanghen wegen 't legaet van mijn suster Sara Nicquet tot 31 december 1633
Item een ditto extract uyt het testament van mijn suster Sara Nicquet was huysvrouw van Dominicus Heemskerck voor soo veel het toucheert aen mij Jacques Nicquet
Item een pacquet met seghelgaren gebonden daerop geschreven stondt : pampieren aengaent het sterffhuys van vader Jan Nicquet off Jan Niq ende van mijne moeder Margareta Boschmans
… van de erffgenamen van Margarita Nicquet
Item een ditto cladde van scheijdinghe van 't sterffhuys van Margaretha Boschmans
Item een ditto reeckeningh courant van Hans Rombouts
Item een ditto reeckeningh courant van Dominicus Heemskerck
Item een ditto reeckeningh courant van Bartholomeus Moor
Item een ditto 1613 kinderen Hans Nicquet
Item een ditto reeckeningh courant van Jacques Nicquet tot primo july 1613
Item een ditto wesende een bilance van den 4 july 1613, beginnende in debit Passchier Lammertijn
Item een ditto 1613 reeckeninghe van interesten van Jacques Nicquet
Item een ditto 1623 testament van mijne suster Sara Nicquet was huysvrouwe van Dominicus van Heemskerck
Item een ditto wesende een copye getrocken uyt een memorie van Jacques Nicquet saliger suster van Sara van Heemskerkck

… rente van f. 100 op gl. 80 gereduceert
Item een geschrift luydende borchtocht van mijne vrienden aenden crediteuren
Item een assignatie van Paul de Willem adij 15 octobris 1615 ter somme van f. 2364.-

   Een doosghen daerinne gevonden :

Twee albasterde ronde balletgens
Item een cooper spaensch schelletgen
Item thien christalijne knoopen
Item eenen gouden signetringh
Item een cleijn blicken doosgen met rieckende pampierkens
Item een cleijn doosgen met twee roode steentgens
Item een cleijn goude kinderringetien
Item eenen ammatiststeen
Item eenen bril met silver

Item een stuck aloeshout
Item een packtgen daer op geschreven twaelff neusdoecken
Item eenen meskoocker met twee messen, eenen priem ende een scheerken
Item een ditto met twee messen ende een forquet

...
Item twee welrieckende paternosters
Item een enckelt braseletgen
Item twee sponsgens
Item een kaertspel
Item een pampierken daer op geschreven staet : 1641 quitantie van een halff jaer huere van 't huys den Oraigneappel, verschenen alderheijlighen 1641
Item een memorie beginnende : Ick Jacques Nicquet van f. 600.– tot behoeff van Sr Gerard Reijnst
Item een memorie van f. 600.– tot behoeff als boven
Item een bundeltgen met hairons
Item een copere plaet van Jan Nicquet den ouden met eenighe affdrucksels van dien
Item een verclaringhe van donatie aen Margareta Bessels van eenighe goude spetien in dato 15 april 1637
Item een trijpen reijssackxen
Item een boeck geintituleert 1642 Amsterdam journael Dd. (monogram) beginnende primo januaryij 1642 van twee bladeren geschreven

... 1641 van fol. 1 tot 132 den 31 decembris 1641
Item een ditto geintituleert grootboeck cc. van Jacques Nicquet begint primo januarij 1636, 1637, 1638, 1639, 1640, 1641, van fol. 1 tot fol. 82 den 31 decembris 1641
Item een ditto geintituleert 1638, 1639 Amsterdam memoriael oncosten van den huyse cleijne reeckeninghen, copyen van brieven 1638, 1639, 1640
Item een ditto geintituleert 1640 in Amsterdam notitie van lijffrenten tot Utrecht, in Den Haghe, notitie voor Anneken Mattheus lijffrenten tot Utrecht
Item een handt clad wit boecksken
Item een pampier daerop geschreven staet : lijste van mijne crediteuren
Item 't origineel accord met sijne crediteuren
Item een copye authenticque van dien

Continuatie gelijck boven op den derthienden dagh der maent may Anno 1642 ter presentie van Johan Hamel ende Daniel le Bleu ...

... langhwerpigh ende een in quarto van de jaeren sesthienhondertsesen-twintich tot sesthienhondertsesendertich
Item een journael ende grootboeck van 't jaer 1621 in folio
Item een journael ende grootboeck van 't jaer 1622, AA in folio
Item een cas ende banckboeck in folio
Item een packsgen daerop geschreven stont : belotene reeckeninghen tusschen Jacques Nicquet ende Gerard Reijnst
Item een packsgen diverschen memorien van 't jaer 1621
Item diversche pacquetten reeckeninghen ende missiven brieven
Item een brieff copyeboeck van 't jaer 1622 in pampier
Item eenighe boecksgens van tractaetjens ende tijdinghen

Item een boeck aengaende de kinderen van Laurens Reael in folio
Item dry cleijne memoriaelen van de jaeren 1636. 1637 ende 1638
Item eenighe pampieren aengaende de kinderen van Jan Nicquet

… (Jacques) Nicket ende Clara de Haze met eeniche andere pampieren aen malcanderen
Item noch enighe tractatgens, comedien ende printen
Item een boeck in folio journaelswijse, daer noch niet ingeschreven en is
Item eenighe printen ende almanach ; voorts andere dinghen
Een kistgen met timmergereedschap
Item een ditto met apothekerije
Item twee evenaers van cleijne balancen ; een groot ende een cleijne met hare coopere schalen
Item een harthouten lessenaer met een slot
Item een chineeschen doos
Item ses led. glase vlesschen
Item een orlogie oft uurwerck met een cooper vergulde casgen
Item een ledigh coffertghen
Item seventhien ledighe bloemladen
Item een incktkoocker met 2 pennemessen ende silvere chachetten met bollen, een scheeve ende een cleijne …

… acht dwersche stocken ende rottinghen
Item een swarte hooren met iser beslaghen
Item een pistool
Item een voghelkouwe van dry huyskens
Item een pijlcooperegewicht van vier pont
Item een kelder van ses vlesschen
Item een ditto van twee vlesschen
Item eenighe ledighe sacken ende vijlen
Item twee groote matstoelen
Item tweeenvijfftich veneetsche copgens ende glasen van verscheijden soorten
Item twaelff romers ende een italiaensche fluyt
Item twee majorcksche peperbuskens
Item twee ditto olie- ende asijnkannekens
Item een majorcksche barbierbecken
Item een majorcksche taillor
Item twee majorcksche schotelen

…
Item een sack met gereetschap om jucht te maecken
Item een out brievenperskens
Item twee groene laeckensche kussens

   Italiaensche boecken op de solder in een kiste

Historie del mondo di Grouan Tareagnota, 5 tomen in o/4	f. 12 :...
Item Plutarcho delle vite o/4	f. 3 :...
Item de Cameron di Boccacioni in o/4	f. 2 :...
Item Le rime del Petrarca in o/4	f. 2 :...
Item Historia della vita et fatti di Bart.° Coglione in o/4	f. 1 :...
Item Historia fiorentina di Piero Buonisegni in o/4	f. 1 :...
Item Del governo di diversi regni di Francisco Sansovino in o/4	f. 1 :...
Item Del compadio dell'historia del regno di Napoli di Thom. Costo in o/4	f. 3 :...
Item Confirmatione delle considerationi del C.N. Paulo contra Bovio in o/4	f. 1 :...
... mondo in o/4	
Item Dell'Historia Venetiana del Bembo in o/4	f. 1 :– :–
Item Consideratione sopra le censure della santita di Papa Paulo V in o/4	f. 1 :– :–
Item Rime del Jeronimo Conestaggio in o/4	f. – :10 :–
Item Descrittione della feste fatte nelle reali nozze del principe di Toscana ets, in o/4	f. – :18 :–
Item Le deche di Tito Livio in o/4	f. 1 :10 :–
Item Dell'Historia universal dell'origine et imperio de Turchi in o/4	f. 1 :10 :–
Item Theatro del cielo et della terra	f. – :5 :–
Item L'Ungeria spiegata ets. in o/4	f. – :10 :–
Item Delle controversie tra il Papa V et la republica di Venetia in o/4	f. – :18 :–
Item Discritt : ne di tutta l'Italia di Leandro Alberti in o/4	f. 3 :– :–
Item Le historie Venetiane di Marco Antonio in o/4	f. 1 :4 :–
Item Della difera della Comedia di Dante in o/4	f. 2 :10 :–
Item Historia della vita et della morte della Sra Giovanna Graja Regina d'Ingliterra in o/8	f. – :12 :–
Item een ditto Lettere volgari di diversi nobillissmi huomini ets. in o/8	f. – :12 :–
...	
Item Il mondo et sue party in 1/8	f. ...
Item Della Republica di Venetiani per Donato Gianotti in o/8	f. ...
Item Il turco vincibile in Ungaria in o/8	f. ...
Item Hadriano tragedia di Groto in o/8 out	f. ...
Item La zucca del Doni fiorentino in o/8	f. – :1 :–
Item Eleganze da Manutio in o/8	f. – :5 :–
Item Quinto Curtio in o/8	f. – :10 :–
Item Dialoghi di Amore di Leone Ebreo in o/8	f. – :15 :–
Item La comedia de Plauto in spaensch in 1/12	f. ...
Item Delle lettere facete et piacevuoli d'atanagi in o/8	f. 1 :4 :–
Item Sylva di varia lettione di Pro Messia in o/8	f. 1 :– :–
Item Le prose di M. Pietro Bembo in 1/12	f. – :1 :–
Item Summario di tutte le scienze in o/8	f. ...
Item Esortatione del Baronio alla Republica di Venetia in o/8	f. 1 :– :–
Item Filostrato lemnio della vita di Appolonio in o/8	f. – :18 :–
Item Salusto con altre cose ets in o/8	f. – :6 :–
Item Een soneti canzoni ets. di Laura Petrarca in o/8	f. ...
...	
Item Il fuggliozio di Tomazo Costo in o/8	f. – :10 :–
Item Quatro comedie del Aretino in o(8	f. 1 :– :–
Item Concetti di Geronimo Garimberto in o/8	f. – :6 :–

Item Le cose Maravigliose di Roma in o/8	f. −:12:−
Item Ameto comedia delle Nimphe Fiorentine in o/8 (out sonder caffetorie – doorgehaald)	f. −:15:−
Item Dante Aligheri fiorentino in folio	f. 2:10:−
Item Geographia de Tolomeo in folio	f. 5:−:−

Fransche

Nicolai Crassi ad Baronium in o/4	f. −:10:−
Item Suetone Tranquille de la ville des Cesars o/4 in 't latijn	f. 1:−:−
Item De la verité de la religion chrestienne Mornay in o/4	f. 1:10:−
Item Tractatus de immunitate eulesiastica ets. in o/4 in 't Latijn	f. 1:−:−
Item Livre de sainte meditation de Rob. Libole in o/4	f. −:12:−
Item Exposition du cathechisme de Pierre Viret in cleijn ..	f. ...

...

Item Quatriesme tome du mercure francois in 1/8	f. ...
Item Le serviteur fidelle in 1/8	f. ...
Item Recueil de ce qui s'est passé en France de 1614 justques 1615 in 1/8	f. ...
Item Discours politiques de la Noe en 1/16	f. −:15:−
Item La sesmaine ou creation du Monde de Bartas ets. in 1/12	f. −:18:−
Item Epistres de Guevare in 1/8	f. 1:4:−
Item Inventaire de Serres en 1/16, 1 tome	f. −:15:−
Item Exposition sur les epistres aux Romains et Hebreux par Primatins in 1/8 oudt	f. −:6:−
Item Vraye narration des choses passées aux Pays Bas l'an 1566 en 1/8	f. −:3:−
Item La Navarre en dueil de Lostal en 1/8	f. −:10:−
Item Les oeuvres de Rabelais en 1/12	f. −:15:−
Item Memoire des comines en 1/12	f. −:15:−
Item L'exercue de l'a.. vertuluse en 1/12	f. −:2:−
Item Instruction pour tous Estats in o/8	f. −:3:−

...

Item Le marchand couverti tragedie in 1/12	f. −:3:−
Item La conformité des merveilles anciennes et modernes de Herodote in 1/8	f. 1:−:−
Item Le Theatre de Van der Noot in 1/8	f. −:12:−
Item Sermons de Jan Caloir sur les dix commandements etc. in 1/4	f. −:6:−
Item Lerees de Bouchet, 3 tomes in 1/12	f. 1:10:−
Item Vier tractaetgens bey malcanderen gebonden	f. −:3:−
Item Vier boecken de la ligue in o/8	f. −:4:−
Item Le triomphe d'Anvers en la susception du Prince Philips en folio	f. 1:10:−
Item L'entrée de Francois fils de France en Anvers ets. en folio	f. 1:10:−

Nederlandsche

Colloquia van Erasmus in o/4	f. 1:16:−
Item Scherm ende schilt der kinderen Godts in o/4	f. −:18:−
Item 't Huwelijck van jonckvrouwen, 13 boeck in o/4	f. −:−:−
Item Nulenes clachte in 1/4 van den selven	f. −:−:−
... hoff de twee maeghden in o/4	f. ...
Item Thien sermoenen over den avontmael in o/4	f. ...
Item Kracht der godtsalicheijt door Triglandus in o/4	f. ...

Item Troostspiegel in geluck ende ongeluck in 0/4	f. 1 :…
Item Schilt des geloofs door Jacob Laurentius in 0/4	f. 1 :1 :…
Item Conquesten van Indien in 0/4	f. 1 :– :–
Item De oorloghen van Italien van Cuiciardin in 0/4	f. 1 :1 :…
Item Historie van Indien in 't langh 0/4	f. 2 :1 :…
Item Ampt der kerckendienaren in 0/4	f. – :1 :…
Item Placaet tegen converticulen ende verhoedinghe van die vande roomsch religie in 0/4	f. …
Item Spiegel der spaenschen tiranye in quarto	f. – :6 :–
Item Een singhenden voorlooper ets. in 0/4	f. – :…
Item Het schilderboeck in 0/4	f. 3 :– :–
… Corte historie … verde Angola ets in 't langh 1/4	f. …
Item Een heijlighe A.B.C. voor sijne scholieren in 0/4	f. – :12 :–
Item Brievenboeck van Coornhart in 0/4	f. 2 :– :–
Item Enghe poorte offe de wech der salicheijt in 0/4	f. 1 :4 :–
Item Poppius Eurijpilus in 0/4	f. – :12 :–
Item De enghe poorte Poppij in 0/4	f. 1 :– :–
Item Opusculen van Erasmus in 0/4	f. 2 :10 :–
Item Wonderlijck schatboeck der historien in 0/8	f. – :6 :–
Item Subtil ondersoeck over het avontmael in cleijn 1/8	f. – :2 :–
Item Cort verhael der insettinghen van de roomsche kercken in cleijn 1/8	f. – :1 :–
Item Navolginghe tegens de idelheijt in 1/8	f. – :12 :–
Item Fondamenten tegens de Jezuiten in 1/8	f. – :15 :–
Item Den bijenkorff der roomsche kercke in 1/8	f. – :6 :–
Item De practijcque tot God salicheijt in 0/8	f. – :10 :–
Item Een grieckscher princessen seijntbrieven in 0/8	f. – :6 :–
…ion de Portugal a la couronne de Castille en 1/8 in out fransch	f. …
Item Cronycke van Hollandt in 1/8 cleijn ende out	f. …
Item Een als boven t'samen	f. …
Item Een synodus van Coornhert in 1/8	f. …
Item Grangrena theologia anababtista in 1/8	f. …
Item Politica Lypsi in 1/8	f. …
Item Livres de chansons out sonder couvert in 1/12 in 't fransch	f. …
Item Spelen van de sinnen in 1/8	f. …
Item Troostspieghel der siecken in 1/8	f. …
Item Costuymen van Antwerpen in 1/4	f. …
Item Goede tijdinghen uyt Canairen in 1/8	f. …
Item Valschen roem des pausdoms in 1/8	f. …
Item Een deel der wercken van Coornhert in folio	f. 2 :…
Item Guiciardin der nederlandsche oorloghen in folio	f. 6 :…
Item Amsterdamsche seecaerten door Albert Haen in folio	f. …
Item Placcaten op 't stuck van de wildernissen door Merula in folio	f. …
…	
Item een ditto Itinerare van Linschoten in folio	f. 4 :…
Item een ditto Josephus van de joodsche oorloghen in folio	f. 3 :– :–
Item een ditto Het sesde deel Carionis door Pieter Bor in folio	f. – :– :–
Item een ditto Tresoor door Merula in folio	f. 4 :– :–
Item een ditto ses tomen van Pieter Bor der Nederlandschen oorloghen in folio met register	f. 48 :– :–

Item een ditto 't Register van der selver boecken (in marge : bij die (vorige) begrepen)	f. –:–:–
Item een ditto Historische betrachtingen van Camerarij in o/4	f. 2 :–:–
Item een ditto De seevaert van Kerckman in o/8	f. 2 :–:–
Item een ditto Emanuel van Meteren in folio	f. 6 :–:–
Item Een caertboeck in folio	f. 8 :–:–
Item een Bijbel in folio	f. 18 :–:–
Item een ditto Paraphrasis Erasmi in folio	f. 6 :–:–
Item een ditto Toneel van de hertoghen ende graven in folio	f. 3 :–:–
Item een ditto Bijbel in o/4	f. 2 :–:–
Item een ditto Cronyck van Everard van Rijck in o/4	f. 2 :10 :–
Item een ditto De nassausche oorloghen in o/4	f. 4 :–:–
Item een ditto Valschen roem des pausdoms in o/8	f. –:15 :–
Item een ditto Vergulde annotatien in o/8	f. –:15 :–
…	
Item dry ditto psalmboecken in o/8	f. …
Item een ditto 't Bewijs van den waeren godtsdienst in o/4 van Grothius	f. …
Item een ditto Oratie Barlei over 't lijden Christi in 1/12	f. …
Item een ditto Vrijmoedigh ondersoeck van verscheijden placaten tegens de remonstranten in o/4	f. …
Item een ditto Standvasticheijt Lipsi in o/8	f. …
Item een ditto Oorsprongh der nederlandsche beroerten in o/4	f. 1 :–:–
Item een ditto De hooffsche .litie in 1/12	f. …
Item een ditto Epithetens handboecken	f. …
Item een ditto Beschrijvinghe van 's-Hertoghenbosch in o/4	f. 1 :…
Item een ditto 't Nederlandsch bestant in o/4	f. 1 :–:–
Item een ditto Oostendes belegh in o/4	f. …
Item een ditto Hooffts Hendrick de Groot in folio	f. 1 :1 …
Item een ditto Plutarchus der doorluchtighe mannen in folio	f. 10 :–:–
Item een ditto Parvus mundi in 't fransch in o/4	f. 1 :–:–
… uyt Astree in 1/12	f. …
Italiaensche boecken	
Item een ditto Delle navigationi 3 tomen per Ramuggio in folio	f. 18 :–:–
Item een ditto Tavolo delle opere di Machiavel in o/4	f. 5 :–:–
Item een ditto Gli annali di Cornelio Tacito in o/4	f. 3 :–:–
Item een ditto Cento favuole morali di Gio Mario in o/4	f. 1 :10 :–
Item een ditto Della guerra di Fiandra van Strada in o/4	f. 3 :–:–
Item een ditto Paolo Giovio in o/4 twee tomen	f. 3 :–:–
Item een ditto L'Origine di Molte citta del mondo in o/4	f. –:6 :–
Item een ditto Memorie historiche dell'Italia in o/4	f. –:15 :–
Item een ditto Historico politico di Monfredino in o/4 2 deelen in o/4	f. –:12 :–
Item een ditto Della guerra di Fiandra di Bentivoghia 3 deelen in 1/8	f. 5 :–:–
Item een ditto Historia della guerra della Germania inferiore di Conestaggio in o/8	f. 2 :–:–
Item een ditto Il filocolo di Boccacio in 1/8	f. –:18 :–
Item een ditto De beschrijvinghe van Virginien in folio in 't fransch	f. 1 :10 :–
… boeck in 't hooghduytsch in 't folio van Theodore de Bry	f. …
Item een ditto La phisica in o/8	f. …

INVENTORY OF JACQUET NICQUET 265

Item een ditto Il seleno historico et politico in o/8   f. ...
Item een ditto La regina fortunata in o/8   f. ...
Item een ditto Syndicato di Thiberio in o/8   f. ...
Item een ditto J. Lucci del genio execrabilé in 1/12   f. ...
Item een ditto Pordigi d'amreo in 1/12   f. −:10:−
Item een ditto Ladamo di Lordano in 1/12   f. ...
Item een ditto Historia catalana in 1/12   f. ...
Item een ditto Antiqua urbis Romae simulachrum van Calvo in folio   f. ...
Item een ditto Stratonica di Luca azarono in 1/12   f. ...
Item een ditto Orlando furioso dell'Ariosto in 1/16   f. 1:...
Item een ditto L'Alcibiade in 1/12   f. ...
Item een ditto Laldemiro di Gio Battista Raggi in 1/12   f. ...
Item een ditto L'ambasciatore individiato in 1/12   f. ...

  Schrijff offt reeckenboecken

    Een ditto journael oft grootboeck B. B. in folio

Item een deel reeckenboeck geligneerde pampier onbeschreven in folio

...

Item Hantvesten van Amsterdam 1613 in folio   f. 3:−:−
Item Philosophische t'samenspreeckinghe van de bije ende spinne in 1/8 cleijn   f. −:2:−
Item Historie van de Meteren in folio   f. 7:−:−
Item Een boecksgen in quarto vol printen   f. −:15:−

  Italiaensche

    Dialoghi di Benevento in o/8   f. −:12:−

Item Tratta delle pene del purgatorio in 1/32   f. −:2:−
Item Trattato dei colori nelle arme in 1/8   f. −:2:−
Item Il gentilhomo del fausto in 1/8   f. −:2:−
Item Gl'inganni comedia in 1/8   f. −:2:−
Item La guerra contra Backari ets. in 1/8   f. −:8:−
Item Il sergio comedia in 1/12   f. −:2:−
Item Rime et prose della casa 1/12   f. −:2:−
Item Lettere amorose di diversi ets. in 1/8   f. −:5:−
Item Clarice comedia ende andere in't selve boeck in 1/12   f. −:6:−
Item Le rime del Burchiello in 1/12   f. −:2:−

...

Item Adventure admirable des siecles passez de la bataille de don Sebastian ets. in o/8   f. ...
Item Sermon joyeux, out in o/8   f. ...
Item Beze sur la vie et mort de Calvin 1/8   f. ...
Item L'arangne de Barleus sur les merveilles del'ame en 1/10   f. ...
Item Discours sur l'estat de fame in 1/8   f. ...
Item Dialogue de choses advenues aux Lutheriens et Hugenots de la France en 1/8   f. ...
Item Copie d'une lettre de Nicolas Tringant ets. in 1/16   f. ...
Item Les fontaines de Spa in 1/12   f. ...
Item Seven fransche oude tractaetgens aen malcanderen gebonden   f. ...

Nederlandsche

    De kercken ordeninghen in 0/4   f. ...

Item Morghenwecker in 0/8   f. ...
Item Een ditto in 0/8   f. ...

...

    journael ende ... Bredenrode in Sweden ets. in 1/4 langh   f. ...

Item Historie van Lazarus in 0/8   f. ...
Item Christelijck gebedtboeck in 1/32   f. –:2:–
Item Christelijcke bekeeringhe van Jaspar Martini in 1/12   f. –:1:–
Item Vijffthien tractaetgens van verscheijden soorten   f. 2:10:–
Item Twee bondels van verscheijden tractaetgens   f. 5:–:–

  Gecontinueert op de xxiijen date junij 164(2) boven op een solder in een groote oostersche kiste

    Een groot schultboeck geteeckent C van de jaeren 1611:1612:1613
Item Een journael van 't selve
Item Een groot schultboeck geteeckent D van de jaeren 1614:1615:1616:1617
Item Een journael van 't selve
Item Een groot schultboeck geteeckent E van de jaeren 1618:1619:1620
Item Een ditto journael van 't selve
Item Een grootboeck geteeckent F beginnende primo januarij 1621 a folio 5
Item Een journael van 't selve beginnende a foli 1 per Willem van Weli ets.
Item Een journael cleijner beginnende a folij 1 primo januarij 1621 per Willem van Weli ets.
Item Een grootboeck van asseurantie beginnende primo januarij 1612
Item Een journael van 't selve
Item Een grootboeck van asseurantie beginnende 24 februarij 1619
Item Een journael van 't selve

Item twee schultboecken van asseurantie beijde beginnende primo martij 1607
Item Een groot- off schultboeck geteeckent A.A. beginnende 23 novembris 1609 van 't capitael van Jacques Nicket
Item Een journael van 't selve
Item Een missieven copijeboeck beginnende primo januarij 1621
Item Een boeck getituleert: anno 1621 in Amsterdam aen banco per Tonneman
Item 12 boecken soo groot als cleijn daerinne begrepen eenen alschabeth inventaris ets. met 2 bundels pampieren raeckende Jan Nicquet den ouden
Item 6 boecken soo groot als cleijn met eenighen schriften van Francois Palgher raeckende Jan Pluquet de jonghe

Op den dertighsten dage der maen ... des jaers 1642 sijn alle de lees-boe(cken op) versoeck der E. Samuel Blomm(art), Adam Bessels door de eers. Dirck Pietersz (Pers) ende Jan Marcussen, beijde b(oeck)vercoopers deser voorss. stede getaxeert ten overstaen van de ... ende ter presentie van Jochum Thielmans ende Jan Swart als getuygh(en).

INVENTORY OF JACQUES NICQUET

des jaers sesthienhondert (tweeendeveertig) is ten sterffhuyse van (Jacques) Nicquet saliger ter presentatie (van) de gesamentlijcke vrienden aldaer (na) de begraeffenisse inde groote ca(mer) achter noortwaerts van de gangh ... thuyn comende vergadert zijnde voor Benedict Baddel, keyserlijcken en bij(den) Hove van Hollandt geadmitteerden o(penbaer) notaris t' Amsterdam residerende ende de naergenoemde getuyghen persoon(lijck) gecompareert den E.Sr Gerardt Reijnst ingeseten coopman deser v(oorss) stede mij notario voorss. seer wel bekent, dewelcke aldaer op de taef(fel) producerende ende mij exhiberende een handtschrift met een memorie seeckere constboecken onder aen gestelt van den voornoemden overledene Jaecques Nicquet, waer van de copije hier naer wel ende getrouwelijck geinsereert is, luydende als volght namelijck :

   Volgt de transcriptie van de   handschrift ende memorie

Ick Jaques Nicket bekenne mits desen schuldi(ch te) wesen aen mijn cosijn Gerart Reijnst de s(om)me van seshondert gulden, te weten dryhondert g(ulden) over eenige maenden ontfangen ende dryhondert gu(lden) nu present, welck seshondert gulden ick hem (heb) te betalen t'allen tijden als hem sal gelieven daer t(oe) hebbe ick gedaen aen mijn verschreven cosijn Gera(rt) Reijnst diversche kunstboecken als bij de memorie (daer) aff sijnde, welcke voorschreve boecken hij gehouden (sal) wesen mij weder te leveren als ick hem deselve s(eshon)dert gulden sal betalen. Doch off geviele dat dese (ses)hondert gulden onder mij bleven tot dat ick des(e...

...

legatere ...ken der waerheijt, desen on(derteeckent) adio 24 septembris 1641 in Amsterdam ende was (ge)teeckent Jacques Nicket.
Een boeck van Albert Durer copersnes
Een boeck van Aldegraeff
Een boeck van hubsen Merten
Een boeck antiquiteijt van Romen
Een boeck printen van Sadelers
Een medalieboeck van Abram Ortelius
Een boeck prospectiva corporum regularum
La nature et diversite des poisons
Een boeck deorum dearumque capita
Een boeck spiegel der weereldt door Pieter Heijns
Een boeck antiquarum statuarum urbis Romae van Lodovico Madruzzi
Een boeck Les genealogies et anciens decens des forestiers et comtes de Flandres
Een boeck Hadriani Baslandi Hollandia comitum historia
Een boeck bijbelfiguren des alten und nieuwen testament houdtsne
Een boeck figure del vechio testamento per Damiano Maratti in Lion 1554
Verclaerde opentlijck, alhoewel den overleden den voorss Jacques Nicket in cas hij deser werelt quame t'overlijden aen hem comparant bij deselve hantschrift hadde gelegateert de constboeck(en), daerinne gespecificeert dat hij evenwel, omme redenen hem daer toe moverende niet alleenlijck absolutelijck desiseerde van 't selve legaet maer oock voor hem ende sijnen broeder Jan Reijnst, wonende tot Venetien, van alsulcke erffenisse, legaet, mits bate off voordeel oock graedt, titule, qualiteijt als hen beijde t'samen ende elck apart uyt cracht

...
mogen werden toegevoegt, ... int geheel ende deel tot behoef ... dispositie
van d'andere sijne v... des voorss Nickets erffge(namen) inde voorss
qualite wel exp(res) renuncierende bij desen, behouden(s) ende uytgeno-
men nochtans sijn aenpart (van) twintich duysent guldens spruytende (uyt)
den sterffhuyse van wijlen Sa(ra) Nicket des voorss Jaecques Nic(quet)
saliger suster die hij niet en vers... in dit desistement ende renuntiati(e)
begrepen te sijn, versoeckende den voo(rss) comparant in de voorrss
qualite h(em) daer aff gelevert te werden openbare acte een off meer in
debita form(e) die ick sijne E. oock hebbe verlee(nt) omme hem alomme
te dienen ende str(ec)ken als naer behooren ter goeder trouwen.
t'Orconde hebb'ick notaris voorss dese met mijn eijghene handt onder-
teeckent op den 28en dagh martij des jaers 1642.
Ende op den derden dagh der maent aprilis daer aen volghende heeft den
E. Gerart Rheijnst voor m(ij) notaris voorss ende de naegenoe(mde) getuyghen comparerende bij de v(oorss) acte van verclaringhe desistem(ent)
ende renuntiatie in qualite als bij deselve gepersisteert alle 't selve con-
firmerende bij dese. Aldus gedaen binnen deser voorss stadt Amsterdam
(ten) huyse ende schrijffcomptoir
...

               w.g.   G. Reijnst
                           J. Hendrixen
                           J. van Wijckersloot

A.E. D'Ailly, ed., 'De Keizersgracht', in *Zeven Eeuwen Amsterdam*, III, Amsterdam (s.a.), pp. 211–29.

Lieuwe van Aitzema, *Saken van Staet en Oorlogh*, 6 parts, The Hague, 1669–72.

Kasper van Baerle, *Blyde Inkomst der allerdoorluchtighste Koninginne, Maria de Medicis, t' Amsterdam*, Amsterdam, 1639.

C. H. Collins Baker, *Catalogue of the Pictures at Hampton Court*, Glasgow, 1929.

Barbara Jeanne Balsiger, *An Analytical Concensus of Early Printed Catalogs of Post-Medieval Collections Kunst- und Wunderkammern*, M.A.Thesis, State University of Iowa, 1955.

Eadem, *The Kunst- und Wunderkammern: A Catalogue Raisonné of Collecting in Germany, France and England, 1565–1750*, Diss. Pittsburgh, 1970.

Violet Barbour, *Capitalism in Amsterdam in the 17th Century*, Ann Arbor, 1963.

W. Bathoe, *A Catalogue of the Collection of Pictures, &c. Belonging to King James the Second; ...*, London, 1758.

W.G. Becker, *Augusteum. Dresdens antike Denkmäler*, 3 vols., Dresden, 1804–11.

Lorenz Beger, *Thesauri regii et electoralis Brandenburgici, volumen tertium: continens antiquorum numismatum et gemmarum, quae cimelarchio Regio-Electorali Brandenburgico nuper accessire, Rariora: Ut & supellectilem antiquariam uberrimam, id est Statuas, Thoraces, Clypeos, Imagines tàm Deorum, quàm Regum & Illustrium ...*, Coloniae Marchicae (1696).

Bernard Berenson, *Italian Pictures of the Renaissance, Venetian Schools*, 2 vols., London, 1957.
Idem, *Italian Pictures of the Renaissance, Central Italian and North Italian Schools*, 2 vols., London, 1968.

Clara Bille, *De Tempel der Kunst of het kabinet van den Heer Braacamp*, Amsterdam, 1961.

Roy Bishop, *Paintings of the Royal Collections*, London/Bombay/Sidney, 1937.

B. Bijtelaar, 'Het Huis Keizersgracht 209,' *Ons Amsterdam*, XIII, no. 1, Jan. 1961, pp. 6–11; no. 2, Feb. 1961, pp. 52–57; no. 4, April 1961, pp. 104–11.

E. Maurice Bloch, 'Rembrandt and the Lopez Collection,' *Gazette des Beaux-Arts*, 6th series, XXIX, 1946, pp. 175–86.

J.C. Block, *Jeremias Falck, sein Leben und seine Werke*, Danzig, 1890.

Petrus Johannes Blok, *History of the People of the Netherlands*, IV, New York/London, 1907.
Idem, *Relazioni Veneziane. Venetiaansche Berichten over de Vereenigde Nederlanden van 1600–1795 (Rijks Geschiedkundige Publicatiën, 7)*, 's-Gravenhage, 1909.

Carl Blümel, *Staatliche Museen zu Berlin, Römische Kopien griechischer Skulpturen des fünften Jahrhunderts v. Chr.*, Berlin, 1931.
Idem, *Staatliche Museen zu Berlin, Römische Bildnisse*, Berlin, 1933.
Idem, *Staatliche Museen zu Berlin, Römische Kopien griechischer Skulpturen des vierten Jahrhunderts v.Chr.*, Berlin, 1938.

Hans Bontemantel, *De Regeeringe van Amsterdam...*, see Kernkamp.

Tancred Borenius, *The Picture Gallery of Andrea Vendramin*, London, 1923.
Idem, 'More about the Andrea Vendramin collection', *Burlington Magazine*, LX, 1932, pp. 140–45.

Marco Boschini, *La Carta del Navegar Pittoresco*, Venice (1660).

C.R. Boxer, *The Dutch Seaborne Empire: 1600–1800*, New York, 1965.

J.P.J. Brants, *Description of the Ancient Sculpture...of the Museum of Archaeology of Leiden*, The Hague, 1927.

A. Bredius, 'Italiaansche Schilderijen in 1672 door Amsterdamsche en Haagsche schilders beoordeeld,' *Oud Holland*, 4, 1886, pp. 41–46; 278–80; 34, 1916, pp. 88–93.
Idem, 'De kunsthandel te Amsterdam in de XVIIe eeuw,' *Amsterdamsch Jaarboekje*, 1891, pp. 54–71.
Idem, *Künstler-Inventare*, (Quellenstudien zur Holländischen Kunstgeschichte), 8 vols., The Hague, 1915–22.
Idem, *Rembrandt, The Complete Edition of the Paintings*, revised by H. Gerson, London, 1969.

J. C. Breen, 'Geschiedenis van het huis 'in 't derde vredejaer', Keizersgracht 604,' *Jaarboek van het Genootschap Amstelodamum*, XVII, 1919, pp. 73–79.

H. Brunsting, 'Twee gouden eeuwen,' *Archeologie en Historie*, Bussum, 1973, pp. 179–90.

J. Bruyn and Oliver Millar, 'Notes on the Royal Collection-III: The 'Dutch Gift' to Charles I,' *Burlington Magazine*, CIV, 1962, pp. 291–94.

Peter Burke, *Venice and Amsterdam, A study of seventeenth-century élites*, London, 1974.

J.W.C. van Campen, ed., *Notae Quotidianae van Aernout van Buchell* (Werken uitgegeven door het historisch genootschap gevestigd te Utrecht), III, no. 70, 1940.

G. Campori, *Raccolta di Cataloghi ed Inventari inediti*, Modena, 1870.

Ralph C.H. Catterall, 'Anglo-Dutch Relations, 1654–1660,' *Annual Report of the American Historical Association for the Year 1910*, Washington, 1912, pp. 103–21.

F. de Clarac, *Musée de Sculpture antique et moderne*, III, Paris, 1850.

Kenneth Clark, *Rembrandt and the Italian Renaissance*, New York, 1966 (paperback edition, New York, 1968).

(A. Conze), *Königliche Museen zu Berlin, Beschreibung der antiken Skulpturen mit Ausschluss der pergamenischen Fundstücke*, Berlin, 1891.

Johan E. Elias, *De Vroedschap van Amsterdam 1578–1795*, Haarlem 1903/05.

Keith Feiling, *British Foreign Policy, 1660–1672*, London, 1930.

Oskar Fischel, review of Borenius, Picture Gallery of Vendramin, in *Zeitschrift für bildende Kunst*, 58, 1924, *Monatsrundschau*, pp. 28–29.

Hanns Floerke, *Studien zur niederländischen Kunst- und Kulturgeschichte*, Munich/Leipzig, 1905.

Melchior Fokkens, *Beschrijvinge der wijdt- vermaarde Koop-stadt Amstelredam*, Amsterdam, 1662.

T. von Frimmel, ed., 'L'anonimo Morelliano, notizia d'opere del disegno,' (*Wiener Quellenschrift zur Kunstgeschichte*, N.S. 1), Vienna, 1888.

Fuchs, 'Christian Knorr von Rosenroth,' *Zeitschrift für Kirchengeschichte*, XXXV, 1914, pp. 548–83.
Idem, 'Aus dem 'Itinerarium' des Christian Knorr von Rosenroth,' *Jaarboek van het Genootschap Amstelodamum*, XIV, 1916, pp. 239–45.

J.J. de Gelder, *Bartholomeus van der Helst*, Rotterdam, 1921.

H.E. van Gelder, 'Een praatje over de kunsthandel en het verzamelen in Nederland,' (*Catalogue of the* VIII.) *Oude Kunst- en Antiekbeurs Delft*, (Museum 'Het Prinsenhof'), 1956.

J.G. van Gelder, 'Notes on the Royal Collection-IV: The 'Dutch Gift' of 1610 to Henry, Prince of 'Whalis', and Some Other Presents,' *Burlington Magazine*, CV, 1963, pp. 541–44.
Idem, 'Jan de Bisschop's Drawings after Antique Sculpture,' *Studies in Western Art, Acts of the Twentieth International Congress of the History of Art*, III, Princeton, N.J. 1963, pp. 51–58.
Idem, 'Jan de Bisschop,' *Oud Holland*, LXXXVI, 1971, pp. 1–88.

Eduard Gerhard, *Berlin's antike Bildwerke*, I, Berlin, 1836.

Horst Gerson, *Rembrandt Paintings*, Amsterdam, 1968

A. Hagen, 'Ueber den Kupferstecher Jeremias Falck,' *Kunstblatt*, no. 16, 30 March 1848, pp. 61–64.

Sigebert Havercamp, *Museum Uilenbroekianum*, (Amsterdam, 1741).

Denys E. Haynes, 'The Arundel Marbles, I,' *Archaeology*, 21, Jan. 1968, pp. 85–91; idem, II, June 1968, pp. 206–11.
Idem, 'The Fawley Court Relief,' *Apollo*, XCVI, July 1972, pp. 6–11.

William Hazlitt, *Criticism on Art*, London, 1843.

R. Hecquet, *Catalogue des estampes gravées d'après Rubens... et de Visscher*, Paris, 1751.

K. Heeringa, *Bronnen tot de Geschiedenis van den Levantschen handel* (Rijks Geschiedkundige Publikatiën, 9/10), 's-Gravenhage, 1910.

K.H. von Heinecken, *Idée générale d'une collection complette d'estampes ...*, Leipzig/Vienna, 1771.

A. Heppner, in A.E. d'Ailly, ed., 'Amsterdamsche Verzamelaars,' in *Zeven Eeuwen Amsterdam*, Amsterdam (s.a.), pp. 227–64.

Frank Herrmann, *The English as Collectors*, New York, 1972.

Mary F.S. Hervey, *The Life, Correspondence and Collections of Thomas Howard, Earl of Arundel*, Cambridge, 1921.

Herman Hettner, *Die Bildwerke der königlichen Antikensammlung zu Dresden*, Dresden 1869.

G. Hoet, *Catalogus of naamlyst van schilderyen met derzelver pryzen...* 's Gravenhage, 1752.

C. Hofstede de Groot *Die Urkunden über Rembrandt (1575–1721)*, (Quellenstudien zur Hollandischen Kunstgeschichte, III), The Hague, 1906.

F.W.H. Hollstein, *Dutch and Flemish Etchings, Engravings and Woodcuts, ca. 1450–1700*, Amsterdam, 1949 ff.

Niels von Holst, *Künstler, Sammler, Publikum*, Darmstadt, 1960.

C.G. 'T Hooft, 'Een geschenk van Amsterdam aan Karel II van Engeland in 1660,' *Jaarboek van het genootschap Amstelodamum*, XIX, 1921, pp. 1–13.

G.J. Hoogewerff, *De twee reizen van Cosimo de' Medici... door de Nederlanden 1667–1669*, Utrecht, 1919.

Idem and J.Q. van Regteren Altena, ed., *Arnoldus Buchelius 'Res Pictoriae'*, (*Quellenstudien zur Holländischen Kunstgeschichte*, XV), 's-Gravenhage, 1928.

Arnold Houbraken, *De groote Schouburgh der Nederlantsche Konstschilders en Schilderessen*, ed. P.T.A. Swillens, 3 vols., Maastricht, 1943–53.

F.W. Hudig, *Frederik Hendrik en de Kunst van zijn tijd*, Amsterdam, 1928.

Emil Jacobs, 'Das Museo Vendramin und die Sammlung Reynst,' *Repertorium für Kunstwissenschaft*, XLVI, 1925, pp. 15–38.

Axel Janeck, *Untersuchung über den holländischen Maler Pieter van Laer*, Diss., Würzburg, 1968.

L.F.J. Janssen, *De grieksche, romeinsche en etrurische Monumenten van het museum van oudheden te Leyden*, Leiden (1848).

N. Japikse, *Resolutiën der Staten-Generaal van 1576 tot 1609*, XI, 1600–1601 (*Rijks Geschiedkundige Publikatiën...*,85), 's-Gravenhage, 1941.

J.N. Jacobsen Jensen, *Reizigers te Amsterdam; beschrijvende lijst van reizen in Nederland door vreemdelingen voor 1850*, Amsterdam, 1919.

J.A. Jochems, *Amsterdams Oude Burgervendels (Schutterij), 1580–1795*, Amsterdam, 1888.

J.C. de Jonge, *Nederland en Venetië*, 's Gravenhage, 1852.

G.W. Kernkamp. *De Regeeringe van Amsterdam soo in 't civiel als crimineel en militare (1653–1672), ontworpen door Hans Bontemantel*, 's Gravenhage, 1897.

Christian Knorr von Rosenroth, '*Itinerarium*', see Fuchs.

Christiaan Kramm, *De levens en werken der hollandsche en vlaamsche kunstschilders, beeldhouwers, graveurs en bouwmeesters, van den vroegsten tot op onzen tijd*, Amsterdam, 1857.

Ernest Law, *A Historical Catalogue of the Pictures in the Royal Collection at Hampton Court*, London, 1881 (2nd edition, 1898).

B. Leplat, *Suite de divers marbres modernes, ...*, (no location), 1733.

Leupe, 'Schilderijen en statuen voor Karel de Tweede, Koning van Engeland 1660,' *De Nederlandsche Spectator*, 1876, pp. 184–86; 1878, pp. 82–83.

Michael Levey, *The Later Italian Pictures in the Collection of Her Majesty the Queen*, Greenwich, Conn. (1964).

Oliver Logan, *Culture and Society in Venice 1470–1790*, New York, 1972.

Frits Lugt, *Mit Rembrandt in Amsterdam*, Berlin, 1920.

Idem, 'Italiaansche Kunstwerken in Nederlandsche verzamelingen van vroeger tijden,' *Oud Holland*, LIII, 1936, pp. 97–135.

Idem, *Répertoire des catalogues de ventes publiques*, I, The Hague, 1938.

Th.H. Lunsingh Scheurleer, *Sprekend Verleden*, Amsterdam, 1959.

Alessandro Luzio, *La Galleria dei Gonzaga venduta all' Inghilterra nel' 1627–28*, Milan, 1913.

Denis Mahon, 'Notes on the 'Dutch Gift' to Charles II,' *Burlington Magazine*, XCI, 1949, pp. 303–05; 349–50; XCII, 1950, pp. 12–18.

Carlo Cesare Malvasia, *Felsina Pittrice, vite de Pittori Bolognesi,...* Bologna, 1678.

Adolf Michaelis, *Ancient Marbles in Great Britain*, Cambridge, 1882.

W.L.F. Nuttall, 'King Charles I's Pictures and the Commonwealth Sale,' *Apollo*, LXXXII, October 1965, pp. 302–09.

David Ogg, *England in the Reign of Charles II*, London/Oxford/New York, 1972.

Franciscus Oudendorp, *Brevis veterum monumentorum ab amplissimo viro Gerard Papenbroekio Academiae Lugduno-Batavae legatorum descriptio*, Lugduno Batavae, 1746.

Caspar Philips, *Verzameling van alle de huizen...langs de Keizers- en Heere- Grachten...*, Amsterdam, 1768 ff. (reprint Amsterdam, 1967).

J.Q. van Regteren Altena and P.J.J. van Thiel, *De portret-galerij van de Universiteit van Amsterdam en haar stichter Gerard van Papenbroeck 1673–1743*, Amsterdam, 1964.

Salomon Reinach, *Répertoire de la statuaire grecque et romaine*, 6 vols., Paris, 1897–1930.

G.J. Renier, *The Dutch Nation*, London, 1944.

M.A. van Rhede van der Kloot, *De Gouverneurs-Generaal en Commissarissen-Generaal van Nederlandsch-Indië 1610–1688,* 's-Gravenhage, 1891.

Carlo Ridolfi, *Le Maraviglie dell'Arte*, Venice, 1648; ed. D. von Hadeln, Berlin, 1914 and 1924.

Cesare Ripa, *Iconologia*, Rome, 1603 (reprint New York, 1970).

N. de Roever, 'Een vorstelijk geschenk,' *Oud Holland*, I, 1883, pp. 169–88.

Hannelore Sachs, *Sammler und Mäzene*, Leipzig, 1971 (with extensive bibliography).

Kurt Salecker, *Christian Knorr von Rosenroth 1636–1689*, Leipzig, 1931.

Luigi Salerno, 'The Picture Gallery of Vincenzo Giustiniani, I,' *Burlington Magazine*, CII, 1960, pp. 21–27.
Idem, 'Arte, Scienza e Collezioni nel Manierismo,' *Scritti di Storia dell' Arte in onore di Mario Salmi*, III, Rome, 1963, pp. 193–214.

Joachim von Sandrart, *Academie der Bau-, Bild- und Mahlerey-Künste von 1675*, ed. A.R. Peltzer, Munich, 1925.

Francesco Sansovino, *Venetia città nobilissima ...*, (with additions by Giustiniano Martinioni), Venice, 1663.

Simona Savini-Branca, *Il Collezionismo veneziano nel '600* (Università di Padova, Pubblicazioni della Facoltà di lettere e filosofia, XLI), Padua, 1965.

R.W. Scheller, 'Rembrandt en de encyclopedische verzameling,' *Oud Holland*, LXXXIV, 1969, pp. 81–147.

Julius von Schlosser, *Die Kunst- und Wunderkammern der Spätrenaissance*, (*Monographien des Kunstgewerbes*, N.F. XI), Leipzig, 1908.

J. Six, 'La famosa Accademia di Eeulenborg,' *Jaarboek der Koninklijke Akademie van Wetenschappen te Amsterdam*, 1925–26, 1926, pp. 229–41.

Seymour Slive, *Rembrandt and His Critics, 1630–1730*, The Hague, 1953.

F.W. Stapel, *De Gouverneurs-generaal van Nederlandsch-Indië...*, The Hague, 1941.
Idem, ed., *Pieter van Dam, Beschryvinge van de Oostindische Compagnie*, II, (*Rijks Geschiedkundige Publicatiën*, 63), 's-Gravenhage, 1927; I^II, (RGP, 68), 's-Gravenhage, 1929; III, (RGP, 87), 's-Gravenhage, 1943.

Johann Gottlob Stimmel, *Catalogue raisonné du cabinet d'estampes de feu Monsieur (Gottfried) Winckler*, V, Leipzig, 1810, pp. 309–17.

B.H. Stricker, 'De verzameling Reynst. Egyptische Antiquiteiten,' *Vooraziatisch-egyptisch Gezelschap 'Ex Oriente Lux', Mededeelingen en Verhandelingen*, 7, 1947, pp. 255–67.

Francis Henry Taylor, *The Taste of Angels, A History of Art Collecting*, Boston/London, 1948.

Gisela Thieme, *Der Kunsthandel in den Niederlanden im siebzehnten Jahrhundert*, Cologne, 1959.

J.J.M. Timmers, *Gérard Lairesse*, Amsterdam, 1942.

Jacobi Tollii Epistolae itinerariae : ex Auctoris Schedis Postumis, cura & studio Henrici Christiani Hennini, Amsterdam, 1700.

Gustav Friedrich Waagen, *Treasures of Art in Great Britain*, 3 vols., London, 1854.

Jan Wagenaar, *Amsterdam in zyne opkomst*, 3 vols., Amsterdam, 1760–67.

E.K. Waterhouse, 'Painting from Venice for seventeenth-century England: some records of a forgotten transaction,' *Italian Studies*, VII, 1952, pp. 1–23.
Idem, 'A note on British collecting of Italian Pictures in the later seventeenth century,' *Burlington Magazine*, CII, 1960, pp. 54–57.

Harold E. Wethey, *The Paintings of Titian*, I, *The Religious Paintings*, London, 1969.
Idem, *The Paintings of Titian*, II, *The Portraits*, London, 1971.

J.A. Worp, *De Briefwisseling van Constantijn Huygens (1608–1687)*, 6 vols., (*Rijks Geschiedkundige Publicatiën*, 15, 19, 21, 24, 28, 32), 's-Gravenhage, 1911/17.

A. von Wurzbach, *Niederländisches Künstler-Lexikon*, 3 vols., Vienna/Leipzig, 1906/11 (reprint 1963).

Johann Wussin, *Cornel Visscher*, Leipzig, 1865.

Filip von Zesen, *Beschreibung der Stadt Amsterdam*, Amsterdam, 1664.

EXHIBITIONS

London 1946–47
Exhibition of the King's Pictures, London (Royal Academy of Arts), 1946–47.

London 1960
Italian Art and Britain, London (Royal Academy of Arts), 1960.

Amsterdam 1965
Het Nederlandse Geschenk aan Koning Karel II van Engeland 1660, Amsterdam (Rijksmuseum), 1965.

INDEX

The general index is printed in roman type. Artists' names are indexed in small capitals. Proper names and names of cities printed in italics refer to former as well as present owners of works of art or archival documents.

AELST, WILLEM VAN 94 (n. 19)

AERSSEN VAN WERNHOUT, JOHANNA VAN 94 (n. 22)

Ahmed I, Sultan of Turkey 18

Albrecht V, Duke of Bavaria 98

ALDEGREVER, HEINRICH 16

Aldrovandi, Ulisse 100

Aleppo (Syria) 29

Alexandretta 29

Alkmaar : Arundel collection 102

Amersfoort : Arundel collection 102

Amory, van de :
Liss (cat. 14) 126
Veronese (cat. 37) 157

Amsterdam 11, 12, 17–20, 22–30, 33, 35–38, 43, 45, 50, 71, 74–75, 87, 93, 102, 104–106

Amsterdam, Athenaeum 24

Amsterdam, Civic Guard (burgervendel) 26

Amsterdam, Dam 13

Amsterdam, Exchange Bank (wisselbank) 25

Amsterdam, Gemeentelijke Archiefdienst 13 (n. 2, 4), 14 (n. 10), 15 (n. 12, 15), 17 (n. 18–19), 18 (n. 23–26), 19 (n. 30), 20 (n. 35, 40), 22 (n. 42), 27 (n. 66–67), 28 (n. 68–71), 29 (n. 78, 82, 84), 32 (n. 95), 35 (n. 98, 101), 39 (n. 3), 58 (n. 49), 92 (n. 17), 93 (n. 19)

Amsterdam, Hartenstraat 20

Amsterdam, Herengracht 18 (n. 25), 20 (n. 39)

Amsterdam, Herengracht 498 27

Amsterdam, Kalverstraat (Keijserskroon) 92, 106 (n. 20)

Amsterdam, Keizersgracht 20 (n. 39), 26 (n. 64), 27 (n. 66), 29 (n. 81)

Amsterdam, Keizersgracht 209 (De Hoop) 9, 10 (n. 2), 19–20, 22, 28, 30, 37, 55, 63, 72, 104, 106, 118 (cat. 9)

Amsterdam, Keizersgracht 211 (De Liefde) 20

Amsterdam, Keizersgracht 213 (Het Geloof) 20

Amsterdam, Keizersgracht 343 18

Amsterdam, Keizersgracht 345 (De Oranjeappel) 18

Amsterdam, Keizersgracht 604 (Het derde vreede jaer) 28

Amsterdam, Kloveniersdoelen 26, 79

Amsterdam, Leidse gracht 20 (n. 39)

Amsterdam, Nederlands Persmuseum (Dutch Press museum) 29 (n. 74)

Amsterdam, Nieuwe Brugsteeg 9 13 (n. 2)

Amsterdam, Nieuwe Brugsteeg 11 (Lybaert) 13

Amsterdam, Nieuwe Heerengracht 27 (n. 66)

Amsterdam, Nieuwe Kerk 15, 18 (n. 23, 25), 19 (n. 30)

Amsterdam, Nieuwendijk 13 (n. 2)

Amsterdam, Oude Hoogstraat 18 (n. 23)

Amsterdam, Oude Kerk 19 (n. 30), 28

Amsterdam, O.Z. Achterburgwal, Westside 13 (n. 2)

Amsterdam, Prinsengracht 20 (n. 39)

Amsterdam, Reestraat 22 (n. 44)

Amsterdam, Rijksmuseum :
Dujardin, K. 28
Helst, B. v. d. 26 (n. 62)
Reynst, Gerrit (portrait of by unidentified artist) 14 (n. 11)
Vroom, H.C. 78 (n. 86)

Amsterdam, Rijksprentenkabinet :
Caelaturae 39 (n. 4)
Goltzius drawing 15 (n. 17)
Icones 45 (n. 17)

Amsterdam, Singel 172 (Bloempot) 13

Amsterdam, Six collection :
Palma Giovane 49, 96 (n. 34)

Amsterdam, Theater 18

Amsterdam, Town council (vroedschap) 24

Amsterdam, Vijzelstraat 27 (n.66)

Amsterdam, Wester Kerk 26–28

Anne, Queen of England :
Bonifazio de' Pitati (cat. 7) 115
Cariani (cat. 9) 117
Girogione, attr. to (cat. 11) 120
Giulio Romano (cat. 12) 122
Lotto (cat. 16–17) 130–134
Oggiono (cat. 19) 135
Parmigianino (cat. 20) 136–137
Reni, attr. to? (cat. 26) 143–144
Schiavone (cat. 28–29) 144–146
Tintoretto (cat. 32) 150
Titian, school of (cat. 36) 155
Veronese (cat. 38) 157–158

Antwerp 15, 17, 30, 33, 54, 90

Appelboom, Harald 50 (n. 30)

INDEX

APPELMAN, BARTHOLOMEUS   94 (n. 19)

Artemisia   136 (cat. 20)

Arundel, Earl of   74, 102

Askew, Pamela   119 (cat. 10)

Athens   23

BACKER, ADRIAEN   93 (n. 18)

Backer, Family archive   27 (n. 67), 28 (n. 68–69), 29 (n. 78)

Backer, J.   39 (n. 3)

Backer, Joores   27

Baddel, B.   17 (n. 19)

BAEN, JOHAN DE   94 (n. 22)

Baerle, Caspar van   20, 24, 30, 36

Baers, Michiel   27 (n. 67), 28 (n. 70)

Baltic States   15

BANDINELLI   130 (cat. 16)

BARENDSZ, DIRCK   112 (cat. 4)

BAROCCI, FEDERICO:
Woman with Dog (cat. 1)   82, 110, 236

Basil, Viscount Feilding   104 (n. 10)

BASSANO, JACOPO   103

BASSANO, JACOPO:
Abraham Leaving Haran (cat. 2)   40, 110
Annunciation to the Shepherds (cat. 3)   40, 111
Christ Carrying the Cross (cat. 4)   39 (n. 3), 40, 82, 84, 111, 236
Entombment of Christ (cat. 5)   40, 105, 112, 147, 228
Virgin and Child and Infant St. John (cat. 6)   40–41, 114

Beger, Laurentius   49 (n. 28)

BEGEYN, ABRAHAM   92

BELJAMBE, P.   141 (cat. 23)

BELLINI, GENTILE:
Madonna and Saints (cat. 43)   165, 173 (no. 102)

BELLINI, GIOVANNI   11, 63, 70, 171 (nos. 1–2, 4, 17, 29, 31, 40), 172 (nos. 66, 69, 73, 85)

Bellori, G. P.   50 (n. 28)

Bembo, Pietro   74, 99 (n. 44)

Bentes, Albert   69 (no. 70), 71 (n. 75)

Bentes, Hillebrant   71 (n. 75)

Berlin-Dahlem, Staatliche Museen:
Liss, St. Paul in Ecstasy   40, 126
Palma Vecchio   105

Berlin (East), Deutsche Staatsbibliothek:
De Antiquorum Tumulis   70 (n. 70)

Berlin (East), Staatliche Museen:
Sculptures   47, 49 (n. 28), 68, 181 (cat. 10), 187 (cat. 25), 194 (cat. 43), 197 (cat. 53), 211 (cat. 94), 218 (cat. 111)

Bernaerts, Isaack   27 (n. 67)

Bernard, Anna   19 (n. 34)

Bicker, Cornelis   25 (n. 51)

Bicker, Roelof   26 (n. 62)

Biesum, Quirijn van:
Van Laer (cat. 41)   161

BISSCHOP, JAN DE   50 (n. 29), 54, 90 (n. 12–13)

Block, J. C.   42 (n. 9), 111–12 (cat. 4–5), 115 (cat. 7), 118 (cat. 10), 120 (cat. 11), 124 (cat. 13), 128 (cat. 15), 134 (cat. 18), 138 (cat. 21A), 144 (cat. 27), 146 (cat. 30)

BLOEMAERT, ABRAHAM:
St. John the Baptist Preaching (cat. A1)   42, 228

BLOEMAERT, CORNELIS   51

Bloemaert, Samuel   18, 89

BLOM, JAN   93 (n. 18)

Boccaccio   152 (cat. 34)

Boelensz, Henrick Dirck   22

Bologna   44

BOLSWERT, S. A.   42, 230 (cat. A5)

BONIFAZIO DE' PITATI   67 (n. 65), 172 (no. 86)

BONIFAZIO DE' PITATI:
Adoration of the Magi (cat. 44)   165
Virgin and Child with Tobias and Saints (cat. 7)   40, 83–84, 115

Bontemantel, Hans   23

BORDONE, PARIS   117 (cat. 9), 152 (cat. 33), 173 (nos. 92, 94)

BORDONE, PARIS:
Male Figure in Armour   105 (n. 16), 173 (no. 92)
Man Holding a Document (cat. 8)   82, 117 (cat. 9), 236
Venetian Lady   71 (n. 74), 93 (n. 17), 94 (n. 23), 96

Bosch, Jeronimo de I   51, 52 (n. 35)

Bosch, Jeronimo de I:
Sculpture   193 (cat. 41)

Bosmans, Margaretha   16

Boston, Museum of Fine Arts:
Guercino, Queen Semiramis (cat. 13)   40, 83, 124

Bradford, Earl of: Bassano, Christ Carrying the Cross (cat. 4)   111

Braganza, Catherine of
see Catherine of Braganza

Brandenburg, Great Elector of   47, 49 (n. 28), 91, 92 (n. 17), 93, 94 (n. 23–24), 95–96, 169 (cat. 58)

Brandenburg, Great Elector of:
Sculptures 181 (cat. 10), 185 (cat. 21), 187 (cat. 25), 191 (cat. 34), 194 (cat. 43), 197 (cat. 53), 201 (cat. 65), 204 (cat. 71), 211 (cat. 94), 218 (cat. 111)

Bredius, A.   101, 103

Broeckhorst, van der (de Wimmenum)   75 (n. 82), 77 (n. 83)

INDEX

Brussels, Bibliothèque royale de Belgique, Cabinet des estampes:
Icones 45 (n. 17)

Bruyningh, J. F. 15 (n. 12), 17 (n. 18), 18 (n. 24), 19 (n. 30)

Buchell, Aernout van 10 (n. 2), 12, 22, 55–58, 65–66, 74, 87, 132 (cat. 17)

Buckingham, Duchess of 119 (cat. 10)

Buckingham, Duke of 53 (n. 38)

Bute, Marquess of:
Saenredam 79

Bye, Johan de 82 (n. 93)

Bijtelaar, B. 20

Calceolari, Francesco, the Younger 100

CAMPAGNOLA, DOMENICO 173 (no. 104)

Campen, Laurens van 51

Campen, Laurens van:
Sculptures 184 (cat. 15)

CARAVAGGIO 42, 114 (cat. 5)

CARAVAGGIO, ATTR. TO:
Four Cyclops in a Forge (cat. A11) 42, 234

CARIANI 172 (no. 57)
Reclining Venus (cat. 9) 38, 70, 72 (n. 76), 82, 84, 85 (n. 102) 117, 172 (no. 57), 236

CARPACCIO, VITTORE 172 (no. 84)

CARRACCI 53 (n. 38), 87, 90 (n. 12), 170 (cat. 59)

CARRACCI, ANNIBALE 103

CARRACCI, ANNIBALE:
Christ and the Woman of Samaria 54

Cassel, Staatliche Kunstsammlungen:
Liss, Brothel (cat. 15) 40, 128

CATENA, VINCENZO 172 (no. 82)

Catherine of Braganza 85 (n. 103)

Catherine of Braganza:
Raphael school (cat. 22) 138
Titian, attr. to (cat. 35) 154

Ceruti, Benedetto 100

Charles I 74, 78 (n. 86), 85 (n. 103), 118 (cat. 10), 120 (cat. 11), 124 (cat. 12)

Charles II 9, 37–38, 41 (n. 6), 48, 63, 65, 75 (n. 82), 77 (n. 83–85), 78 (n. 86, 88), 79, 81, 82 (n. 93–94), 83 (n. 96), 84–85 (n. 103), 102, 109, 124 (cat. 13)

Charles II:
Barocci (cat. 1) 110
Bassano (cat. 4) 111
Bonifazio de' Pitati (cat. 7) 115
Bordone (cat. 8) 117
Cariani (cat. 9) 117
Giorgione, attr. to (cat. 11) 120
Giulio Romano (cat. 12) 122
Guercino (cat. 13) 124
Lotto (cat. 16) 130
Lotto (cat. 17) 132
Oggiono (cat. 19) 135
Parmigianino (cat. 20) 136
Raphael school (cat. 22) 138
Schiavone (cat. 28) 144
Schiavone (cat. 29) 145
Tintoretto (cat. 32) 150
Titian (cat. 34) 152
Titian, attr. to (cat. 35) 154
Titian, school of (cat. 36) 155
Veronese (cat. 38) 157
Unidentified artist (cat. 39) 159
Sculptures: 214–217 (cat. 99–110)

Christian IV, King of Denmark 29 (n. 81)

Christina, Queen of Sweden 50, 119

Cincinnati, Public Library 39 (n. 4), 43, 231 (cat. A6, 7)

Clarac, F. de 47–48

Clark, Sir Kenneth 105 (n. 16), 106

CODDE, PIETER 92

Coen, Jan Pietersz 14

COLLENIUS, HARMAN 94 (n. 20)

Congo (Africa) 14

Contarini, Michele 74

Copes, Henrico 32 (n. 94)

Copes, Johan 32 (n. 93–94)

Copes, Rizzardo 32 (n. 94)

Cornaro, Cardinal 125 (cat. 13)

CORNELISZ, CORNELIS VAN HAARLEM:
The Duet (cat. A2) 42, 228

CORREGGIO 11, 57, 122 (cat. 12), 130 (cat. 16), 136 (cat. 19)

Cospi, Ferdinando 100

Coymans, Constantia 35 (n. 98)

Cust, L. 124 (cat. 12)

DALEN, CORNELIS II VAN 38 (n. 1), 40, 44 (n. 15), 53 (n. 38), 132 (cat. 17), 150 (cat. 32), 150 (cat. 33), 152 (cat. 34)

DALENS, DIRCK 92

Delft 93

Diemen, van: Van Laer (cat. 40) 160

DIEST, JERONIMUS VAN 94 (n. 22)

Djakarta 15 (n. 15)

DODIJNS, WILLEM 71 (n. 74), 92 (n. 17)

DOOMER, LAMBERT 92

DOSSO 103

DOU, GERRIT:
Young Mother
79 (n. 88), 81 (n. 91–92), 82 (n. 93), 85 (n. 102)

Drakenstein, Slot 27

INDEX

Dresden, Gemäldesammlungen :
Guercino (cat. 13)   125
Veronese (cat. 56)   168

Dresden, Skulpturensammlung   47, 49, 68
Cleopatra   58 (n. 51)
Sculptures :   191 (cat. 34), 201 (cat. 65), 204 (cat. 71), 211 (cat. 94)

Dresden, Staatliche Kunstsammlungen, Kupferstichkabinett : Caelaturae   39 (n. 4)

DUCQ, JOHAN LE   94 (n. 22)

DUJARDIN, KAREL   28 (n. 73), 71 (n. 74), 87, 92 (n. 17), 94 (n. 21), 95

DURER, ALBRECHT   16

Dutch East India Company   13–14, 82 (n. 93)

Dutch East Indies   14

Dutch Gift   9, 26 (n. 65), 37, 38–39, 41 (n. 6), 48, 63, 65 (n. 62), 72 (n. 76), 75–86, 90, 103, 115 (cat. 7), 117 (cat. 8), 122 (cat. 12), 124 (cat. 13), 135 (cat. 19), 136 (cat. 20), 144 (cat. 28), 145 (cat. 29), 150 (cat. 33)

DYCK, ANTHONY VAN   90 (n. 7, 11)

DYCK, ANTHONY VAN :
Christ Carrying the Cross   42 (n.9)
Christ on the Cross (cat. A3)   23

EECKHOUT, GERBRAND VAN DEN   92

EGMONT, JUSTUS VAN :
Virgin Mary (cat. A4)   42 (n. 9), 230

Egyptian art   59–63, 99, 102

Elias, J. E.   28 (n. 68)

ELSHEIMER, ADAM :
Mocking of Ceres   79, 81, 82 (n. 94)

England   15, 63

English Royal Collections :
Bonifazio de' Pitati (cat. 7)   40, 115

Bordone (cat. 8)   40, 117
Cariani (cat. 9)   40, 117
Giorgione, attr. to (cat. 11)   40, 120
Giulio Romano (cat. 12)   40, 122
Lotto (cat. 16, 17)   40, 130, 132
Oggiono (cat. 19)   40, 135
Parmigianino (cat. 20)   40, 136
Schiavone (cat. 28, 29)   40, 144–145
Tintoretto (cat. 32)   40, 150
Titian (cat. 34, 36)   40, 152, 155
Veronese (cat. 38)   40, 157
Sculpture : (cat. 108)   48, 216

Este, Cardinal Alessandro d'   119 (cat. 10)

EVERSDIJK, DAVID   94 (n. 20)

FALCK, JEREMIAS   38 (n. 2), 40, 42 (n. 9) 43, 44 (n. 14–15), 105, 111 (cat. 4), 112 (cat. 5), 115 (cat. 7), 118 (cat. 10), 120 (cat. 11), 124 (cat. 13), 126 (cat. 14), 128 (cat. 15), 134 (cat. 18), 138 (cat. 21 A), 144 (cat. 27), 146 (cat. 30), 147 (cat. 31), 228 (cat. A1, 2), 230 (cat. A3, 4), 232 (cat. A9)

Father Time   10, 47 (n. 19)

FERRERIS, DIRCK   94 (n. 20)

FETTI, DOMENICO   89

FETTI, DOMENICO :
Vision of St. Peter (cat. 10)   40, 118

Fischel, O.   105 (n. 16)

Flinck, Nicolaas Antoni   51

Flinck, Nicolaas Antoni :
Titian (cat. 33)   150
Icones 1 (cat. 1)   175
Icones 6 (cat. 6)   179
Icones 37 (cat. 37)   192
Icones 45 (cat. 45)   195
Icones 49 (cat. 49)   196
Icones 61 (cat. 61)   200

Florence, Museo Archeologico :
Sleeping Ariadne (replica)   58 (n.50)

Florence, Uffizi :
Reni, Susannah (cat. 23)   139

Fokkens, Melchior   22

Foscarini, Antonio   99 (n. 44)

France   15, 30–32, 63

Frankfurt, Stadt- und Universitäts-bibliothek :
Icones   45 (n. 17)

Frederick Henry, Prince   9, 29 (n. 81), 35, 58 (n. 51)

Freedberg, S. J.   124 (cat. 12)

Frey, A. von :
Liss (cat. 14)   126

FROMANTIOU, HENDRIK VAN   91, 92 (n. 17), 93

GALLE, CORNELIS   44 (n. 14)

GARDYN, NICOLAES DU   28 (n. 73)

GEERTGEN TOT SINT JANS :
Lamentation; Legend of the Relics of St. John the Baptist   78 (n. 86)

Geesdorp, Franchoys   25 (n. 56)

Gerson, Horst   105–106

Ghent, St. Michael church :
Van Dyck (cat. A3)   230

GIORGIONE   70, 103 (n. 2), 132 (cat. 17), 171 (nos. 3, 11, 13, 15, 18, 24–27, 36), 172 (nos. 67, 79, 83)

GIORGIONE :
A Picture of Music   120 (cat. 11)
Ceres   70, 93 (n. 17), 94 (n. 23), 95, 171 (no. 18)
David with the Head of Goliath   68 (n. 69), 171 (no. 3)
Man and Woman (Two Lovers)   71 (n. 74), 171 (no. 11)

GIORGIONE, ATTR. TO :
The Concert (cat. 11)   40, 82, 84, 85 (n. 102), 120, 236

Giraud collection   147 (cat. 30)

GIULIO ROMANO :
Portrait of Isabella d'Este (cat. 12)   38 (n. 1), 40, 83–85, 122, 237

Giustiniani, Marchese Vincenzo   51

Göttingen, Kunstsammlungen der Georg-August-Universität:
Cornelisz, Cornelis (cat. A2) 42, 228

Göttingen, Niedersächsische Staats- und Universitätsbibliothek:
Caelaturae 39 (n. 4)
Icones 45 (n. 17)

GOLTZIUS, HENDRICK 15–16

Gonzaga see Mantua, Duke of

Goorle, Abraham van 59

GRAAT, BARENT 94 (n. 19)

Graeff, Andries de 78 (n. 88), 79, 103

Graeff, Cornelis de 23

Graefland, Joan 20 (n. 35), 28

Grafton, Dukes of 83

Grafton, Duke of:
Guercino (cat. 13) 124

GREBBER, ANTHONY DE 93 (n. 18)

Greece 57, 87

Grimani, Cardinal Domenico 88

Grimani, Giovanni, Patriarch of Aquileia 88

Gronau, Jacob 59

Gronovius, Fredericus 36

GUERCINO 89, 136 (cat. 20)

GUERCINO:
Queen Semiramis (cat. 13) 40, 83–84, 87, 124, 237

Haarlem 19

HAENSBERGEN, JOHANNES VAN 94 (n. 22)

Huygens, A. 41 (n. 9), 111 (cat. 4), 115 (cat. 7), 120 (cat. 11), 124 (cat. 13), 126 (cat. 15), 134 (cat. 18), 138 (cat. 21A), 146 (cat. 30). 230 (cat. A3, 4)

Hamburg 38 (n. 2), 43

Hamilton, Marquess of 74, 104 (n. 10)

Hampton Court 37, 81

Hampton Court:
Mabuse 78 (n. 86)

Hanmer, Sir Thomas:
Guercino (cat. 13) 124

Hannover, Niedersächsische Landesgalerie: Reni, Susannah (cat. 23) 139

HARINGH, DANIEL 94 (n. 22)

Haze, Clara de 17, 18 (n. 23), 19 (n. 34), 29 (n. 81)

Haze, Constantia de 18 (n. 23), 19 (n. 34), 29 (n. 81)

Haze, Sara de 18 (n. 23), 29 (n.81)

Hecquet, R. 41, 43

Heimbach, Chr. von 49 (n. 28)

Heinecken, K. H. von 41–43, 138 (cat. 21A), 228 (cat. A1), 230 (cat. A5), 232 (cat. A9)

Heinsius, Nicolaas 36

HELST, BARTHOLOMEUS VAN DER 26, 28, 79

Henry, Prince of Wales 78 (n. 86)

Heijrmans, Abraham 29, 31 (n. 84), 32 (n. 95)

Hinlopen, Geertruyd Jacobs 17, 18 (n. 25)

Historia 10

HOLBEIN, HANS:
Portrait 93 (n. 17), 94 (n. 23), 103 (n. 2)

HOLLAR, WENZEL 54

HOLSTEYN, CORNELIS 122 (cat. 12)

HOLSTEYN, PIETER II 38 (n. 1), 40, 44 (n. 15), 122 (cat. 12)

HONDECOETER, MELCHIOR DE 94 (n. 19)

Honselaarsdijk 59 (n. 52)

HONTHORST, GERARD VAN 232 (cat. A9)

Hooftman, Eva 28 (n. 68)

Hoorn, Barent van 20

Hoorn, Lucretia van 22 (n. 46a)

Hoorn, Simon van 22 (n. 46a), 26 (n. 63, 65), 27, 79

Houbraken, Jacob 27–28, 53 (n. 38), 82 (n. 93), 91

Huizen (near Naarden), Kommerrust 22–23

Huygens, Constantia 35 (n. 101)

Huygens, Constantijn, the Elder 18, 30, 35 (n. 101), 36, 93, 104, 169 (cat. 58)

Huygens, Constantijn, the Younger 90

Huygens, Lodewijk 90

Jacobs, E. 39 (n. 2), 48 (n. 24), 49 (n. 28), 68–69, 71 (n. 71), 95 (n. 30), 110 (cat. 2), 111 (cat. 4), 117 (cat. 9), 122 (cat. 12), 132 (cat. 17), 138 (cat. 22), 143 (cat. 26), 150 (cat. 32), 152 (cat. 34), 155 (cat. 36), 157 (cat. 38), 167 (cat. 52)

James II:
Bassano (cat. 4) 111
Bonifazio de' Pitati (cat. 7) 115
Bordone (cat. 8) 117
Cariani (cat. 9) 117
Girogione, attr. to (cat. 11) 120
Giulio Romano (cat. 12) 122
Lotto (cat. 16, 17) 130–32
Oggiono (cat. 19) 135
Parmigianino (cat. 20) 136
Raphael school (cat. 22) 138
Reni, attr. to (cat. 26) 143
Schiavone (cat. 28, 29) 144–45
Tintoretto (cat. 32) 150
Titian (cat. 34) 152
Titian, attr. to (cat. 35) 154
Titian, school of (cat. 36) 155

Java 19

ICONES see Reynst, Gerard

Jeude, Marinus de:
Van Laer (cat. 41)   161

Imperiale, Vincenzo:
Strozzi (cat. 30)   147

Indonesia (East Indies)   14–15, 17

Inferville, D'   31 (n. 91)

Jonghe, Clement de   39, 111 (cat. 4), 130 (cat. 16), 134 (cat. 18), 136 (cat. 20), 138 (cat. 21A), 139 (cat. 23), 141 (cat. 24), 143 (cat. 26), 144 (cat. 27), 146 (cat. 30), 152 (cat. 34), 157 (cat. 37), 160 (cat. 40), 161 (cat. 41), 162 (cat. 42)

JORDAEN, JOHANNES   93

JORDAENS, JACOB:
Flora, Silenus, and Zephyrus (cat. A5)   42 (n. 9), 230

Italy   15, 19, 29 (n. 81), 36, 57, 87, 89, 104

KALF, WILLEM   92

Kate, Lambert ten   199 (cat. 58)

Kemp, Gerrit   18 (n. 29)

Kernkamp, G. W.   23

Kircher, Athanasius   59 (n. 53), 60, 63, 99

Klinkenberg, G. J. van   59

Knorr von Rosenroth, Christian   47 (n. 17), 48, 55, 57, 63–66, 72, 78 (n. 85), 86–87

KONINCK, PHILIPS   92

Kramm, Christian   160 (cat. 40)

Kretzer, Maerten   103

La Bosse   136 (cat. 20)

LAER, PIETER VAN   19 (n. 33), 43, 87

LAER, PIETER VAN:
The Ambush (cat. 40)   39 (n. 3), 40, 64, 160
The Horse Stable (cat. A8)   42 (n. 8), 231
The Large Limekiln (cat. 41)   39 (n. 3), 40, 161

Marauders Setting Fire (cat. A6)   231
Resting Herd (cat. A7)   231
The Shot with the Pistol (cat. 42)   39 (n. 3), 40, 162

LAIRESSE, GERARD DE   45, 47 (n. 17, 19, 22), 48, 92

Law, Ernest   110 (cat. 2), 120 (cat. 11), 122 (cat. 12), 130 (cat. 16), 132 (cat. 17), 135 (cat. 19), 144 (cat. 28–29), 150 (cat. 32), 152 (cat. 34), 155 (cat. 36), 157 (cat. 38)

Le Blanc, Ch.   48

LE BLON, J. CHR.   136 (cat. 19), 155 (cat. 36)

Le Blon, Michel   50 (n. 30), 77 (n. 84), 89–90 (n. 7)

Leiden, Gemeente Archief   49 (n. 27)

Leiden, Orangerie of the Botanical Garden   49

Leiden, Rijksmuseum van Oudheden   48 (n. 26), 49, 51, 68, 70

Leiden, Rijksmuseum van Oudheden, Library:
Icones   45 (n. 17)

Leiden, Rijksmuseum van Oudheden:
Scarab   49, 60

Leiden, Rijksmuseum van Oudheden:
Sculptures   175–180 (cat. 1–7), 189 (cat. 31–32), 196 (cat. 49), 200 (cat. 63), 203 (cat. 70), 208 (cat. 85), 209 (cat. 88), 211–213 (cat. 95–96, 98), 219–221 (cat. 112–118, 120–121)

Leiden, University of   49

Leipzig, Museum der bildenden Künste:
Guido Reni, attr. to (cat. 25)   143

Leningrad, Hermitage:
Van Laer (cat. 42)   40, 162
Stomer, attr. to (cat. A9)   42, 232

LEONARDO DA VINCI   135 (cat. 19)

Leopold Wilhelm, Archduke   54

Levant Company (Levantsche handel)   31 (n. 89, 93), 32

LEYDEN, LUCAS VAN   171 (no. 41), 172 (no. 74)

LEYDEN, LUCAS VAN, ATTR. TO:
Adam and Eve   78 (n. 86)
St. Jerome   78 (n. 86)

LICINIO   172 (no. 63), 173 (no. 98)

Lier, Willem van (Heer van Oosterwijck)   31 (n. 92), 32, 36

LIEVENS, JAN ANDRE   94 (n. 19)

Ligne, Abraham de   29 (n. 81)

LINGELBACH, JOHAN   92

Lisbon, Gulbenkian Foundation:
Rembrandt, Alexander   105 (n. 16)

LISS, JAN   89, 118 (cat. 10), 146 (cat. 30), 228 (cat. A2)

LISS, JAN:
The Brothel (cat. 15)   40, 63, 96 (n. 38), 128
St. Paul in Ecstasy (cat. 14)   40, 126

Listingh, Nicolaas   28 (n. 68)

London, British Museum:
De Annulis, et Sigillis   60 (n. 56), 69 (n. 70)
De Mineralibus   70 (n. 70)
De Picturis   69 (n. 70)
De Rebus Naturalibus   69 (n. 70)

London, British Museum, Dept. of Prints:
Caelaturae   41 (n. 4)

London, National Gallery:
Reni, Susannah (cat. 23)   139

London, The Surveyor's Office:
Caelaturae   39 (n. 4)

London, Victoria and Albert Museum:
Caelaturae   41 (n. 4)

London, Whitehall   37, 81, 83 (n. 96), 85–86

Lopez, Alfonso   54, 103–104

Loredan, Bernardino 99 (n. 44)

Lormier, Willem:
Van Laer (cat. 41) 161

LOTTO, LORENZO:
Andrea Odoni (cat. 16) 38 (n. 1), 39 (n. 3), 40, 57, 83–85, 87, 105 (n. 16), 130, 237
Portrait of a Gentleman (cat. 17) 40, 57, 82, 84, 85 (n. 102), 132, 236

LOTTO, LORENZO, ATTR. TO:
Adoration of the Shepherds (cat.18) 40, 67 (n. 65), 134

Louis XIV 31 (n. 91), 32–33

LUDICK, LODEWIJK VAN 94 (n. 20)

Lugt, Frits 53 (n. 38), 97, 102, 105 (n. 16), 111 (cat. 4), 112 (cat. 5), 115 (cat. 7), 117 (cat. 9), 120 (cat. 11), 124 (cat. 13), 130 (cat. 16), 132 (cat. 17), 134 (cat. 18), 138 (cat. 21A), 139 (cat. 23), 144 (cat. 28–29), 147 (cat. 31), 150 (cat. 32–33), 152 (cat. 34), 155 (cat. 36), 157 (cat. 38)

LUTMA, JAN II 40, 44 (n. 15), 143 (cat. 25)

Luzio, A. 124 (cat. 12)

Lyons 33

MABUSE:
Adam and Eve 78 (n. 86)

Madrid, Prado:
Raphael, Holy Family (cat. 22) 138

Mahon, Denis 79, 81, 109, 110–11 (cat. 1–4), 115 (cat. 7), 117 (cat. 8–9), 120 (cat. 11), 122 (cat. 12), 124 (cat. 13), 130 (cat. 16), 132 (cat. 17), 134 (cat. 18), 135 (cat. 19), 136 (cat. 20), 138 (cat. 22), 139 (cat.23), 143 (cat. 26), 144 (cat. 28–29), 150 (cat. 32–33), 152 (cat. 34), 154 (cat. 35), 155 (cat. 36), 157 (cat. 38), 159 (cat. 39)

Malvasia, Carlo Cesare 11 (n. 4), 87, 90 (n. 12)

MANDER, CAREL VAN 16

Mansi, Marchese:
Van Laer (cat. 40) 160

Mantua, Duke of 74, 122 (cat. 12)

Mantua, Duke of, Vincenzo 119 (cat. 10)

MARCONI, ROCCO 171 (no. 8)

Mariette, P. J. 122 (cat. 12), 124 (cat. 13)

MARSEUS, OTTO 94 (n. 19)

MATHAM, JACOB 118 (cat. 10), 126 (cat. 14), 138 (cat. 21A), 157 (cat. 38)

MATHAM, THEODOOR 40, 41, 44 (n. 15), 51, 114 (cat. 6), 118 (cat. 10), 126 (cat. 14), 141 (cat. 24), 143 (cat. 26), 146 (cat. 30), 154 (cat. 35), 157 (cat. 38)

Maurits, Prince 95 (n. 27)

Mazarin, Cardinal 31–32

Medici, Prince Cosimo de' 66–67

Medicis, Marie de, Queen of France 20, 58

MICHELANGELO 104

MICHELANGELO:
Venus and Cupid 92 (n. 17), 94 (n. 23)

Michiel, Marcantonio 74, 87–88, 99 (n. 44), 130 (cat. 16)

Milan 27 (n. 67)

Minerva 23

MOMPER, PHILIPS DE 92

MONINCKS, JOHAN 94 (n. 22)

MONINCX, PIETER 94 (n. 22)

Moscardo, Lodovico 100

Moscow 27 (n. 67), 44

Moscow, Pushkin Museum:
Strozzi, Old Courtesan (cat. 30) 40, 146

Moscow, Museum Rumianzov:
Strozzi (cat. 30) 146

Muselli, studio of 137 (cat. 20)

Musson, Matthijs 90 (n. 7)

MIJTENS, MARTINUS 94 (n. 22)

Nantes, Musée des Beaux-Arts:
Reni, Susannah (cat. 23) 139

Naples, Banca Sannitica:
Van Laer, Ambush (cat. 40) 40, 160

Naples, Museo Nazionale:
Flora Farnese 50

Nassau, L. van 79

Nave, Bartolommeo della 74–75, 104 (n. 10)

Near East 35

NETSCHER, CASPAR 92

Neufville, Robert de:
Sculpture 203 (cat. 70)

Neuhauser, Barthold 59–63

New or Brabant Company (Nieuwe/Brabantsche Compagnie) 13

New York, Metropolitan Museum of Art:
Rembrandt, Aristotle 106
Rembrandt, St. James 105 (n. 16)
Caelaturae 41 (n. 4)

Nichetti, Giovanni 17 (n. 22)

Nicquet, Jacques 15–18, 19 (n. 34), 29, 32 (n. 94), 89

Nicquet, Jacques, Inventory 248–68

Nicquet, Jan (b. 1539) 15–17, 89

Nicquet, Jan (b. 1565) 18 (n. 23), 19 (n. 34), 29 (n. 81), 32 (n. 94)

Nicquet, Margaretha 15, 19

NIEDECK, PIETER PZ 94 (n. 20)

Nieuwburg, Huis ter 58

Noack, A. 112 (cat. 4)

Northbrook collection:
Guercino (cat. 13)   124

Nuremberg, Germanisches Nationalmuseum:
Liss (cat. 15)   129

Nijs, Daniel   74–75, 88–89, 104 (n. 10)

Odoni, Andrea   57, 74, 87, 99 (n. 44)

Odoni, Andrea:
Lotto (cat. 16)   130
Titian, attr. to (cat. 35)   154

Odoni, Alvise:
Lotto (cat. 16)   130

OGGIONO, MARCO D': Christ Child and St. John (cat. 19)   82, 135, 236

Orléans, Duc d':
Reni, Susannah (cat. 23)   139

ORSI, LELIO   103

Orville, Philippe d'   221 (cat.122)

Oudendorp, Franciscus   48 (n. 26)

Oxford, Bodleian Library:
De Sacrificiorum..Vasculis   60 (n. 55), 69 (n. 70)

Padua, Sta. Maria in Vanzo:
Bassano   105

Paets, Adriaen:
Van Laer (cat. 40)   160

PALMA GIOVANE, JACOPO:
Dance of Naked Children   49, 92 (n. 17), 94 (n. 23), 96, 169 (cat. 58)

PALMA VECCHIO, JACOPO   53 (n. 38), 70, 103, 105, 115 (cat. 7), 117 (cat. 9), 141 (cat. 24), 154 (cat. 35), 171 (nos. 5, 14, 19, 30, 42), 172 (nos. 61, 80)

PALMA VECCHIO:
St. Paul   93 (n. 17), 94 (n. 23)

Palthe, J.: Van Laer (cat. 41)   161

Papenbroek, Gerard van   48 (n. 26), 49, 51, 52 (n. 35), 70, 103

Papenbroek, Gerard van:

Icones 1–7   175–180
Icones 31–32   189
Icones 41   193
Icones 48–49   196
Icones 55   198
Icones 63   200
Icones 70   203
Icones 85   208
Icones 89   209
Icones 95–96   211–213
Icones 98   213
(cat. 112–118, 120–121)   219–221

Paris   31–33, 36, 43

Paris, Bibliothèque Nationale:
Aelia Flaccilla   50
Caelaturae   41 (n. 4), 45 (n. 17)
Icones   45 (n. 17)

Paris, Louvre:
Sleeping Ariadne   58 (n. 50)

PARMIGIANINO   11, 53 (n. 38), 122 (cat. 12), 135 (cat. 19)

PARMIGIANINO:
Athena (cat. 20)   39 (n. 3), 40, 82, 84, 136, 236

Pater, Albert Dircksz   25 (n. 51)

PELLECUM, MATTHEUS VAN   94 (n. 19)

PERRIER, FRANÇOIS   47 (n. 19)

PERSIJN, REIJNIER A   51, 54

Pesser, Maria Dircxdr   103

PICINUS, JACOB   11 (n. 5)

Pickfatt, Richard:
Van Laer (cat. 41)   161

PIOMBO, SEBASTIANO DEL   122 (cat. 12), 132 (cat. 17), 150 (cat. 32), 172 (no. 48)

Pölnitz, Gerardt Bernard von   93, 95 (n. 27), 169 (cat. 58)

PORCELLIS, JAN:
Seastorm   78 (n. 86)

PORDENONE   172 (nos. 52, 59)

Portugal   45

PIJNACKER, ADAM   92

Provence   33

Prudence   47

Prussia   59 (n. 52)

QUELLINUS, ARTUS   75, 77 (n. 84), 90

Quichelberg (Quiccheberg), Samuel von   98

Ram, Giovanni   74, 99 (n. 44)

Ranst   90 (n. 7)

RAPHAEL   11, 99 (n. 44), 103, 122 (cat. 12), 138 (cat. 21A, B), 171 (no. 10)

RAPHAEL:
Castiglione   54
Holy Family with the Lamb   139 (cat. 22)
Old Man   93 (n. 17), 94 (n. 23)

RAPHAEL SCHOOL:
Christ on a Lamb (cat. 22)   82, 138, 236
Virgin and Child (cat. 21A, B)   39 (n. 3), 40, 42 (n. 6), 44 (n. 15), 138

Rauwart, Claes   18

REMBRANDT   97, 104, 143 (cat. 25), 152 (cat. 34)

REMBRANDT:
Adriaen Banck   78 (n. 85)
Alexander the Great   105 (n. 16)
Aristotle with Bust of Homer   106
Descent from the Cross   104–105
Hendrickje Stoffels   106
St. James   105 (n. 16)
Susannah   78 (n. 85)

Renard (Renouard), Louis   45 (n. 17)

RENI, GUIDO   53 (n. 38), 99 (n. 44)

RENI, GUIDO:
Susannah (cat. 23)   39 (n. 3), 40, 44, 63, 139

RENI, GUIDO, ATTR. TO:
Allegory of Painting (cat. 26)   39 (n. 3), 40, 83–84, 143, 237
St. Bartholomew (cat. 24)   39 (n. 3), 40, 63, 141
St. John (cat. 25)   40, 143

Renialme, J. de   103

RENIERI, NICOLO   35, 67 (n. 65), 87, 89, 161 (cat. 40), 168 (cat. 57)

Renieri, Nicolò:
Bonifazio de' Pitati (cat. 44)   165
Schiavone (cat. 47–48)   165–66
Sustris (?) (cat. 51)   166
Tintoretto (cat. 52)   167
Titian (cat. 53)   167
Veronese (cat. 57)   168

Reynst, Abraham   27

Reynst, Albert   27, 28 (n. 67)

Reynst, Catharina   19 (n. 30)

Reynst, Constancia (b. 1638)   27–28, 29 (n. 75)

Reynst, Constantia (b. 1603)   19 (n. 30), 24

Reynst, Gerard (b. 1599)   9–13, 16–17, 19, 22–27, 29–30, 35–39, 41–43, 45, 47, 51, 54–55, 57–58, 63, 66–68, 70–72, 74–75, 81, 84–85, 87–90, 93, 95–102, 104–106

Reynst, Gerard:
Caelaturae   11, 37, 38–45

Reynst, Gerard : Paintings :
Barocci (cat. 1)   110
Bassano (cat. 2–6)   110–114
Bonifazio de' Pitati (cat. 7)   115
Bordone (cat. 8)   117
Cariani (cat. 9)   117
Carracci (cat. 59)   170
Fetti (cat. 10)   118
Giorgione, attr. to (cat. 11)   120
Giulio Romano (cat. 12)   122
Guercino (cat. 13)   124
Laer, van (cat. 40–42)   160–164
Liss (cat. 14, 15)   126–129
Lotto (cat. 16, 17)   130–134
Lotto, attr. to (cat. 18)   134–135
Oggiono, attr. to (cat. 19)   135

Parmigianino (cat. 20)   136–137
Raphael school (cat. 21, 22)   138–139
Reni (cat. 23)   139–141
Reni, attr. to (cat. 24–26)   141–144
Schiavone (cat. 27–29)   144–146
Strozzi (cat. 30)   146–147
Tintoretto (cat. 31, 32)   147–149, 150
Titian (cat. 33, 34)   150–154
Titian, attr. to (cat. 35)   154
Titian, school of (cat. 36)   155–157
Veronese (cat. 37, 38)   157–158

Reynst, Gerard : Sculptures :
Aristotle, Funerary Chest of   55, 65, 68, 106
Cleopatra   9, 11 (n. 2), 58
Icones   11, 37, 45–54, 64 (n. 1), 88, 102, 238–240
Icones 1   47 (n. 21), 48 (n. 26), 49–50, 51 (n. 31), 54 (n. 38), 175 (cat. 1)
Icones 2   47 (n. 21), 48 (n. 26), 49–50, 51 (n. 31), 176 (cat. 2)
Icones 3   47 (n. 21), 49–50, 51 (n. 31), 177 (cat. 3)
Icones 4   47 (n. 21), 48 (n. 26), 49, 51 (n. 31), 177 (cat. 4)
Icones 5   47 (n. 21), 48 (n. 26), 49–50, 51 (n. 31), 178 (cat. 5)
Icones 6   47 (n. 21), 48 (n. 26), 49–50, 51 (n. 31), 54 (n. 38), 179 (cat. 6)
Icones 7   47 (n. 21), 48 (n. 26), 49, 51 (n. 31), 180 (cat. 7)
Icones 8   47 (n. 21), 181 (cat. 8)
Icones 9   47 (n. 21), 181 (cat. 9)
Icones 10   47 (n. 20–21), 49 (n. 28), 181 (cat. 10)
Icones 11   47 (n. 21), 51 (n. 31), 183 (cat. 11)
Icones 12   47 (n. 21), 183 (cat. 12)
Icones 13   47 (n. 21), 51 (n. 31), 183 (cat. 13)
Icones 14   47 (n. 21), 48, 50, 184 (cat. 14)
Icones 15   47 (n. 21), 48, 52 (n. 33), 184 (cat. 15)
Icones 16   47 (n. 21), 48, 50, 51 (n. 31), 52 (n. 33), 184 (cat. 16)

Icones 17   47 (n. 21), 48, 184 (cat. 17)
Icones 18   47 (n. 21), 48, 51 (n. 31), 185 (cat. 18)
Icones 19–20   185 (cat. 19–20)
Icones 21   47 (n. 20), 49, 185 (cat. 21)
Icones 22   51 (n. 31), 187 (cat. 22)
Icones 23–24   187 (cat. 23–24)
Icones 25   47 (n. 20), 49, 187 (cat. 25)
Icones 26   188 (cat. 26)
Icones 27   50, 188 (cat. 27)
Icones 28–30   188–89 (cat. 28–30)
Icones 31   48 (n. 26), 49, 51 (n. 31), 189 (cat. 31)
Icones 32   48 (n. 26), 49, 52 (n. 33), 58, 189 (cat. 32)
Icones 33   51 (n. 31), 191 (cat. 33)
Icones 34   47 (n. 20), 49, 191 (cat. 34)
Icones 35–36   191–92 (cat. 35–36)
Icones 37   54 (n. 38), 192 (cat. 37)
Icones 38   192 (cat. 38)
Icones 39   51 (n. 31), 192 (cat. 39)
Icones 40   51 (n. 31), 193 (cat. 40)
Icones 41–42   193 (cat. 41–42)
Icones 43   47 (n. 20), 49 (n. 28), 194 (cat. 43)
Icones 44   195 (cat. 44)
Icones 45   54 (n. 38), 195 (cat. 45)
Icones 46–48   195–96 (cat. 46–48)
Icones 49   48 (n. 28), 49, 51 (n. 31), 54 (n. 38), 196 (cat. 49)
Icones 50–52   197 (cat. 50–52)
Icones 53   47 (n. 20), 49 (n. 28), 197 (cat. 53)
Icones 54   52 (n. 33), 198 (cat. 54)
Icones 55   51 (n. 31), 198 (cat. 55)
Icones 56   199 (cat. 56)
Icones 57   48, 199 (cat. 57)
Icones 58–60   199–200 (cat. 58–60)
Icones 61   54 (n. 38), 200 (cat. 61)
Icones 62   200 (cat. 62)
Icones 63   48 (n. 26), 49, 51 (n. 31), 200 (cat. 63)
Icones 64   201 (cat. 64)
Icones 65–66   47 (n. 20), 49, 201 (cat. 65–66)
Icones 67–70   48, 203 (cat. 66–69)

Icones 71 47 (n. 20), 48–49, 203 (cat. 70)
Icones 72–84 204–08 (cat. 72–84)
Icones 85 48, 52 (n. 33), 208 (cat. 84)
Icones 86 48–49, 208 (cat. 86)
Icones 87–88 48, 209 (cat. 86–87)
Icones 89 48–49, 51 (n. 31), 52 (n. 33), 209 (cat. 88)
Icones 90–91 48, 209–10 (cat. 90–91)
Icones 92 48, 52 (n. 33), 65, 209 (cat. 88), 210 (cat. 91)
Icones 93 47 (n. 21), 48, 51 (n. 31), 210 (cat. 92)
Icones 94 47 (n. 20), 48, 210 (cat. 93)
Icones 95 48–49, 211 (cat. 94)
Icones 96 48–49, 211 (cat. 95)
Icones 97 48–49, 213 (cat. 96–97)
Icones 98 48–49, 213 (cat. 98)
Icones A–M 47 (n. 21), 65, 85–86, 214–217 (cat. 100–110)

Reynst, Gerard (b. 1630–40) 27
Reynst, Gerrit (Gerard) (d. 1615) 13–15, 17–19, 29 (n. 73)
Reynst, Jacobsz 13
Reynst, Jacobus 106 (n. 20)
Reynst, Jan (b. 1591) 19 (n. 30)
Reynst, Jan (b. 1601) 9–13, 16–17, 19, 22, 29, 30–33, 35–37, 39, 57, 66–67, 71–72, 84, 87–89, 97, 99–102, 104–106

Reynst, Jan :
Bellini (cat. 43) 165
Bonifazio de' Pitati (cat. 44) 165
Cariani (cat. 9) 117
Carracci drawings (cat. 59) 170
Lotto, attr. to (cat. 18) 134
Rottenhammer (cat. 45) 165
Schiavone (cat. 28–29, 46–50) 144–145, 165–166
Sustris (cat. 51) 166
Tintoretto (cat. 52) 167
Titian (cat. 53–54) 167
Titian, attr. to (cat. 35) 154
Titian, school of (cat. 36) 155
Veronese (cat. 37–38, 55–57) 157, 167–168

Reynst, Joan 27–28, 66
Reynst, Margaretha 19 (n. 30)
Reynst, Margryt 19 (n. 30)
Reynst, Maria 29 (n. 73)
Reynst, Pieter (b. ca. 1510) 13
Reynst, Pieter (b. 1592) 19
Reynst, Wijntje 19 (n. 30, 32)

RIBERA:
Head of St. John 93, 94 (n. 24)

Ricci, Daniele 87

Ricci, Daniele :
Guercino (cat. 13) 124

Richardson, J. R. 199 (cat. 58)

Ridolfi, Carlo 9–11, 24, 30, 31 (n. 85), 32, 35, 37, 66, 67 (n. 65), 68 (n. 69), 84, 89 (n. 4), 134 (cat. 18), 144 (cat. 28), 145 (cat. 29), 152 (cat. 34), 155 (cat. 36), 157 (cat. 38), 165 (cat. 43–47), 166 (cat. 48–51), 167 (cat. 52–56), 168 (cat. 57)

Ripa, Cesare 10, 47

ROGHMAN, ROELANDT 92

Rome 23, 28, 59 (n. 53)

ROTTENHAMMER, HANS:
Madonna in Adoration (cat. 45) 165

Rotterdam 79

RUBENS, P. P. 45 (n. 16), 128 (cat. 15)

RUBENS, P. P.:
Large Hunt 78 (n. 85)
After Elsheimer 83 (n. 94)

Ruttens, S. 29 (n. 84)

Ruzzini, Carlo 75

Rijswijk Palace 55 (n. 45)

SADELER 16

SAENREDAM, PIETER:
View of the Groote Kerk at Haarlem 79, 81

SANDRART, JAN VAN 94 (n. 22)
SANDRART, JOACHIM VON 54, 128 (cat. 15), 160 (cat. 40)

Sandrart, Joachim von : Van Laer (cat. 41) 161

Sansovino, Francesco 74

SANTVOORT, DIRCK 93 (n. 18)

SANTWIJCK, FRANÇOIS VAN 94 (n. 22)

SARTO, ANDREA DEL 138 (cat. 21A)

Scamozzi, Vincenzo 67, 74

Schaep, Pieter 27–28

Scheller, R. W. 100–101

Schellinger, Bernard 27

Scheveningen 75

SCHIAVONE, ANDREA 70, 111 (cat. 4), 173 (nos. 113–114, 116–117, 119–120, 122–123)

SCHIAVONE, ANDREA:
Battle between Lapiths and Centaurs (cat. 48) 67 (n. 65), 166
Christ Before Pilate (cat. 28) 66, 82, 84, 144, 236
Christ Presented to the People (cat. 46) 165
Judgment of Midas (cat. 29) 66, 82, 84–85 (n. 102), 145, 236
Madonna and Child (cat. 47) 67 (n. 65), 165
Perseus (cat. 49) 166
Philosophers' Heads (cat. 50) 166
Presentation in the Temple (cat. 27) 39 (n. 3), 40, 144

Scholten, Hendrick 54 (n. 39)

SCHONGAUER, MARTIN 16

Schuyt, Albert Gijsbertsz 19, 29 (n. 81), 31 (n. 92)

Schuyt, Anna 17, 19, 27, 39, 66, 75, 77 (n. 84, 85), 81, 228 (cat. A1), 233 (cat. A9)

Sepibus, Georgius de 59 (n. 53)

Settala, Lodovico and Manfredo 100

Six, Jan 41 (n. 5), 50, 51 (n. 31–32), 53 (n. 38), 54, 91, 96 (n. 34, 37–38)

Six, Jan:
Icones 1–7 175–180
Icones 11 183
Icones 13 183
Icones 16 184
Icones 18 185
Icones 20 185
Icones 22–23 187
Icones 31 189
Icones 33 191
Icones 37 192
Icones 39–40 192–193
Icones 49 196
Icones 55 198
Icones 61 200
Icones 63 200
Icones 88 209
Icones 92 210
cat. 120–121 221
Liss (cat. 15) 128–129

Six, Pieter 37, 71 (n. 74), 96

Six, Pieter:
Liss (cat. 15) 128

Six van Hillegom, Jhr. J:
Palma (cat. 58) 49, 169

Smissaert, Joan Carlo (Heer van Niel) 24

Smits, Gerard 29

Solms, Amalia van 9, 11 (n. 2), 36, 58

SOMEREN, HENDRICK VAN 94 (n.19)

Spain 15

Spiegel, Hendrick Dircksz 26

States General 14, 18, 29–33, 78 (n. 86), 79. 82 (n. 93)

States of Holland and West Friesland 9, 38, 48, 63, 75, 83 (n. 96), 84–85

STELLA, JACQUES:
Virgin and Child 42 (n. 9)

Stimmel, J. G. 41 (n. 9), 43, 110 (cat. 2), 111 (cat. 3–4), 112 (cat. 5), 114 (cat. 6), 115 (cat. 7), 118 (cat. 10), 120 (cat. 11), 122 (cat. 12), 124 (cat. 13), 126 (cat. 14), 130 (cat. 16), 132 (cat. 17), 134 (cat. 18), 136 (cat. 20), 138 (cat. 21A), 139 (cat. 23), 141 (cat. 24), 143 (cat. 25), 143 (cat. 26), 144 (cat. 27), 147 (cat. 31), 150 (cat. 32–33), 152 (cat. 34), 155 (cat. 36), 157 (cat. 37–38), 160 (cat. 40), 161 (cat. 41), 162 (cat. 42)

Stockholm, Riksarkivet 50 (n. 30)

STOMER, MATHIAS, ATTR. TO:
Esau (cat. A9) 42, 232

Stricker, B. H. 59–60

STRIEP, CHRISTIAEN 94 (n. 20)

STROZZI, BERNARDO:
Old Courtesan (cat. 30) 39 (n. 3), 40, 44, 146

STRIJCKER, WILLEM 94 (n. 20)

SUSTRIS, LAMBERT 171 (no. 7)

SUSTRIS, LAMBERT:
Death of the Pharao (cat. 51) 67 (n. 65), 166

Swieten, P. van 94 (n. 23)

TEMPEL, ABRAHAM VAN DEN 93 (n. 18)

TENIERS, DAVID 54

The Hague 32

The Hague, Algemeen Rijksarchief 31–33, 75 (n. 82), 77 (n. 83–85), 78 (n. 88), 81 (n. 91), 83 (n. 96)

The Hague, 'Conflerle-kamer' 92

The Hague, Mauritshuis:
Dou 79

The Hague, Museum Meermanno-Westreenianum:
Icones 45 (n. 17)

The Hague, Royal Library:
Caelaturae 41 (n. 4)
Huygens Correspondence 95 (n. 27), 104 (n. 12)

The Hague, Rijksbureau voor kunsthist. Documentatie:
Caelaturae 41 (n. 4)
Six catalogue 96 (n. 34)

Timmers, J. J. M. 47 (n. 17), 48

TINTORETTO, JACOPO 11, 70, 103, 150 (cat. 33), 174 (nos. 138, 143), 232 (cat. A9)

TINTORETTO, JACOPO:
Group Portrait (cat. 52) 67 (n.65), 167
Pietà (cat. 31) 40, 43, 147 (cat. 31)
Portrait of Dominican Friar (cat.32) 40, 82, 84, 85 (n.102), 150, 236
Venetian Senator 71, 96, 174 (no. 138)

TITIAN 11, 38 (n. 1), 53 (n. 38), 57, 70, 103 (n. 2), 117 (cat. 9), 132 (cat. 17), 130 (cat. 32), 150 (cat. 35), 171 (nos. 20, 22), 172 (nos. 49, 54, 65)

TITIAN:
Ariosto 54, 57 (n. 46)
Children 92 (n. 17)
Concert of Five Half-lengths 120 (cat. 11)
Flora 54
Holy Family 57
Landscape with Satyr and Nymph 93 (n. 17), 94 (n. 23)
Portrait of Giorgione 92 (n. 17), 94 (n. 23)
Portrait of Isabella 122 (cat. 12)
Portrait of a Man (Aretino) (cat.33) 40, 53 (n. 38), 150
Portrait of a Man (Sannazaro) (cat. 34) 38 (n. 1), 39 (n. 2), 40, 82, 84–85, 152, 236
Portrait of a Senator (cat. 54) 167
Shepherd and Shepherdess 92 (n. 17), 94 (n. 23)
St. Francis in Adoration (cat. 53) 67 (n. 65), 167
Venetian Lady 93 (n. 17), 94 (n. 23)

TITIAN, ATTR. TO:
Holy Family with St. John (cat. 35) 40, 82, 84, 87, 154, 236

TITIAN, SCHOOL OF:
Virgin and Child with Tobias and Angel (cat. 36) 40, 66, 67 (n. 65), 79, 82, 84, 85 (n. 102), 155, 236

Tixerandet, F. 92 (n. 17), 96 (n. 32)

Tongeren, Johann van:
Van Laer (cat. 41) 161

Torrington, Viscount:
Bassano (cat. 4) 111

Toulon 31 (n. 90)

Turkey 57, 87

Uffelen, Lucas van 57 (n. 46), 89, 103–104, 161 (cat. 40)

Uffenbach, Zacharias Conrad von 53 (n. 38)

Uilenbroek, Gosuin 50, 52 (n. 33), 67, 103

Uilenbroek, Gosuin:
Sculptures 184 (cat. 15, 16), 189 (cat. 32), 198 (cat. 54), 208 (cat. 84), 209 (cat. 88), 210 (cat. 91), 221 (cat. 120, 122)

Uittenbogaert, Jan 22

UNIDENTIFIED ARTIST:
Christ and the Virgin (cat. 39) 82, 159, 236

United Company of Amsterdam (Vereenigde Companie van Amsterdam) 13

United Provinces 29, 31–33, 41 (n. 6), 45, 75 (n. 82), 81

Utrecht, Centraal Museum:
Tomb Relief 70, 221 (cat. 119)

Utrecht, University Library: Van Buchell, Notae 22, 55

Uylenburgh, Gerrit 50 (n. 28), 54 (n. 39), 72 (n. 76), 75, 77 (n. 85), 81 (n. 91), 83 (n. 96), 90–93, 94 (n. 24), 96

Uylenburgh, Hendrick 96–97

Uylenburgh, Saskia van 72 (n.76), 97

Vaduz, Prince of Liechtenstein:
Van Laer (cat. A6) 231

Vaduz, Prince of Liechtenstein (formerly):
Van Laer, Large Limekiln (cat. 41) 40, 161

VAILLANT, JACQUES 92

VAILLANT, WALLERANT 93 (n. 18)

VASARI, GIORGIO 132 (cat. 16)

Vatican, Museums:
Jupiter Verospi 50
Sleeping Ariadne 11 (n. 2), 58

Ven, J. van de 27 (n. 67), 32 (n. 95), 35 (n. 98)

Veerle, Jacob and Johannes van 54, 89

VELDE, WILLEM VAN DE 78 (n. 87)

Vendramin, Andrea 11–12, 37–39, 47–48, 51 (n. 31), 52 (n.34), 55, 57, 59, 63, 65, 67–75, 84, 86 (n. 105), 87, 98–100, 102

Vendramin, Andrea:
Bellini 171 (nos. 1–2, 4, 17, 29, 31, 40), 172 (nos. 66, 69, 73, 85), 173 (no. 102)
Bonifazio de' Pitati 172 (no. 86)
Bordone 173 (nos. 90, 92, 94)
Campagnola 173 (no. 104)
Cariani 117 (cat. 9), 172 (no. 57)
Carpaccio 172 (no. 84)
Catena 172 (no. 82)
Corenzio 174 (no. 132)
Giorgione 171 (nos. 3, 11, 13, 15, 18, 24–27, 36), 172 (nos. 67, 79, 83)
Leyden, van 171 (no. 41), 172 (no. 74)
Licinio 172 (no. 63), 173 (no. 98)
Marconi 171 (no. 8)
Palma Vecchio 171 (nos. 5, 14, 19, 30, 42), 172 (nos. 61, 80)
Piombo, del 172 (no. 48)
Pordenone 172 (nos. 52, 59)
Raphael 171 (no. 10)
Schiavone 173 (nos. 113–114, 116–117, 119–120, 122–123)
Sustris 171 (no. 7)
Tintoretto, Jacopo 174 (nos. 138, 143)
Titian 171 (nos. 20, 22), 172 (nos. 49, 54, 65)
Veronese 171 (no. 32)
Unidentified artists 171 (nos. 6, 9, 12, 16, 21, 23, 28, 33–35, 37–39, 43), 172 (nos. 44–47, 50–51, 53, 55–56, 58, 60, 62, 64, 68, 70–72, 75–78, 81, 87), 173 (nos. 88–89, 91, 93, 95–97, 99–101, 103 105–112, 115, 118, 121, 124–128), 174 (nos. 129–131, 133–137, 139–142, 144–155)

Vendramin, Andrea: MS. catalogues:
De Annulis 60, 63, 69 (n. 70)
De Antiquorum tumulis 48 (n. 26), 55, 65, 68, 70 (n. 70)
De Mineralibus 70 (n. 70)
De Picturis 63, 68, 69 (n. 70), 70–71, 84, 87, 95–96, 105
De Rebus Naturalibus 69 (n. 70)
De Sacrificiorum Vasculis 57 (n. 48), 60, 63, 69 (n. 70)
De Sculpturis 47–49, 68, 70, 71 (n. 70), 72, 86 (n. 105)

Vendramin, Andrea: Sculptures 175–183 (cat. 1–11), 184–185 (cat. 16–20), 187 (cat. 22–23, 25), 188–192 (cat. 26–39), 193–197 (cat. 41–53), 198–200 (cat. 55–56, 58–62), 206 (cat. 78), 210 (cat. 91–92), 213 (cat. 96), 214–221 (cat. 99–101, 103–107, 110–122)

Vendramin, Andrea:
Sculptures, Concordance with Icones 244–247

Vendramin, Gabriele 74

VENNECOL, JACOB 94 (n. 19)

Venice 11–15, 17–19, 29–30, 32–33, 36, 55, 57, 63, 72, 74, 104–105

Venice, Museo archaeologico:
Vitellius 50

Venne, Lucas van de 14

Vereenigde Compagnie van Amsterdam,
see United Company

VERMEER 92, 94 (n. 23)

VERONESE, PAOLO 11, 103 (n. 2)

VERONESE, PAOLO:
Good Samaritan (cat. 56) 167
Leda 119 (cat. 10)
Marriage of St. Catherine (cat. 38) 40, 66, 67 (n. 65), 82, 84, 85 (n. 102), 157, 236
Portraits of Soranza Family (cat. 57) 67 (n. 65), 168
Resurrection of Christ (cat. 37) 39 (n. 3), 40, 64, 67 (n. 65), 157
Sacrifice of Abraham (cat. 55) 167

VERSCHUYR, THEODOOR 92

Vertue, George 85 (n. 103)

Vienna, Kunsthistorisches Museum:
Fetti 119 (cat. 10)
Geertgen tot Sint Jans 78 (n. 86)
Vos, Simon de 129 (cat. 15)

Villiers, Barbara 83

Villiers, Barbara:
Guercino (cat. 13) 124

VISSCHER, CORNELIS 38 (n. 1), 40–44, 110 (cat. 2), 111 (cat. 3), 122 (cat. 12), 130 (cat. 16), 136 (cat. 20), 139 (cat. 23), 146 (cat. 30), 147 (cat. 31), 154 (cat. 35), 155 (cat. 36), 157 (cat. 37), 160 (cat. 40), 161 (cat. 41), 162 (cat. 42)

Visscher, Nicolaes 45

Vlaeming van Outshoorn, Cornelis de 75, 77 (n. 84), 90

Vondel, Joost van den 27 (n. 65), 79

VORSTERMAN, LUCAS 152 (cat. 34)

Vos, Jan 22, 28

VOS, SIMON DE:
Musical Company 129 (cat. 15)

Vossius, Isaac 36

Vries, Johan de 50, 51 (n. 32)

Vries, Johan de:
Sculptures 192–193 (cat. 39–40)

VROOM, H. C.:
Battle at Gibraltar 78 (n. 86)

Wagenaar, Jan 23

Warnaertsz, J. 29 (n. 82), 31 (n. 84), 33, 35 (n. 98)

Weigel, Rudolf 42 (n. 8), 232 (cat. A8)

Weltevreden (Java), Rijswijk Palace 14 (n. 11)

WERVEN, JACOB VAN 49

Weston Park, Earl of Bradford:
Bassano (cat. 4) 40, 111

Wildt, Eva de 28 (n. 68)

Wilhem, David Le Leu de 29 (n. 81), 35–36

Wilhem, Paulo de 18 (n. 23), 19 (n. 34), 29–30, 36

Willem II 58 (n. 51), 79 (n. 89)

Willem III 28, 53 (n. 38), 82 (n. 93)

William III:
Titian (cat. 33) 150

Wilton House:
Sleeping Ariadne 58 (n. 50)

WILS, JAN 93, 94 (n. 19)

Wimmenum, de see Broeckhorst

Winckler 41, 43, 228 (cat. A2), 232 (cat. A9)

Windsor Castle, Royal Library:
Busts & Statues in Whitehall Gardens 86

Winterthur, O. Reinhart Coll.:
Tintoretto 96 (n. 37), 174 (no. 138)

Witsen, Cornelis 26, 29

Witsen, Nicolaes 50

Witsen, Nicolaes:
Sculptures 211 (cat. 95), 213 (cat. 98), 221 (cat. 117)

Witt, Jan de Jr. 103

Witt, Jan de:
Sculpture 203 (cat. 70)

Wolfenbüttel, Library:
Knorr, Itinerarium 63

Wolff, Rejnier van der 79 (n. 88), 103

WOLFRAEDT, DANIEL 94 (n. 19)

Wurzbach, A. von 42, 110 (cat. 2), 111 (cat. 3–4), 112 (cat. 6), 115 (cat. 7), 118 (cat. 10), 120 (cat. 11), 122 (cat. 12), 124 (cat. 13), 126 (cat. 14), 128 (cat. 15), 130 (cat. 16), 132 (cat. 17), 136 (cat. 20), 138 (cat. 21A), 139 (cat. 23), 146 (cat. 30), 147 (cat. 31), 150 (cat. 32–33), 152 (cat. 34), 154 (cat. 35), 155 (cat. 36), 157 (cat. 37–38) 160 (cat. 40), 161 (cat. 41), 162 (cat. 42)

Wussin, Johann 42 (n. 8–9), 43, 110 (cat. 2), 111 (cat. 3–4), 112 (cat. 5), 114 (cat. 6), 115 (cat. 7), 118 (cat. 10), 120 (cat. 11), 122 (cat. 12), 124 (cat. 13), 126 (cat. 14–15), 130 (cat. 16), 132 (cat. 17), 134 (cat. 18), 136 (cat. 20), 138 (cat. 21A), 139 (cat. 23), 141 (cat. 24), 143 (cat. 25–26), 144 (cat. 27), 146 (cat. 30), 147 (cat. 31), 150 (cat. 32–33), 152 (cat. 34), 154 (cat. 35), 155 (cat. 36), 157 (cat. 37–38), 160 (cat. 40), 161 (cat. 41), 162 (cat. 42)

WIJNANTS, JAN 92

York, Art Gallery:
Bassano (after) 111 (cat. 4)

Zesen, Filip von 22

Zwieten, Heer van:
Van Laer (cat. 41) 161

LIST OF TEXT FIGURES

Vignette on title page:
Michel Le Blon,
Coat of Arms of the Reynst family (photo after Van der Kellen by A. Frequin).

Fig. 1. *Detail of Map of Amsterdam* by Balthasar Florisz van Berckenrode, 1625, Gemeentelijke Archiefdienst, Amsterdam (photo).

Fig. 2. *The House 'De Hoop'* at top with reconstruction of the original house at bottom, based on Philips' *Grachtenboek*, 1772 (photo Gemeentelijke Archiefdienst, Amsterdam).

Fig. 3. *The House 'De Hoop', Keizersgracht 209*, Amsterdam (photo Gemeentelijke Archiefdienst).

Fig. 4. Bartholomeus van der Helst, *The Four Governors of the Kloveniersdoelen*, 1655. Historisch Museum, Amsterdam.

Fig. 5. *Signature of Gerard Reynst* (D.T.B. 946, 10.5.1654), Gemeentelijke Archiefdienst, Amsterdam (photo).

Fig. 6. *Signature of Jan Reynst* under Marine Treaty, signed April 18, 1646, Algemeen Rijksarchief, The Hague (photo).

Fig. 7, 7a-d. *Title page and introduction to Carlo Ridolfi*, LE MARAVIGLIE DELL ARTE, Venice, 1648 (photo Beinecke Rare Book Library, Yale University, New Haven).

Fig. 8. *Title page for the* VARIARUM IMAGINUM…CAELATURAE, Rijksprentenkabinet, Amsterdam.

Fig. 9. Gerard de Lairesse, *Title page for* SIGNORUM VETERUM ICONES, Rijksprentenkabinet, Amsterdam.

Fig. 10. '*Sepultura di Aristotile*', illustration from Andrea Vendramin's DE ANTIQUORUM TUMULIS, 1627, fol. 9r, Deutsche Staatsbibliothek, East Berlin (photo).

Fig. 11. I. v. Werven, *Marmora Papenbroekiana*, proposed installation in Leiden. Gemeentelijke Archiefdienst, Leiden (photo N. v. d. Horst).

Fig. 12. *Illustration from Andrea Vendramin's* DE SACRIFICIORUM… VASCULIS, Bodleian Library, Oxford (photo).

Fig. 13, 13a. *Drawings after Egyptian Vase in Andrea Vendramin's* DE SACRIFICIORUM …VASCULIS, fol. 34, Bodleian Library, Oxford, and illustration of the same Vase in Kircher.

Fig. 14-18, 22. *Drawings after Pieces of Egyptian Art in the Reynst collection, illustrated in Kircher.*

Fig. 19, 19a. *Drawing after Scarab in Andrea Vendramin's* DE ANNULIS.., 1627, fol. 12, British Museum, London and illustration of the same scarab in Kircher.

Fig. 20, 20a. *Drawing after Scarab in Andrea Vendramin's* DE ANNULIS…, 1627, fol. 13, British Museum, London, and illustration of the same scarab in Kircher.

Fig. 21-21a,b.
The Scarab Klinkenberg, Rijksmuseum van Oudheden, Leiden. *Drawing after the same Scarab in Andrea Vendramin's* DE ANNULIS…, 1627, fol. 14, British Museum, London, and illustration from Kircher.

Fig. 23. Gerrit Dou, *The Young Mother*, Mauritshuis, The Hague (photo A. Dingjan, The Hague).

Fig. 24. J. van den Vondel, *De Kunstkroon van den koning van Groot-Britannie.* (October 1660).

Fig. 25. P. J. Saenredam, *The Groote Kerk at Haarlem*, Collection Marquess of Bute, Muont Stuart, England.

PAINTINGS

Plate P2. Cornelis Visscher, *Abraham Leaving Haran*, after Jacopo Bassano.

Plate P3. Cornelis Visscher, *The Annunciation to the Shepherds*, after Jacopo Bassano.

Plate P4. Jeremias Falck, *Christ Carrying the Cross*, after Jacopo Bassano.

Plate P4a. Jacopo Bassano, *Christ Carrying the Cross*, the Earl of Bradford, Weston Park, England (photo Royal Academy of Arts, London).

Plate P5. Jeremias Falck, *The Entombment of Christ*, after Jacopo Bassano.

Plate P6. Theodor Matham, *The Virgin and Child and the Infant St. John*, after Jacopo Bassano.

Plate P7. Jeremias Falck, *The Virgin and Child with Tobias and Saints*, after Bonifazio de' Pitati, called Veronese.

Plate P7a. Bonifazio de' Pitati, called Veronese, *The Virgin and Child with Tobias and Saints*, Hampton Court.

Plate P8. Paris Bordone, *Portrait of a Man Holding a Document*, Hampton Court.

Plate P9. Cariani, *Reclining Venus*, Hampton Court.

Plate P10. Jeremias Falck, *The Vision of St. Peter*, after Domenico Fetti.

Plate P11. Jeremias Falck, *The Concert*, after Giorgione (attributed to).

Plate P11a. Giorgione (attributed to), *The Concert*, Hampton Court.

Plate P12. Pieter Holsteyn II, *Portrait of Isabella d'Este*, after Giulio Romano.

Plate P12a. Giulio Romano, *Portrait of Isabella d'Este*, Hampton Court.

Plate P13. Jeremias Falck, *Queen Semiramis Receiving News of the Revolt of Babylon*, after Guercino.

Plate P13a. Guercino, *Queen Semiramis Receiving News of the Revolt of Babylon*, Museum of Fine Arts, Boston (photo).

Plate P14. Jeremias Falck, *St. Paul in Ecstasy*, after Jan Liss.

Plate P14a. Jan Liss, *St. Paul in Ecstasy*, Staatliche Museen, Berlin-Dahlem (photo Walter Steinkopf, Berlin).

Plate P15. Jeremias Falck, *The Brothel*, after Jan Liss.

Plate P15a. Jan Liss, *The Brothel*, Gemäldegalerie, Cassel (photo).

Plate P16. Cornelis Visscher, *Portrait of Andrea Odoni*, after Lorenzo Lotto,

Plate P16a. Lorenzo Lotto, *Portrait of Andrea Odoni*, Hampton Court.

Plate P17. Cornelis van Dalen II, *Portrait of a Gentleman*, after Lorenzo Lotto.

Plate P17a. Lorenzo Lotto, *Portrait of a Gentleman*, Hampton Court.

Plate P18. Jeremias Falck, *The Adoration of the Shepherds*, after Lorenzo Lotto (attributed to).

Plate P19. Marco d'Oggiono (attributed to), *The Infant Christ and St. John Embracing*, Hampton Court.

Plate P20. Cornelis Visscher, *Athena*, after Parmigianino.

Plate P20a. Parmigianino, *Athena*, Windsor Castle.

Plate P21A. Jeremias Falck, *The Virgin and Child and St. Anne*, after Raphael school.

Plate P21B. Unidentified engraver, *The Virgin and Child and St. Anne*, after Raphael school.

Plate P23. Cornelis Visscher, *Susannah and the Elders*, after Guido Reni.

Plate P23a. Guido Reni (studio?), *Susannah and the Elders*, Musée des Beaux-Arts, Nantes (photo studio Madec, Nantes).

Plate P24. Theodor Matham, *St. Bartholomew*, after Guido Reni (attributed to).

Plate P25. Jan Lutma II, *St. John the Evangelist*, after Guido Reni (attributed to).

Plate P25a. Guido Reni (attrib. to), *St. John the Evangelist*, Museum der bildenden Künste, Leipzig (photo).

Plate P26. Theodor Matham, *Allegory of Painting*, after Guido Reni (attributed to).

Plate P27. Jeremias Falck, *Presentation in the Temple*, after Andrea Schiavone.

Plate P28. Andrea Schiavone, *Christ Before Pilate*, Hampton Court.

Plate P29. Andrea Schiavone, *The Judgment of Midas*, Hampton Court.

Plate P30. Jeremias Falck, *The Old Courtesan*, after Strozzi.

Plate P30a. Bernardo Strozzi, *The Old Courtesan*, Pushkin Museum, Moscow.

Plate P31. Cornelis Visscher, *Pietà*, after Tintoretto.

Plate P32. Cornelis van Dalen II, *Portrait of a Dominican Friar*, after Tintoretto.

Plate P32a. Jacopo Tintoretto, *Portrait of a Dominican Friar*, Hampton Court.

LIST OF PLATES

Plate P33. Cornelis van Dalen II, *Portrait of a Man* (so-called *Pietro Aretino*), after Titian.

Plate P34. Cornelis van Dalen II, *Portrait of a Man* (so-called *Sannazaro*), after Titian.

Plate P34a. Titian, *Portrait of a Man* (so-called *Sannazaro*), Hampton Court.

Plate P35. Cornelis Visscher, *Holy Family with St. John and St. Elizabeth*, after Titian (attributed to).

Plate P36. Cornelis Visscher, *The Virgin and Child with Tobias and the Angel*, after school of Titian.

Plate P36a. Titian (school of), *The Virgin and Child with Tobias and the Angel*, Hampton Court.

Plate P37. Cornelis Visscher, *The Resurrection of Christ*, after Veronese.

Plate P38. Theodor Matham, *The Marriage of St. Catherine*, after Veronese.

Plate P38a. Paolo Veronese, *The Marriage of St. Catherine*, Hampton Court.

Plate P40. Cornelis Visscher, *The Ambush*, after Pieter van Laer.

Plate P40a. Pieter van Laer. *The Ambush*, Banca Sannitica, Naples (photo).

Plate P41. Cornelis Visscher, *The Large Linnekiln*, after Pieter van Laer.

Plate P42. Cornelis Visscher, *The Shot with the Pistol*, after Pieter van Laer.

Plate P42a. Pieter van Laer, *The Shot with the Pistol*, Hermitage, Leningrad.

Plate P58. Jacopo Palma the Younger, *Dance of Naked Children*, Collection Jhr. Six van Hillegom, Amsterdam (photo Art Promotion, Amsterdam).

SCULPTURES

Plate S1. '*Consul*'.

Plate S1a,b. *Torso and Head of Male Figure Draped in Himation*, Rijksmuseum van Oudheden, Leiden.

Plate S2. '*T. Hostilius*'.

Plate S2a. *Torso of a Roman Military Statue*, Rijksmuseum van Oudheden, Leiden.

Plate S3. '*Venus*'.

Plate S3a,b. *Statuette of Venus*, Rijksmuseum van Oudheden, Leiden.

Plate S4. '*Flora*'.

Plate S4a, b. *Statue of Fortune*, Rijksmuseum van Oudheden, Leiden.

Plate S5. '*Venus*'.

Plate S5a, b. *Statuette of Venus with a Dolphin*, Rijksmuseum van Oudheden, Leiden.

Plate S6. '*Abundantia*'.

Plate S6a. *Statue of a Muse*, Rijksmuseum van Oudheden, Leiden.

Plate S7. '*Hercules*'.

Plate S7a. *Statuette of the Hercules Farnese*, Rijksmuseum van Oudheden, Leiden.

Plate S8. '*Iulia*'.

Plate S9. '*Adonis*'.

Plate S10. '*Gladiator*'.

Plate S10a. *Statuette of Trajan*, Staatliche Museen, East Berlin.

Plate S11. '*Apollo*'.

Plate S12. '*Cupid*'.

Plate S13. '*Hermaphroditus*'.

Plate S14. *Draped Female Figure Standing*.

Plate S15. *Statuette of Male Nude Wearing a Chlamys*.

Plate S16. *Draped Female Figure Standing, with a Pitcher at her Feet*.

Plate S17. *Partly Draped Female Figure Standing*.

Plate S18. *Draped Female Figure Standing*.

Plate S19. '*Hercules*'.

Plate S20. '*I. Caesar*'.

Plate S21. '*Agrippina Major*'.

Plate S21a. *Female Portrait Bust*, Staatliche Museen, East Berlin.

Plate S22. '*Poppea Sabina*'.

Plate S23. '*Augustus*'.

Plate S24. '*Tiberius*'.

Plate S25. '*Calphurnia*'.

Plate S25a. *Female Bust*, Staatliche Museen, East Berlin.

Plate S26. '*Octavia*'.

Plate S27. '*Vitellius*'.

Plate S28. '*Domitia*'.

Plate S29. '*Pallas*'.

Plate S30. '*Milenus*'.

Plate S31. '*Livia*'.

Plate S31a. *Female Portrait Bust*, Rijksmuseum van Oudheden, Leiden.

Plate S32. '*Cleopatra*'.

Plate S32a. *Female Bust*, Rijksmuseum van Oudheden, Leiden.

Plate s33. '*Trajanus*'.

Plate s34. '*Gordianus*'.

Plate s34a. *Bust of Gordian III*, Skulpturensammlung, Dresden.

Plate s35. '*Helena*'.

Plate s36. '*Octavia*'.

Plate s37. '*Hadrianus*'.

Plate s38. '*Caracalla*'.

Plate s39. '*Terentia*'.

Plate s40. '*Iulia Mammea*'.

Plate s41. '*Agrippina*'.

Plate s42. '*Iupiter Ammon*'.

Plate s43. '*Aristea*'.

Plate s43a. *Bust of a Youth*, Staatliche Museen, East Berlin.

Plate s44. '*Pallas*'.

Plate s45. '*Geta*'.

Plate s46. '*Flavia*'.

Plate s47. '*Faunus*'.

Plate s48. '*Antoninus*'.

Plate s49. '*Hadrianus*'.

Plate s49a. *Portrait Bust of Hadrian*, Rijksmuseum van Oudheden, Leiden.

Plate s50. '*Cupido*'.

Plate s51. '*Faunus*'.

Plate s52. '*Cijrus*'.

Plate s53. '*Flavia*'.

Plate s53a. *Portrait of a Woman*, Staatliche Museen, East Berlin.

Plate s54. '*Lucilla*'.

Plate s55. '*Faustina*'.

Plate s56. '*Silvius Postumius*'.

Plate s57. *Bust of a Young Boy*.

Plate s58. '*Germanicus*'.

Plate s59. '*Papirius*'.

Plate s60. '*Antonia Major*'.

Plate s61. '*Hadrianus*'.

Plate s62. '*Claudia*'.

Plate s63. '*Bachus*'.

Plate s63a. *Bust of a Youthful Pan*, Rijksmuseum van Oudheden, Leiden.

Plate s64. '*Severus*'.

Plate s65. '*Priapus*' (front).

Plate s65b. '*Priapus*' (back).

Plate s65a. *Herm of Priap*, Skulpturensammlung, Dresden.

Plate s66. *Male Portrait Bust*.

Plate s67. *Male Portrait Bust*.

Plate s68. *Female Portrait Bust*.

Plate s69. *Bust of Septimius Severus*.

Plate s70. *Female Portrait Bust*.

Plate s70a. *Bust of Helena (?)*, Rijksmuseum van Oudheden, Leiden.

Plate s71. *Bust of Man with Priestly Crown*.

Plate s71a. *Bust of Man with Priestly Crown (Arsaces XX?)*, Skulpturensammlung, Dresden.

Plate s72. *Male Portrait Bust*.

Plate s73. *Male Portrait Bust*.

Plate s74. *Male Bust*.

Plate s75. *Male Bust*.

Plate s76. *Male Bust*.

Plate s77. *Bust of a Philosopher*.

Plate s78. *Ideal Portrait of a Poet (?)*.

Plate s79. *Female Bust*.

Plate s80. *Laureate Male Bust*.

Plate s81. *Laureate Male Bust*

Plate s82. *Female Portrait Bust*.

Plate s83. *Female Bust as Oratress(?)*.

Plate s84. *Male Portrait Bust*.

Plate s85. *Male Portrait Bust*.

Plate s85a. *Male Portrait Bust*, Rijksmuseum van Oudheden, Leiden.

Plate s86. *Female Bust as Oratress(?)*.

Plate s87. *Female Bust*.

Plate s88. *Julius Caesar (?)*.

Plate s88a. *Julius Caesar (?)*, Rijksmuseum van Oudheden, Leiden.

Plate s89. *Male Portrait Bust*.

Plate s90. *Bust of a Youth (Caracalla?)*.

Plate s91. *Female Bust*.

Plate s92. *Faun*.

Plate s93. *Bust of a Poet*.

Plate s94. *Herm of a Bearded God*.

Plate s94a. *Herm of a Bearded God*, Skulpturensammlung, Dresden.

Plate s95. *Bust of Apollo (?) or Diana (?)*.

Plate s95a. *Bust of Apollo (?) or Diana (?)*, Rijksmuseum van Oudheden, Leiden.

Plate s96, 97. *Female Head* (top); *Male Head* (bottom).

Plate s96a. *Statuette of Hygieia*, Rijksmuseum van Oudheden, Leiden.

Plate s98. *Male Portrait Head*.

Plate s98a. *Male Portrait Head*, Rijksmuseum van Oudheden, Leiden.

Plate s99. '*Sabina*'.

Plate s100. '*Caracalla*'.

Plate s101. '*Aesculapius*'.

Plate s102. '*Cupido*'.

LIST OF PLATES

Plate S103. *'Vesta'*.

Plate S104. *'Cijbele'*.

Plate S104a. *Busts & Statues in Whitehall Gardens*, ms. Windsor Castle.

Plate S105. *'Tiberius'*.

Plate S106. *'Domitianus'*.

Plate S106a. *Busts & Statues in Whitehall Gardens*, ms. Windsor Castle.

Plate S107. *'Commodus'*.

Plate S108. *'Faustina'*.

Plate S108a. *Busts & Statues in Whitehall Gardens*, ms. Windsor Castle.

Plate S108b. *Bust of a Woman*, Hampton Court.

Plate S109. *'Scipio Africanus'*.

Plate S110. *'M Brutus'*.

Plate S111. *Statuette of a Youth*, Staatliche Museen, East Berlin.

Plate S112. *Statuette of Venus with Cupid*, Rijksmuseum van Oudheden, Leiden.

Plate S114. *Herm of Bacchus Indicus*, Rijksmuseum van Oudheden, Leiden.

Plate S115. *Standing Figure of a Draped Woman*, Rijksmuseum van Oudheden, Leiden.

Plate S116. *Large Right Foot*, Rijksmuseum van Oudheden, Leiden.

Plate S117. *'Sepolcro antiquo'*, illustration from Andrea Vendramin's DE ANTIQUORUM TUMULIS, 1627, fol. 7r, Deutsche Staatsbibliothek, East Berlin.

Plate S117a. *Funerary Chest*, Rijksmuseum van Oudheden, Leiden.

Plate S118. *'Sepolcro'*, illustration from Andrea Vendramin's DE ANTIQUORUM TUMULIS, 1627, fol. 11r, Deutsche Staatsbibliothek, East Berlin.

Plate S118a. *Tomb Monument*, Rijksmuseum van Oudheden, Leiden.

Plate S119. *'Sepoltura'*, illustration from Andrea Vendramin's DE ANTIQUORUM TUMULIS, 1627, fol. 14 r, Deutsche Staatsbibliothek, East Berlin.

Plate S119a. *Tomb Relief of Two Couples*, Centraal Museum, Utrecht.

Plate S120. *Tomb Relief*, illustration from Andrea Vendramin's DE ANTIQUORUM TUMULIS, 1627, fol. 22 r, Deutsche Staatsbibliothek, East Berlin.

Plate S120a. *Tomb Relief*, Rijksmuseum van Oudheden, Leiden.

Plate S121. *Tomb Relief*, illustration from Andrea Vendramin's DE ANTIQUORUM TUMULIS, 1627, fol. 23 r, Deutsche Staatsbibliothek, East Berlin.

Plate S121a. *Tomb Relief*, Rijksmuseum van Oudheden, Leiden.

APPENDIX

Plate A1. Jeremias Falck, *St. John the Baptist Preaching*, after Abraham Bloemaert.

Plate A2. Jeremias Falck, *The Duet*, after Cornelis Cornelisz van Haarlem.

Plate A2a. Cornelis Cornelisz van Haarlem, *The Duet*, Kunstsammlungen der Georg-August-Universität, Göttingen, (photo).

Plate A5. Schelte A. Bolswert, *Flora, Silenus, and Zephyrus*, after Jordaens.

Plate A9. Jeremias Falck, *Esau Selling his Birthright to Jacob*, after Stomer.

Plate A9a. Matthias Stomer, *Esau Selling his Birthright to Jacob*, Hermitage, Leningrad.

Plate A11. Jeremias Falck, *Cyclops Forging the Arms of Achilles*, after Caravaggio (?).

CREDIT FOR PHOTOGRAPHS

Amsterdam, Gemeentelijke Archiefdienst: text figs. 1–3, 5.

Amsterdam, Rijksmuseum: text figs. 4, 8–9: plates P 2–P 7, P 10–P 18, P 20–P 21A, B, P 23–P 27, P 30, P 32–P 38, P 40–P 42; S 1–S 18; A 1–A 2, A 5, A 9, A 11.

Amsterdam, Jhr. Six van Hillegom: plate P 58 (photo Art Promotion).

Berlin-Dahlem, Staatliche Museen: plate P 14a (photo Walter Steinkopf).

Berlin (East), Deutsche Staatsbibliothek: text fig. 10; plates S 117–121.

Berlin (East), Staatliche Museen: plates S 10a, S 21a, S 25a, S 43a, S 53a, S 111.

Boston, Museum of Fine Arts: plate P 13a.

Cassel, Staatliche Kunstsammlungen: plate P 15a.

Cincinnati, Public Library: plate P 31 (photo C. W. Bostain).

Dresden, Skulpturensammlung: plates S 34a, S 65a, S 71a, S 94a.

Göttingen, Kunstsammlungen der Georg-August Universität: plate A 2a.

Leiden, Gemeentelijke Archiefdienst: text fig. 11 (photo N. v. d. Horst).

Leiden, Rijksmuseum van Oudheden: text fig. 21b; plates S 1a, b S 2a, S 3a, b, S 4a, b, S 5a, b, S 6a, S 7a, S 19–S 31, S 31a, S 32, S 32a, S 33–S 49, S 49a, S 50–S 63, S 63a, S 64–S 65, S 65 b, S 66–S 70, S 70a, S 71–S 85, S 85a, S 86–S 88, S 88a, S 89–S 95, S 95a, S 96, S 96a, S 97–S 98, S 98a, S 99–S 112, S 114–S 116, S 117a, S 118a, S 120a, S 121a.

Leipzig, Museum der bildenden Künste: plate P 25a.

Leningrad, Hermitage: plates P 42a, A 9a.

London, British Museum: text figs. 19–21.

London, English royal collections (copyright reserved); plates P 7a, P 8–P 9, P 11a, P 12a, P 16a, P 17a, P 19, P 20a, P 28–P 29, P 32a, P 34a, P 36a, P 38a; S 104a, S 106a, S 108a, b.

Moscow, Pushkin Museum: plate P 30a.

Nantes, Musée des Beaux-Arts: plate 23a.

Naples, Banca Sannitica: plate P 40a.

New Haven, Yale University, Beinecke Rare Book Library: text figs. 7 a–d.

Oxford, Bodleian Library: text figs. 12–13.

The Hague, Mauritshuis: text fig. 23 (photo A. Ding jan).

The Hague, Algemeen Rijksarchief: text fig. 6.

Utrecht, Centraal museum: plate S 119a.

Weston Park, Earl of Bradford: plate P 4a.

COLOPHON

Typography: Frederik Bos,
Editorial Department Koninklijke
Nederlandse Akademie van
Wetenschappen, Amsterdam

Typesetting and printing:
Casparie Alkmaar bv

Lithography: RCO Van Setten,
Amsterdam

Binding: Wörmann b.v.
boekbinders, Zutphen